GNSS Systems and Engineering

GNSS Systems and Engineering

The Chinese Beidou Navigation and Position Location Satellite

Shusen Tan
Beijing Satellite Navigation Center, Beijing, China

WILEY

国防工業出版社
National Defense Industry Press

This edition first published 2018 by John Wiley & Sons Singapore Pte. Ltd under exclusive licence granted by National Defense Industry Press for all media and languages (excluding simplified and traditional Chinese) throughout the world (excluding Mainland China), and with non-exclusive license for electronic versions in Mainland China.

Registered Offices
John Wiley & Sons, Inc., 111 River Street, Hoboken, NJ 07030, USA
John Wiley & Sons Singapore Pte. Ltd, 1 Fusionopolis Walk, #07-01 Solaris South Tower, Singapore 138628

Editorial Office
1 Fusionopolis Walk, #07-01 Solaris South Tower, Singapore 138628

For details of our global editorial offices, customer services, and more information about Wiley products visit us at www.wiley.com.

Wiley also publishes its books in a variety of electronic formats and by print-on-demand. Some content that appears in standard print versions of this book may not be available in other formats.

Library of Congress Cataloging-in-Publication Data applied for

ISBN: 9781118897034

Cover design by Wiley
Cover image: © lixuyao/Gettyimages

Set in 10/12pt WarnockPro by SPi Global, Chennai, India

10 9 8 7 6 5 4 3 2 1

Contents

Preface

The contemporary satellite navigation system has become an important infrastructure in the main developed countries. The US Global Positioning System (GPS) and the Russian Global Navigation Satellite System (GLONASS), established in the middle 1990s, are not only important military equipment but also widely used in the fields of global navigation and positioning, high precision time transfer, spacecraft tracking telemetry and control, and so on. The European Union started building the Galileo Satellite Navigation System at the end of the twentieth century. China has started building the China Compass Navigation Satellite System since the mid-1990s. The Compass Satellite Navigation and Positioning Reporting System was successfully built on December 27, 2012, realizing the service capability in Asia-Pacific Region. The author, starting from the Compass Satellite Navigation and Positioning Reporting System and engineering practice and taking the Radio Determination Satellite System (RDSS) and Radio Navigation Satellite Service (RNSS) as the important content, wrote *GNSS Systems and Engineering: The China Compass Navigation and Position Location Satellite.*

This book consists of three parts, 15 chapters in total. The first part includes four chapters introducing the theory and engineering design of Radio Determination Satellite Service (RDSS) mainly including basic concepts, fundamental principle, engineering design, user equipment technology, and so on. Chapter 5, Theory and Engineering Design of Navigation and Determination Satellite Service, serves as a link between the previous and following chapters introducing the theory and design of Radio Navigation and Determination Satellite Service integrating RDSS and RNSS. The third part includes the next nine chapters that introduce the theory and engineering design of Radio Navigation Satellite Service (RNSS), mainly including basic concepts, fundamental principle, engineering design, application examples, user equipment testing, and so on.

During writing and compiling of this book, departments at all different levels and relevant experts provided care and support. I would like to thank Xu Qifeng and Xie Youcai for their valuable revision comments; Wang Gang, researcher, for providing the analog computation result of the relative motion features of the static-state user; and to thank the China Aerospace Science and Technology Corporation (CASC) for providing the GLONASS document. I express my sincere thanks to Dr. Chen Xiangdong, Senior Engineer Zhao Wenjun, Zhang Aiyong, Mo Zhongqiu, Qu Wenke, Shen Fei, Yang Hua, Wei Gang, Li Yanyan, Jiao Cheng, Zhu Weigang, Zhao Jing, Hou Li, Yan Jianhua, Luo Keke; they made great effort in manuscript preparation, printing, and illustration preparation. At the same time, I express special thanks to Dr. Chen Xiangdong, Dr. Li Lin, Dr. He Feng, Dr. Sun Li, Dr. Ma Min, Dr. Li Min, Dr. Hu Caibo, Dr. Hu Zhirui, Dr. Tang Bin, Dr. Yang Huafeng, Tang Yun, Wang Sen, Shi Xin, Wang Chuhan, Zhang Linfeng, Zhang Hui, Dou Changjiang, Sha Hai, Dr. Wang Dongxia, Xin Jie, and Li Yanjie, for their great effort in translating the book.

1
Overview

The term "satellite navigation" is a short version of Radio Navigation Satellite Service (RNSS). The service that obtains the position vector $\overrightarrow{R} = (X, Y, Z, \dot{X}, \dot{Y}, \dot{Z}, T)$ under the fixing time of the Earth coordinate system by determining the traveling distance and rate of distance change of the radio signal transmitted by satellite is called the "RNSS," where (X, Y, Z) is the user's coordinate in the Earth-fixed coordinate system; $(\dot{X}, \dot{Y}, \dot{Z})$ is the user's velocity component; and T is the moment when the user is located at the coordinate (X, Y, Z). The satellite navigation system consists of the satellite that realizes the radio navigation target, the ground operation control system, and the application terminal.

Transit and цикaдa, built by USA and the Soviet Union successively in the 1960s, can only obtain the position coordinate (X, Y, Z) of the user under static state conditions. Movement velocity of the ship must be given to obtain the position coordinates in motion. The second generation of satellite navigation positioning systems, Global Positioning System (GPS) and Global Navigation Satellite System (GLONASS), built by the USA and Russia, respectively, can realize high-accuracy timing, spatial 3D positioning, and measurement of kinematic velocity for aerospace, aviation, navigation, and ground users at any point on Earth or in terrestrial space. Compared with the first generation of satellite navigation systems, the difference is significant in terms of the name of the system, positioning principle, positioning method, and so on. The first generation of Chinese satellite navigation positioning system utilized a positioning principle totally different from the one mentioned here. The confirmation of radio navigation parameter and position calculation required by user positioning is accomplished by ground control center other than the user, in fact, both positioning of the user terminal and positioning report from the user terminal to ground control center are completed, therefore it is called the Radio Determination Satellite Service (RDSS). RNSS is a subset of RDSS. China Compass is a multi-functional aerospace application system integrating satellite radio positioning, navigation, communication, location reporting, and identification.

In the corresponding chapters of this book, the system that can only confirm the position coordinate of the user is called the satellite positioning system, and the system that can completely confirm position vector of the user is called the satellite navigation system. However, RDSS is a complete system that accomplishes navigation, positioning, location reporting, and short message communication. This system was created originally by China Compass. Its function of location reporting has the features of lower

GNSS Systems and Engineering: The Chinese Beidou Navigation and Position Location Satellite, First Edition. Shusen Tan.

systematic complexity and cost as well as shorter responding time of positioning reporting and higher security than RNSS positioning, which are the advantages of Compass and also the future directions for GLONASS development.

1.1 Origin of GLONASS

GLONASS, which originated from satellite positioning, has been through the development of satellite navigation. On December 27, 2012 regional deployment of Compass Satellite Navigation System positioning was accomplished; and the system integrating satellite navigation and positioning reporting pushed the GLONASS to its zenith. The development process can be divided into the three stages that follow.

1.1.1 Stage 1: Satellite Radio Positioning

This originated in the 1960s. Typical systems were the Transit (the US Navy Satellite Navigation System) and Russian цикада. Their principles originated from Professor Shekbusavidge's team of former Soviet Union Leningrad Marjongski Airforce Engineering Academy of Science and Frank T. McClure, George C. Wesabah, and William H. Jill of the Institute of Applied Physics at Johns Hopkins University, USA. In principle, the space-borne transmitting signal is the navigation signal and the receiver of the Earth station receives the satellite signal. Through measurement of the Doppler frequency of electricity generated by satellite signal under relative movement, as well as the orbital parameter given by the satellite navigation signal, the location of the receiver, the user's positioning can be confirmed and accomplished. Determination of orbital parameters is realized through the reverse operation of the Doppler measurement after the satellite navigation signal is received by the fixed station on the Earth. This kind of system can only achieve the user's point positioning. Continuous navigation requires phase assistance. Therefore, the early stages of satellite navigation development only accomplished positioning.

1.1.2 Stage 2: RNSS

The RNSS originated in the 1970s. Typical systems are the US GPS and the Russian GLONASS. The principle roots can be found in the US Navy Research Council's (NRC) "Timation," the EZIB Program by the Air Force's SAMSO Organization, and the former Soviet Union's research program. The fundamental principle is that the user uses the radio signal transmitted by satellite to determine the position vector $R = (X, Y, Z, \dot{X}, \dot{Y}, \dot{Z}, T)$ under the Earth's fixed time coordinate system through measurement of the distance and rate of distance change: this is referred to as the RNSS, where (X, Y, Z) is the position coordinate of the Earth-fixed coordinate system of the user; $(\dot{X}, \dot{Y}, \dot{Z})$ is the velocity component of the user; and T is the moment when the user is located at coordinate (X, Y, Z). RNSS is the English terminology.

The satellite navigation system consists of the satellite that realizes the radio navigation target, the ground operation control system, and the application terminal. The application terminal provides not only the user's location but also the velocity, heading, distance to the destination, and trip time of the user. Therefore, the system that provides the RNSS signal service is also called the satellite navigation system. It can continuously provide the positioning velocity determination parameter.

1.1.3 Stage 3: Satellite Navigation Positioning Reporting

This originated in the 1990s. A typical system is China Compass Satellite Navigation Positioning System. As the information era progresses, the user not only needs to know his own location, velocity, and time, but also needs to share location, velocity, and duration of voyage with other users. In 1994, a Compass testing system was started to realize positioning, location reporting, and short message communication. In 1994, the "Double Satellite Navigation Positioning System" was approved. By taking two satellites as the space segment and taking the ground user electron height as the third material, it constitutes two spherical surfaces with the two satellites as the center of sphere and distance of the user as the radius, and the third spherical surface where the Earth's core is the center of sphere and the user's distance from surface of referenced ellipse sphere to the Earth's core plus height is the radius. The cross point of the three spherical surfaces is the location of the user; therefore, its principle is similar to the three-sphere positioning principle. The location reporting to the central control system is accomplished during the positioning of the user and it can realize short message communication between users. Compass provides the complete RNSS service at the same time; it is an integrated GLONASS system integrating navigation, positioning, location reporting, and short message communication. It has a greater number of outstanding advantages in achieving double satellite positioning reporting and triple satellite positioning reporting so as to avoid positioning reporting failure caused by conditions where four positioning satellites cannot be received, such as in the city among high-rise buildings and in cliff/remote gorge environments.

1.2 Development and Future Plans for the GPS System

In 1973, the United States Department of Defense approved the proposal of GPS and named it the NAVSTAR Global Positioning System, [1] which aimed to (1) deliver weapons accurately and (2) provide uniform navigation positioning to reverse the booming kinds of military navigation. The proposed constellation was a total of 24 satellites distributed on three circular orbital planes, with eight satellites on each orbital plane with a dip angle of 63°. The orbital planes were distributed with uniform space along the equator with an orbit height of 10,980 sea miles. Such an orbit height was a semi synchronous orbit, which was able to produce the repeated ground track. The large-scale antenna with an uplink injection station set in the mainland of the USA could accomplish the injection to the satellite according to the plan orientation safely and reliably. This selection of the three orbital planes not only features in good coverage but also facilitates backup satellite distribution with one backup satellite on each orbital plane and accomplishes the backup of the failure satellite. This constellation allows users to observe 6–11 satellites anytime in any place.

Two L-frequency band radio navigation signals are transmitted by each satellite, with L1 of 1575.42MHz and L2 of 1227.6MHz. L1 modulates two orthorhombic spreading code signals. I-subcircuit is a C/A code, named the coarse code or acquisition code with a code rate of 1.023 Mbps and used for civilian navigation; the Q-subcircuit is a precise distance measurement code, also called the P code with a code rate of 10.23 Mbps and used for authorized users. L2 only modulates P code, and is only used for authorized users. During the implementation of the engineering, due to concern about the prospect of GPS, financial expenditure is a constraint. In order to guarantee the effective experiment set up in Yuma Proving Grounds, three orbital planes were changed to six orbital planes to reflect the limited experimental satellite on the proving ground. There were four satellites on each plane and this had a good effect on the experiment but made the constellation layout difficult. In order to not waste the quantity of satellites or form a blank area in coverage, six orbital planes remain as they currently are. However, the three-orbital-plane scheme is still recommended by the GPS Modernization Plan, and a new civilian navigation signal has been added to L2.

From the proposal approved in 1973, GPS was mainly used for military purposes. Two navigation frequencies can be used for calibrating ionosphere propagation delay. However, there is only one navigation frequency L1 for civilian use. In order to meet civilian needs, calibration parameters of the ionosphere are provided in the L1 navigation message. Under the support of the model, 70% of ionosphere propagation deflection can be calibrated through these ionosphere calibration parameters. Therefore, for civilian users, they can only achieve positioning accuracy to a few tens of meters. Early US government protocols only allowed provision of positioning with an accuracy of 100 m (95%) to the civilian user and added SA measurement that reduced the accuracy on the civilian navigation signal. Due to the increase in civilian demand, various difference navigation positioning technologies have appeared that might improve accuracy to the meter-level. The Wide Area Augmentation System (WAAS) scheme provided by the Federal Aviation Administration (FAA) was typical of this. The US President released the SA Policy of reducing accuracy in 2000, and increased civilian accuracy to 25 m. In order to further expand the civilian market, the civilian navigation frequency was actively advocated and it was decided to modulate the civilian signal on L2 too, in order to carry out the I/Q reused QPSK modulating method. However, as for this proposal, military and civilian users were still on the same frequency for reuse, which brought about difficulties to war navigation implementation. Therefore, the US government took the lead in proposing to use the L5 frequency, that is, 1176.45 MHz, as the working frequency, which was registered in the ITU. At the World Radio Conference in 2000 (WRC-2000), it was proposed that satellite radio navigation frequency should include downlink frequency 1260∼1300 MHz, 5010∼5030 MHz, and the newly added uplink frequency included 1300∼1350 MHz, 5000∼5010 MHz, 1164∼1215 MHz; downlink navigation frequency lump power flux-density should be no greater than -121.5 dBW/(m^2•MHz). Now, GPS actually occupies three frequency bands; 1575.42±12 MHz, 1227.6±12 MHz, and 1176.45±12 MHz. Although the navigation frequency is so abundant it cannot implement effective war navigation, the increase in military signal in a war zone inevitably affects the civilian one. The interference with the civilian signal will also affect the military one. Therefore, BOC modulation emerges. Reuse of navigation frequency was increased by utilizing space spectrum frequency division. The ideal solution is that military and civilian users, respectively, have two independent navigation frequencies, which is favorable to civilians. However, the current frequency source is limited and it is difficult to find another frequency source on the L-frequency band.

Another technological problem in GPS with any big change is encryption of the military P Code. The initial phase is precise pseudo-code of P code that has a relatively high security property. In order to increase the difficulty of theft, it was changed to a Y Code after adding scrambling code and then M Code.

The latest development in GPS is to continuously improve the navigation performance of GPS on the original track and frequency resource, and the navigation war performance is mainly improved. For instance, a reinforcing measure of the military code war zone is adopted, and control of civilian code is enhanced. The time Keeping System (TKS) and autonomous navigation (Auto-NAV) are added on the BLOCK-IIR satellite. This has autonomous working ability for 180 days without requiring ground system intervention. Autonomous navigation performs distance measurement and information exchange through UHF inter-satellite links, autonomously updating ephemeris on board, and accomplishing the calibration of the satellite clock. The time system is maintained by using a highly stable cesium atomic clock and rubidium atomic clock. It also has the reprogrammable performance of satellite space; namely, that an operative flight program could be completely newly programmed according to ground command. After cold start, the processor will execute the program stored in PROM (Programmable Read-Only Memory), which will maintain the working routine through beam uplink.

The GPS III Program was launched in 2000. Its fundamental target up until 2020 is: (1) to realize the navigation signal's capability of penetrating vegetation; (2) to reach positioning accuracy of 1 m; (3) to realize full ILS capability through a wide area differential; (4) to improve timing accuracy to 1 ns; (5) to possess excellent war navigation performance – in a war zone at least, the navigation signal could be enhanced by 30 db (1000 times); and (6) to fully realize an automatic early warning navigation signal.

1.3 Development and Future Plans for GLONASS

Based on the Soviet Union's successful low orbit system, цикаαа, which began in the 1970s, Russia's GLONASS improved the orbit forecast accuracy and long-term stability of an atomic clock in space and digital signal processing techniques of propagation time delay correction accuracy of the troposphere and ionosphere. The purpose of study was to provide continuous and high-accuracy timing, spatial (three-dimensional) positioning, as well as movement velocity vectors for spaceflight, aviation, and ground users at any point on Earth or in near Earth space. GLONASS also consists of three parts:

1. Satellite system, consisting of 24 satellites on three orbital planes.
2. Monitoring control system, consisting of a ground monitoring station and a control station.
3. User equipment.

Just like as GPS, navigation positioning of GLONASS automatically performs location and velocity calculations through measurement of non-inquiring pseudo-range and the radial pseudo-velocity is finished by user equipment, which is a typical RNSS positioning system. The pre-conditions for this positioning system had been accomplished in ΛВВИА research led by B. C. Sebusa in the period of 1955–1957 before the first man-made Earth satellite was launched.

The rationality and feasibility of a radio navigation system based on the conformance to the requirement of high accuracy for the ground user, offshore user, and air and space user was demonstrated 1 year after the launch into orbit of the first satellite цикаαа used by a non-dynamic user. After the first GLONASS (TLOHACC) satellite, namely

KOCMOC 1413, was launched on October 12, 1982, 65 satellites had been launched up until the middle of 1995. As of December 14, 1995, one rocket with three satellites was successfully launched by Russia that brought the total number of satellites in the GLONASS constellation to over 24. On January 18, 1990, the establishment of GLONASS was announced. The suggestion made in the ICAO 10th Airborne Navigation Conference was approved by the Russian government, which proposed to provide a standard accuracy channel of GLONASS to world aviation users and promised that the positioning accuracy would meet the parameters in Table 1.1. From the successful launch of the first GLONASS satellite to the establishment of the system, 73 GLONASS satellites in total had been launched, 67 of which were successful. Due to the conflict between short lifetime of satellite and long networking period, the satellites had been in orbit for 12 years. GLONASS promises not reduce accuracy.

A GLONASS satellite has a lifetime of at least 3 years. Since the year 2000, satellite lifetime has been anticipated to be 5 years.

GLONASS adopts the frequency division (FDMA) identification system. Each satellite uses a fixed frequency bandwidth. The purpose of the design is to enhance jamming (intentional interference) rejection capability. However, due to the small frequency space of each satellite, which is only 0.5625 MH_2, the resistance ability to modern bandwidth disturbance is very limited, and burdens the design of receivers more to provide a different reference frequency to each satellite. Since L_1 frequency of GLONASS uses the radio astronomy operating frequency band, it has been suggested by the World Radio Conference this must change, which also makes the design of GLONASS satellite frequency more difficult. The only solution is sharing one frequency between two satellites corresponding to Earth, which undoubtedly will cause difficulty in the identification of low elevation of satellites. Is FDMA suitable for the navigation system? Many people disagree with this viewpoint. The Galileo System in Europe is still using a CDMA system. GLONASS frequency assignment is listed in Table 1.2.

It was decided in WRC-87 in 1987 and WRC-92 in 1992 that 1610–1626 MHz was assigned to the satellite mobile mss ground to spatial service. By 1998, GLONASS had conceded 1610.6–1613.8 MHz to radio astronomy. Therefore, from 1998 to 2005,

Table 1.1 GLONASS accuracy properties.

Parameters	Measurement Accuracy	
	GPS (P = 0.95)	**GLONASS (P = 0.997)**
Horizontal Plane, m	100 (72/18) (C/A Code) 300 (P = 0.9999) (C/A Code) 18 (P Code. Y Code.)	60 (CT Code) (39)
Vertical Plane, m	<200 (C/A Code) 20 (P Code. Y Code)	75 (CT Code) (39)
Velocity, cm/s	<200 (C/A Code) 20 (P Code. Y Code)	15 (CT Code)
Accelerated Velocity, cm/s^2	8 (C/A Code) <19 (C/A Code)	–
Time, μs	0.34 (C/A Code) 0.18 (C/A Code)	1 (CT Code)

Table 1.2 GLONASS frequency assignment.

Channel No.	Frequency (MH_2) L_1	L_2
00	1602.0	1246.0
01	1602.5625	1246.4375
02	1603.125	1246.875
03	1603.6825	1247.3125
04	1604.25	1247.75
05	1604.8125	1248.1875
06	1605.325	1248.625
07	1605.9325	1249.0625
08	1606.5	1249.5
09	1607.0625	1249.9375
10	1607.625	1250.375
11	1608.1875	1250.8125
12	1608. 75	1251.25
13	1609.3125	1251.6875
14	1609.875	1252.125
15	1610.4375	1252.5625
16	1611.0	1253.0
17	1614.5625	1253.4375
18	1612.125	1253.875
19	1612.6875	1254.3125
20	1613.25	1254.75
21	1613.8125	1255.1875
22	1614.375	1255.625
23	1614.9375	1256.0625
24	1615.5	1256.5

GLONASS satellites only used Channels 0–12 (1602.0–1608.25, 1246.0–1251.25 MHz). Using Channel 13 is an exception. After the year 2005, GLONASS-M satellite transmitted a channel frequency of K = −7–4, the high-end frequency of which is 1604.25 + 5.11 MHz = 1069.36 MHz. At the same time, the GLONASS-M satellite radiated navigation signals of frequencies L_1 and L_2 for civilian users, in which radiant power of an L_2 civilian user is 12 W. Antenna gain along the direction of axis of a satellite pointing to the ground is 8.8 dB, and corresponding angle of GLONASS relative to this axis is $\pm 15^{\circ}$, 11 dB, $\pm 19^{\circ}$, 9 dB. The civilian ranging code of L2 is the pseudorandom sequence of maximal length the same as L1. The polynomial is $1 + x^5 + x^7$ and the period is 1 m; pseudo-code velocity is 5.11 Mbps. In the message, representation of the difference $\Delta\tau_n$ of L1 and L2 frequencies satellite time delay equipment is added. When L2 signal is lagging behind L1 signal, $\Delta\tau_n > 0$. When L1 signal is lagging behind L2 signal, then $\Delta\tau_n < 0$, and the deflection $\delta_{\tau_n} < 2 \times 10^{-9} s$.

Besides $\Delta\tau_n$ in the navigation message of GLONASS-M satellite, n parameters for improving the user positioning reliability were introduced.

The difference between representation system master clock time scale transmitted by a GLONASS satellite and time UTC (SU) time scale is τ_c. Satellite ranging code phase position is subject to the system master clock, while UTC (SU) is the Russian time frequency standard, and ephemeris of navigation satellite adopts UTC (SU) for calculation.

In order to fit in the astronomical time, UTC (SU) time scale can be corrected once or twice a year, with 1 s for each correction. The correction can be performed at 00:00 of the night from December 31 to January 1, March 31 to April 1, June 30 to July 1, and September 30 to October 1.

The correction features of UTC (SU) time scale introduced in GLONASS-M satellite navigation message are:

10 – Pending time correction
00 – No correction
01 – Correction with +1 second
11 – Correction with −1 second

Under the condition that UTC (SU) time scale is prepared for correction, this information shall be introduced into the navigation message not less than 2 months before correction.

The features of the satellite remodel are prepared to be introduced into the navigation message:

00 – GLONASS Satellite
01 – GLONASS-M Satellite

In a GLONASS-M satellite, the time scale difference between GPS and GLONSS planned to transmit is $\tau_{GPS-GLN}$. Phase position of the ranging code is subject to the time scale of GPS and GLONASS. The maximum numerical range transmitted by this parameter is $\pm 1.9 \times 10^{-3} s$.

GLONASS will contribute to improve the long-term stability of the satellite clock. The GLONASS-M satellite will be installed with a new type of cesium clock, which will improve maximum day-and-night instability from $5 \times 10^{-13} s$ to 1×10^{-13}. The GLONASS-M satellite will adopt two methods to improve reliability and integrity: Method 1: Autonomous monitoring of basic functions is continuously performed on the satellite. When a phenomenon that affects the quality of navigation quality is found, sign Bn, which automatically connects in the downlink navigation message, means that this satellite cannot be used for navigation. Method 2: Monitoring network is setup by the ground control system to inspect the navigation signal. When the navigation signal is found to possess deviation affecting positioning quality, sign Cn introduced in the navigation message means that this satellite cannot be used for navigation.

1.4 Development and Future of the Chinese Navigation Satellite System

Originating from RDSS, the Chinese satellite navigation system has a weakness in continuous and autonomous navigation, but it was full of advancements – integration of positioning and positioning reporting, which means the integration of navigation and communication was accomplished; and the first generation of the Compass Navigation Positioning System, which was built in 2000. From 2000 to 2004, the three-step development planning of the China Compass was completed. The second generation of the satellite navigation system was launched in 2004. On December 27, 2012, Phase 1 of the Compass Second Generation Project covering China and some regions of Asia-Pacific was formally announced to start running, which completed the creation and perfection of the positioning reporting system of China Compass so as to provide RDSS and RNSS services to the user. The third step of development planning was launched in 2011. It aims to complete the global Compass system around 2020 to provide RDSS and RNSS services to users.

The Compass Navigation System, built in 2012, had navigation signals of three frequencies for the ionospheric calibration and improvement in carrier phase service efficiency. At the same time, it also provides calibration parameters in grid and model representations of ionosphere propagation delay of the user of single frequency, which contains eight model parameters of the ionosphere, α_n and β_n, 8 bits each, 64 bits in total; see to Table 1.3. (Refer to Table 1.4 for the ionosphere grid.)

Table 1.3 Calibration parameters of the ionosphere of the compass system.

	Parameter of Ionosphere		
Parameter	Bit	Quantization Unit	Unit
α_0	8	2^{-30}	s
α_1	8	2^{-27}	s/π
α_2	8	2^{-24}	s/π^2
α_3	8	2^{-24}	s/π^3
β_0	8	2^{11}	s
β_1	8	2^{16}	s/π
β_2	8	2^{16}	s/π^2
β_3	8	2^{16}	s/π^3

Note: Parameters marked "1" are the complement of "2." The most significant bit (MSB) is sign bit.

Table 1.4 Ionosphere parameter error tab information arrangement.

Parameter	IGP	$\Delta\tau_i$	GIVE1
Bit	8	9	4

IGP is the grid point number.

$\Delta\tau_i$ is the vertical delay of ionosphere of grid point.

The system distance is 0.125 m.

GIVE1 is the ionosphere vertical delay correction error tab of this grid point.

Table 1.5 Conversion parameters of compass time and GPS time.

Parameter	Bit	Unit
A_0GPS	14[1]	0.1 ns
A_1GPS	16[1]	0.1 ns/s

Note: Parameters marked "1" are the complements of "2." The MSB is sign bit.

Conversion formula:

$t_{GPS} = (t_E - \Delta t_{GPS})$

Wherein $\Delta t_{GPS}=A_{0GPS}+A_{1GPS}\times t_E$

t_E is the BDT of user's calculation

Table 1.6 Conversion parameters of compass time and Galileo time.

Parameter	Bit	Unit
$A_0Galileo$	14[1]	0.1 ns
$A_1Galileo$	16[1]	0.1 ns/s

Note: Parameters marked "1" are the "2." The MSB is a sign bit.

Conversion formula:

$t_{Galileo} = t_E - \Delta t_{Galileo}$

Wherein $\Delta t_{Galileo}=A_{0Galileo}+A_{1Galileo}\times t_E$

t_E is the BDT of user's calculation

Satellite broadcasts the integrity sign of self-test and the ground test to improve the reliability and integrity of positioning.

In order to meet the needs of multi-system compatibility, it successively broadcasts the time conversion parameters of GPS, GLONASS, Galileo systems (see Table 1.5 and Table 1.6), satellite integrity, and precision positioning clock correction parameter as well as satellite position correction parameters (see Table 1.7).

For time synchronization and satellite orbit determination of Compass Navigation System, the unique method for missions accomplished in the station setup in the home country alone that was mentioned previously is considered, which creates conditions for regional high-accuracy navigational positioning.

The Compass Navigation System is an independent satellite navigation system boasting advantages such as application compatibility with the advanced satellite navigation system, which will become a component of international GLONASS.

1.5 Galileo Navigation Satellite System

The Galileo Satellite Navigation System has been in planning for a long time. From March of 2005, it officially learned to use the (27+3) MEO constellation instead of

Table 1.7 Satellite clock errors, ephemeris correction parameter, and GPS error information.

Parameter	T	PRN	Δx	Δy	Δz	A_0	A_1	$\Delta\dot{x}$	$\Delta\dot{y}$	$\Delta\dot{z}$	EPREI	IOD
Bit	9	6	11	11	11	13	8	8	8	8	4	8

Note: T is the corresponding time of the parameters, expressed with integral minute of the same day in units of 3 min; and PRN is the satellite number of GPS

$\Delta\chi$, Δy, Δz are the correction of broadcast ephemeris of the satellite in unit of 0.25 m.

$\Delta\dot{x}$, $\Delta\dot{y}$, $\Delta\dot{z}$ are the change rates of $\Delta\chi$,Δy,Δz in unit of 0.0025 m/s.

A_0, A_1 are the slow varying correction of satellite clock error. A_0 is the clock offset in unit of 0.5 ns; A_1 is the clock rate in unit of 0.005 ns/s. Clock error correction is relative to the satellite clock error correction in GPS navigation message. Broadcasting value of the navigation message shall be added in user's calculation.

IOD is the date of data issuance. EPREI is the state parameter pointer of equivalent range error of the satellite ephemeris correction. (Omitted).

the GEO satellite constellation. A specific frequency plan was announced officially to the consultative country of the frequency (see Section 9.4.1.) The first navigation experimental satellite, Giove A, was launched on December 28, 2005. Giove B, another experimental satellite, was planned for launch in 2006. Before 2011, another two satellites would be launched adopting at least four work stars to perform the experiment in orbit. And then the remaining 26 satellites would be launched to form the full constellation.

Galileo measurement and control adopts the S frequency band. Uplink injection is L-frequency band of 1300~1350 MHz. Signal monitoring is in the C frequency band of 5000~5010 MHz.

1.6 Indian Navigation Satellite System

The Indian government officially approved implementation of the Indian Regional Navigation Satellite System (IRNSS) on May 9, 2006. A space segment consisted of 7–8 satellites including three GEO satellites and four IGSO satellites with an inclination angle of 29°. It was changed to six IGSO satellites in 2012. Therefore, it consisted of nine satellites in total. In September of 2007, Naill, an official of the Indian Space Research Organization said at an International Spaceflight Conference that 16 billion Rupees would be spent on the setup of the IRNSS system. A navigation satellite would transmit navigation signal of three frequencies. One is BPSK (1) with the center frequency of 1575.42 MHz. The other one is BOC (5, 2) with the center frequency of 1176.45 MHz, and BPSK (1). The third one is the signal of S frequency band with the center frequency of 2491.005 MHz, EIRP = 35.5 dBw. The plan was to finish deployment of seven satellites in 2012 with service regions of 35°E–130°E, 45°S–45°N.

At the same time, India implemented the GPS Aided GEO Augmented Navigation based on the GPS satellite based augment, which is similar to US WAAS. The space segment includes GPS satellites and one GEO (INMARSAT-4F1) satellite. The GEO satellite adopts the C and L bands as the working frequencies. The two L band signals are the same as L1 and L5 of GPS. It is fully compatible with the EU's Galileo and the Japanese GPS Augmented System, MSAS, based on MTSAT.

1.7 Japanese Regional Navigation Satellite System

The Japanese Navigation System serves as a regional navigation satellite system (Quasi Zenith Satellite System) and satellite based augmentation system (multi-function satellite augmentation system). The Multi-Function Satellite Augmentation System is an augmentation system of GPS, which is similar to the US WAAS. Its satellite is the multi-purpose satellite (MTSAT) launched by Japan. Two GEO satellites are located at 140°E and 145°E covering the Asia-Pacific Area. It has possessed the initial service ability since 2007. Ground stations are located in Hawaii, USA; Canberra, Australia, and the Philippines, with the broadcasting signal covering the Asia-Pacific Area.

The QZSS (Quasi Zenith Satellite System) was primarily defined as a multi-mission satellite system providing cell phone communication and broadcasting service with an angle measurement (70°) and positioning service with a rotation angle of 30°. Once the GPS system signal is interrupted, it can still possess reliable navigation positioning and communication capabilities.

The space segment consists of three GEP satellites and four asymmetrical IGSO satellites. At least four satellites at the optimum location within 2500 km around the Japanese archipelago can be observed, among which is at least one has an elevation angle of 70°. This is to improve anti-occluding capability in cities among high rises and mountainous regions to maintain smooth communication, broadcasting, and position reporting.

The navigation signal transmitted by QZSS is completely compatible with GPS, including L1: 1575.42 MHz, L2: 1227.6 MHz, L5: 1176.45 MHz. The broadcasting service signal is 2605 MHz-2630MHz (S frequency band). Low velocity communication service signal frequency is 2170–2200MHz (S frequency band) and the LEX signal with a center frequency of 1278.75 MHz and a modulation system of BPSK (5) is compatible with Galileo E6CS. The minimum received power is −155.7 dBw. Centimeter-level surveying and mapping augmentation information is broadcast at a data rate of 2 kbps through the L-frequency band spiral array antenna with the Reed–Solomon error correcting code (255,233). The broadcasting parameter is ionospheric delay using a ruled surface to match actual measurement at each penetration point, which is vastly superior to the accuracy of Klobuchar model ionospheric correction.

2

Concept and Application Prospects of Satellite Positioning Reporting Engineering

2.1 Satellite Positioning Reporting Service

There are two ways of determining a satellite radio service user's location. One is the Radio Navigation Satellite System (RNSS), which is widely known. Positioning and parameter calculation of navigational speed and voyage are accomplished autonomously by the user through the radio satellite navigation signal received. The other is the Radio Determination Satellite System (RDSS). The user's position can be determined through the measured pseudorange as well as change rate of pseudorange by the user, or the user can be informed through the same system after the distance measurement and location calculation performed by external system. Mainly, the latter method aims to provide the location information of user and it's difficult to provide the navigation parameters such as velocity, course error, and prediction time to the destination. The engineering design and equipment manufacturing engineering of this satellite positioning service is named satellite positioning reporting engineering, a feature of which is to complete location reporting upon positioning. Refer to Table 2.1 for the performance of satellite positioning reporting engineering and satellite navigation engineering.

2.2 Type of Service and Frequency Assignment

Satellite positioning engineering adopts two GEO satellites to realize the accurate positioning of all kinds of mobile carrier or fixed carrier. It can also conveniently realize user's bidirectional data communication and precise time synchronization. According to classification method of radio satellite service of International Telecommunication Union (ITU), satellite positioning engineering can operate Radio Determination Satellite Service (RDSS), and Mobile Satellite Service (MSS). It also operates fixed satellite service for the feed link. It is always a tough mission to arrange the RDSS radio frequency spectrum range. The frequency range suitable for mobile user positioning is jammed. RDSS service, being as important as it is, is directly related to life-saving and rescue so that L and S frequency bands are the first choice. Scramble between RDSS frequency spectrum and MSS frequency spectrum is very intense. In 1985, based on the Geostar "Three-star" commercial concept created by O'Neill, the Federal Communication Commission officially affirmed RDSS technique standard based on the spread spectrum code division multiple access modulation and authorized three companies to build the RDSS

GNSS Systems and Engineering: The Chinese Beidou Navigation and Position Location Satellite, First Edition. Shusen Tan.

Table 2.1 Comparison between satellite positioning reporting engineering and satellite navigation engineering.

Name Performance	Satellite Positioning Reporting Engineering	Satellite Navigation Engineering
Principle	User receiver to determine the location of user non-autonomously (informed by MCC)	User receiver to determine the location and velocity autonomously.
Available satellite constellation	Geostationary orbit (GEO) satellite constellation	Geostationary orbit (GEO) satellite constellation, MEO satellite constellation, IGSO satellite constellation.
Service	Positioning, time service, location reporting, communication	Positioning, velocity measurement, time service
User to transmit signal response	Required	Not required
Observed quantity	Distance sum from user to control center system through satellite	Satellite-earth pseudorange, value of Doppler measurement
Complexity of satellite load	Simple	Complicated
Coverage performance	Region + Region ...	Regional and global integrated design
User dynamic adaptability	Medium and low dynamic user single one-time service	Low, medium and high dynamic continuous service
Application	Positioning, location reporting, communication, rescue	Navigation, weapon guidance

Table 2.2 US 1618 MHz frequency spectrum management[a)].

Governmental Frequency Assignment (MHz)	Non-Governmental Frequency Assignment (MHz)
1610~1626.5 Aviation radio navigation	1610~1626.5 Aviation radio navigation

a) 1610~1626.5 MHz frequency is assigned to ground-to-air of RDSS at the same time.

system. Official redistribution 2 of RDSS spectrum is introduced in Tables 2.2, 2.3, and 2.4, respectively.

Under the arrangement of Federal Council, an approach to affirm RDSS service is found. However, it is not in conformance with international radio frequency assignment. Therefore, the effort of fighting for frequency assignment has been made by the International Telecommunication Union (ITU). Through coordination with the ITU, CCIR, IMO, and ICAO, it endeavored to be listed in the agenda of WARC MOB-87 held in

Table 2.3 US 2492 MHz frequency spectrum management[a)].

Governmental Frequency Assignment (MHz)	Non-Governmental Frequency Assignment (MHz)	User designated by FCC	
		Planning Portion	Unused Frequency
2450~2483.5	2450~2483.5 Fixed mobile radio positioning	Auxiliary broadcasting (74) Private operation-fixed (94) Private operation-mobile (90)	2450 ± 50 MHz Industrial frequency, scientific frequency and medical frequency
2483.5~2500	2483.5~2500 Radio positioning reporting (air-to-ground)	Satellite communication (25)	

a) Auxiliary broadcasting services and private radio stations with a license on July 25, 1985 could keep operating on the basis that they are mainly used on RDSS.

Table 2.4 US 5100 MHz frequency spectrum management[a)].

Governmental Frequency Assignment (MHz)	Non-Governmental Frequency Assignment (MHz)	User designated by FCC	
		Planning Portion	Dedicated Frequency (MHz)
5000~5250 Aviation radio navigation	5000~5250 Aviation radio navigation	Aviation (87)	

a) 5117~5183 MHz sub-band is assigned to fixed satellite service air-to-ground transmission, which is connected with the RDSS operated in frequency band of 1610~1626.5 MHz and 2483.5~2500 MHz for use. All of the total power flux density reached to the earth surface shall not be greater than −159 $dBW/m^2/4\,kHz$.

Geneva in October 1987. In July 1985, before the conference, the Executive Committee of ITU decided formally to list the assignment of RDSS frequency into the conference agenda. And in early 1986, some countries formulated unanimous standards for RDSS on the CCIR working team conference. The RDSS frequency spectrum was approved in 1987.

Frequency bands used by RDSS include:

Central station to satellite	6525~6541.5 MHz
Satellite to central station	5150~5200 MHz
Satellite to user	2483.5~2500 MHz
User to satellite	1610~1626.5 MHz

Re-assignment of S and L frequency bands used for RDSS happened for two reasons. One is the GeoStar Company running the RDSS service in USA was declared bankrupt due to management incompetence. The other important reason is the worldwide common voice of developing a synchronous orbit satellite covering a large communication network region. In order to reduce the expense of user terminal, from February to March of 1992, WARC-92, which was responsible for assigning the dedicated frequency band, was convened in Torremolinos, Spain. Mobile satellite service assigned again the L frequency band of 1610~1626.5 MHz (ground-to-air) and S frequency band of 2483.5~2500 MHz (air-to-ground) as the global main service, and RDSS was included in the secondary service. Due to the efforts made by Chinese government, countries in the third zone such as China and 28 other countries have the main service of RDSS in S frequency band of 2483.5~2500 MHz. This means that air-to-ground S frequency band has obtained the status as the main RDSS service of our country. In 2012, it was confirmed again by WARC-2012 that this frequency band was shared by radio navigation satellite frequency and MSS service. However, ground-to-air L frequency band of 1610~1626.5 MHz still took RDSS as secondary service. The radio station of secondary service was not supposed to disturb the designated main service. In this frequency band, the main services included ground-to-air service of MSS, aviation radio navigation, radio astronomy, and so on. Therefore, as for the techniques of satellite network frequency application, MSS is also be applied for while RDSS service is applied to make the coordinated position superior to the satellite network system applied afterward. This complies with the actual condition of RDSS system combining MSS service. Now, the global systems, Globalstar, Odyssey, and Iriaum have these frequency band MSS services. In the implementation of satellite positioning engineering design coordination with the system mentioned here should be accomplished, interference analysis should be performed, and measures of interference elimination should be taken accordingly.

The C extending frequency band determined in 1987 is excellent. The ionosphere propagation error due to the feed links from MCC to satellite and satellite to MCC is extremely small. Due to the lack of a low cost general purpose device, another C or Ku frequency band was adopted, which needs to be brought into the satellite fixed service for coordination. Due to the geosynchronous earth orbit, satellites became very jammed; it was very difficult to coordinate frequencies in the mature frequency band of devices. Because of this, attention needs to be paid in the engineering design stage and a detailed design for interference elimination should be provided.

Although worldwide demand of mobile communication increases greatly, the frequency resource that could be assigned was really inadequate. Relevant organizations of ITU proposed to assign RDSS air-to-ground frequency of 2483.5~2500 MHz and ground-to-air frequency of 1610~1626.5 MHz to the global mobile communication service. In the WARC of 2007, the Chinese delegation made the suggestion that the governments of the countries in the third zone could autonomously decide to use the frequency band mentioned previously in order to protect the achievement of satellite positioning engineering by uniting all available forces. Due to the convenience of device manufacturing and excellent performance index of the device, the struggle for the frequency in the frequency band mentioned here will continue. Mutual interference of multi-system sharing must be solved. The experience of the China Compass using the S frequency band attracted attention from the global navigation industry. In the World Radio Conference 2007, 22 countries, including Germany and Italy, proposed to extend the S frequency band to 2483.5~2500 MHz to a global navigation frequency band, and China expressed its support. WARC-2012 confirmed it as global navigation frequency to provide support for the RDSS service.

2.3 System Interference Analysis and Strategy

Since the frequency bands of 2483.5~2500 MHz and 1610~1626.5 MHz were assigned to be shared by multi-service systems, interference analysis in engineering design became especially important. Since MSS network systems adopted multiple systems such as CDMA, FDMA, and so on, interference analysis became more complicated. Performance deterioration of L and S frequency bands incurred at the receiving demodulation terminal as well as the impact of the system capacity should be specifically analyzed according to the system designs.

2.3.1 L Frequency Band Interference Analysis

Ground-to-air L frequency of 1610~1626.5 MHz inbound link interference not only directly affects the MCC's acquisition, tracking, demodulation, and distance measurement, but also impacts on systematic user capacity. The acquisition ability, accuracy of measurement, demodulation error rate, and systematic user capacity should be reviewed, respectively, according to the factors such as network number, EIRP transmitted by user, user's working frequency, user capacity, modulation mode, and commonly used frequency bandwidth of MSS/RDSS/RNSS in the same airspace. Besides the analysis of interference from other systems, the impact on the system to be coordinated from this system should also be assessed (because bilateral harmful interference should be eliminated for the frequency coordination). To be specific, the mutual interference analysis of Chinese system L frequency band should include GlobalStar, Odyssey, Iriaum, and other systems in Asia. Multiple systems running in the same region cannot work together without coordination. Both parties can only work together under the premise that both parties are taking technology strategy and assigning corresponding service type and user capacity.

Analysis shows that it is beneficial to both the system itself and other systems to properly design the transmitting power of the user to reach a threshold receiving level of MCC according to the receiving power of the user.

The second measure of reducing L-frequency band inbound interference is for the user's receiver to adopt a beam forming scheme. Destination of the signal transmitted in the L-frequency band is determined according to direction of satellite signal in the S-frequency band; by adopting the beam forming scheme, the transmitting antenna beam of main beam pointing to the destination is generated to reduce the possibility of exposing itself.

2.3.2 S Frequency Band Interference Analysis

Air-to-ground S-frequency band of 2483.5~2500 MHz outbound link interference directly affects the performance of acquisition, tracking, and demodulation of the user's received signal. Interference analysis should be performed according to the omni-directional EIRP modulation mode of the satellite systems running at the same time, and RDSS system scheme design indexes that are proposed to be used such as information transmission rate, spread-spectrum code rate, length, and so on to determine the numbers of simultaneous working systems and the strategies. Since multiple systems are often working in the same region, space division of satellite signal beam cannot be adopted. However, it is suitable to use the nulling antenna for the receiving antenna system at the user terminal. The index design of the nulling antenna should be on the basis of meeting the receiving distance measurement of two satellites in the S frequency band signal of the RDSS system, namely, generating two received beams with significant transmission gain, to ensure the receipt of the signal of two satellites and measure the time difference of reaching the user between the two satellite signals so that the beam pointing and power level of transmitting single can be determined according to the minimum receiving power to make the mutual interference of the users minimal. The indexes of an appropriate receiving system nulling antenna are as follows:

Number of received beams: 2
Received beam forming gain: >3.0 dBi

From the analysis here, it can be concluded that it would be a better choice to use multiple beams nulling antenna techniques for the receiving and transmitting antenna of a user's RDSS system device in order to solve the mutual interference and man-made interference of multiple systems.

Certainly, the cost of user receiver will be increased by using multiple beam nulling antenna. Therefore, the development of the multiple beam nulling antenna with low cost is the key to make the measures mentioned previously more widely used.

2.4 Service Optimization of Satellite Positioning Reporting Engineering

Satellite positioning reporting engineering is an extremely costly aspect of engineering for satellite application, which contains the extremely valuable position of geosynchronous orbit and the satellite application frequency bands of L, S, C, or Ku. It is enormously wasteful to only design one service that will not reduce cost to the user. From the angle of system integration, satellite positioning reporting can realize the integration of multiple application system in the system-level and integration of user terminals most easily.

2.4.1 Integration of RDSS with RNSS and MSS

The integration of RDSS and MSS means that Radio Determination Satellite System reporting service can be accomplished in the same channel as the satellite mobile communication service with the same function module of user positioning equipment and communication equipment. The measures that meet the positioning demand of thousands of the random user are also suitable for random access of user's communication to make the communication response of short message finished within one second (under smooth outbound queue).

Due to the high integration of positioning and communication, the positioning and location reporting are also integrated in the system and realized at the same user terminal, which makes positioning and location reporting easier, more economic and more competitive in the market than the satellite navigation system and communication system; meanwhile, position field monitoring is acquired. The RDSS system could be used for the field monitoring in logistics and arduous and hazardous area to provide full real-time understanding of the scene and representation of the failure to the command and dispatch organizations.

2.4.2 Integration of RDSS and WAAS

Since the application of GPS, a wide area augmentation system has been proposed on the basis of GPS in order to improve the performance of user's positioning accuracy and integrity. Its purpose is to not only provide the wide area difference correction number and integrity information but also add an enhancement satellite the same as GPS satellite for the user to improve the user's positioning accuracy and get more constellation selection. In order to meet the requirement mentioned previously, generally, an additional satellite will be launched in the GEO. As for the regions and countries that have already established RDSS, these functions can be accomplished by adopting GEO satellite of RDSS with the shared data link with RDSS. When RDSS satellite carries radio frequency load of same working frequency as GPS satellite, then RDSS satellite becomes the veritable enhancement satellite of GPS, the economy of which is self-evident. In particular, RDSS's bidirectional L/S-frequency band short message working system establishes the foundation for multi-layer interactive on-demand service between MCC and user. Through the higher data rate information of the S-frequency band, users of 1.0 m level could be provided with difference integrity information to realize precise RNSS navigation.

2.4.3 Integration of RDSS and TDRSS

Since the 1980s, many countries have competitively developed aerospace communication TT&C network with the core of data relay system, namely the Aerospace Communication TT&C Space-based Network. For instance, by utilizing TDRSS of two synchronous orbit GEO satellites, LEO and MEO space could be basically covered to realize the real-time observation and control of MEO and LEO satellites and aerospace system. Now that RDSS's GEO satellite possesses the service capability of MSS, observation and control of the near-earth satellite can be tested on the system. Space-based TT&C terminal of near-earth satellite is RDSS's satellite-borne user receiver. One RDSS system has the capability to support space-based TT&C of the near-earth satellite. The coverage area might be small (equivalent to RDSS coverage), but number of supportable satellite and reliability are impressive.

Meanwhile, due to the function requirement of RDSS and WAAS, the orbit determination accuracy of RDSS GEO satellite is much higher than that of relay satellites. The position error of orbit determination accuracy of WAAS loads for extrapolation of 12 hours is better than 15 m and radial direction is better than 1.5 m. The user's distance measurement data acquired through RDSS system is able to provide precise orbit determination resource for user satellite. At the same time, time comparison accuracy of 20 ns could be obtained through two-way time service of the RDSS.

However, it has its own weakness. As RDSS's goal is to serve users of MSS, its data relay capability is limited. Through integration design, the requirement of basic TT&C parameters transmission of spacecraft could be met. It has the potential to relay control instructions at certain important stages and image transmission of partial objectives. In order to resolve conflict between low data transmission rate and a large amount of information relayed about the RDSS system, a code division system could be adopted to divide the satellite resources of the inbound link to improve the inbound data transmission rate of the satellite-borne user to meet basic needs.

2.5 RDSS Application

Satellite positioning reporting system is the proper method to connect navigation and communication, and it is possible it may become the satellite radio short message communication system that is most widely used. In the fields of aviation, aerospace, navigation, and land transportation, the application prospect is wide. It is shown in primary application that the consumption ratio is satisfactory.

2.5.1 Aviation Application

Outstanding contribution of the aviation application is the air traffic control. When combined with pressure altimeter and RNSS, it can complete the missions such as navigation, monitoring, and communication required by air traffic control.

The combination of RDSS and pressure altimeter could get the location reporting of the aviation user to air traffic control center accomplished independently. Currently, the relative pressure altimeter accuracy is superior to 80 m and absolute accuracy is superior to 200 m at a flight height of 12000 m, which completely meets the requirement of flight path layering. By taking the positioning accuracy distribution of land service and sea service as an example, to the north of north latitude 18°, the positioning accuracy is superior to 500 m. Even in the Nansha offshore area, the accuracy is superior to 1200 m, which completely meets the requirement of route location accuracy.

After the combination of RDSS and RNSS, the user's positioning need could be met through RNSS service. RDSS could meet the need of location reporting and communication scheduling. The aviation user's need of management could be met through more convenient method.

Distribution of airplane in the airport area is often approximately 20% and around 20% in the area near the airport and 60% en route. According to the different stages of flight, the rate of positioning and location reporting rate could be investigated. According to conditions below, the interval of each airplane applying positioning or location reporting is 10 s. RNSS positioning or air route calculation could be used for the rest of the time.

1. In China, average flight time is 90 min;
2. Average cruise duration is 20 min, and interval of location reporting is 20 s;
3. Average climbing duration is 20 min, and interval of location reporting is 20 s;
4. Average descending time is 20 min, and interval of location reporting is different, 10 s for 600 m away from an airport and 5 s for 300 m away from an airport.

If one time of RDSS positioning or location reporting is performed according to the time interval of 10 s, six services are applied for each airplane every minute, 360 services applied per hour. Even if 500 airplanes require aviation management service at the same time, during the flight, 1.8×10^4 times the services are applied per hour, and this requirement can easily be met.

The execution of this flight plan requires thorough and close coordination as well as the authorized coordinating agency, which uniformly controls the user coming into the airspace and set up the technical management standard for compulsory execution. This is not only for the safety of voyagers but also for national security.

According to this analysis, RDSS is a complete system of communication, navigation and surveillance (CNS). It can not only accomplish the surveillance of the ground command staff to location of airplane in the air, but also provide surveillance for the person on board to adjacent airplane position. The surveillance system of each airport only needs to be networked with MCC of RDSS. It is more economic, more accurate and safer than the surveillance system relying on the ground radar station altimeter and report of air crew. In particulars, the insufficient surveillance on the marine navigation due to the limited operating distance of radar could be compensated for.

If necessary, collisions could be avoided through providing a ATC situation map of 30~60 nautical miles to the pilot from RDSS ground station, which will show the location of the airplane near the center. This ATC situation map could also show the airport runway, navigation equipment, navigator fix, ATC coasting grade, and flight path of other flight user. The information could be provided through RDSS control center. Thus, RDSS has the potential to become the basic method of ATC and is possible to become the infrastructure of CNS.

RDSS has made reality as a lifesaving tool for the pilot. Urgent positioning button of user's terminal equipment is linked with the lifesaving appliance. During an aircraft accident, by activating the user's terminal to work, the ground center creates the falling track and the positioning accuracy is at least one order of magnitude higher than the regular Emergency Locator Transmitter (ELT). The positioning accuracy of the regular ELT is 10 miles.

2.5.2 Aerospace Application

In the aerospace field, it can provide precise time synchronization to the LEO aerospace user. As two-way time service of RDSS will not be affected basically by the dynamic performance of user and position coordinates accuracy, it only relates to two-way round-trip delay measurement accuracy from central control system via RDSS satellite to user. Therefore, RDSS is an excellent method for providing precise time synchronization to the dynamic user. If the round-trip delay from the central control system via satellite to the user is the same, it only relates to random error of delay measurements. However, the round-trip delay inequality only relates to the one between RDSS satellite outbound transponder and inbound transponder, the one between transmitting system and receiving system of central control system, the one between user's receiving and transmitting and receiving system, and the one between forward and reverse transmission of space ionosphere. Each part of the delay inequality after calibration can be controlled within 20 ns. Thus, such synchronization accuracy is superior to 50~100 ns. The interactive calculation and low accuracy of the complicated orbit of time synchronization performed by GPS for dynamic users are avoided.

The second application is to provide method of orbit determination for the near-earth satellite. When the user satellite enters into the beam of RDSS satellite, its observed quantity is the round-trip distance sum from central control system via RDSS satellite to users. When there are two RDSS satellites, there are two distance sums. Taking near-earth satellite Kepler plus harmonic terms as the model and utilizing the multiple times of long-time observation data, the orbit parameter of the near-earth satellite could be determined. Its accuracy is within hundreds of meters. Orbital relative accuracy of each near-earth satellite is superior to 100 m. It could be used as supplementary for aerospace small satellite orbit determination.

The third application is the transmission of the telemetering and telecontrol information for near-earth satellites. Through RDSS communication link, it could be used as a supplement for telecontrol command transmission.

2.5.3 Navigation Application

Navigation application realizes the mutual location reporting of ship formations to achieve harmonious formation configuration and sailing speed through the working principle of the RDSS commanding user receiver. A wider application is for marine rescue. RDSS location reporting could realize location reporting and scout monitoring of any ship and warship right through to a private canoe to achieve the purpose of fast rescue and even reporting injuries and diagnostic reports for long distance emergency treatment.

2.5.4 Land Transportation Application

The most attractive application is urban traffic information reporting. Through the positioning and location reporting functions of RDSS user receivers on vehicles plus the control mechanism of velocity response, traffic information is reported to the central control system when travel is hindered, and an alarm given to vehicles that are speeding.

Logistics management information reporting: Large logistics and haulage vehicles could report the dynamic conditions of journey time, motion, safe state, and integrity of goods through an RDSS user terminal for full digitalization of logistics management.

2.5.5 Hazardous and Difficult Site Monitoring

The China Compass Satellite Navigation System has realized automatic telemetry of hydrological information for the main sections of the Yangtze River, Yellow River, and Heilongjiang River. The hydrological change data are provided for flood prevention.

For dangerous areas with arduous living conditions and high risks, such as security monitoring and alarms on a petroleum transmission pipeline, the RDSS system is able to realize information reporting and simple security operation helping to avoid loss of life.

Through location reporting and monitoring of helicopters at the scene of fires, the situation map of a fire scene can be reported to the command center accurately and rapidly. The area of fire spread can be determined by the flight path recorded by the helicopter flying along the scene of the fire in order to provide support for high-speed decision making to prevent a forest fire spreading.

3

Principles of Satellite Positioning Reporting

3.1 Theory of Positioning Reporting

Satellite positioning reporting applies to non-autonomous or passive positioning. The distances from the satellite to the user are measured by the Measurement Control Center (MCC), substituted for the user receiver, and then the position is determined by MCC. Figure 3.1 shows the simplest layout for satellite positioning reporting, including one MCC, two GEO satellites, and a certain number of Calibration Stations (CS).

Intersection measurement positioning of three spheres is the fundamental principle of satellite positioning reporting. Since the distances are measured by MCC, and MCC can use a frequency standard above 1×10^{-14}, high accuracy time frequency is not required by the satellite and it can be one order of accuracy lower than MCC.

MCC transmits the time reference signal for inquiring through satellites S_1 and S_2. When the user receives the signal, the reply signal will be transmitted and returns to MCC through S_1 and S_2. The signal time delays returned from S_1 and S_2 are measured by MCC, respectively. Due to the known locations of S_1 and S_2 at every moment, during the process of data processing the distance between the user and the two satellites can be obtained considering the relative movement of S_1 and S_2 as well as the transmission delay of MCC and the transponders of S_1 and S_2, the transmission delay of the user receiver, and influence of the ionosphere and troposphere during the transmission of these signals; the coordinate position of the user can be calculated according to the geodetic elevation data of the user's point of presence (the solution of elevation data will be narrated in latter relevant chapters). In order to analyze and explain this better, Figure 3.1 changes to Figure 3.2.

The following formula shows the constituted basic observation quantity and mathematical model [3].

$$\begin{aligned} S_1 = {} & D_{c1}(t_1) + \mathrm{c}\,\delta t_{s1(t_1)} + D_{1u}(t_1) \\ & + \mathrm{c}\,\delta t_u(t_2) + d_{u1}(t_3) + \mathrm{c}\,\delta \mathrm{t}_{\mathrm{s1}}(t_3) + d_{c1}(t_3) + \delta \mathrm{t}_{\mathrm{c1O}} + \delta t_{\mathrm{c1I}} \end{aligned} \tag{3.1}$$

$$S_2 = D_{c1}(t_1) + \mathrm{c}\,\delta t_{s1(t_1)} + D_{1u}(t_1) + \mathrm{c}\,\delta t_u(t_2) + \mathrm{d}_{\mathrm{u2}}(\mathrm{t}_4) + \mathrm{c}\,\delta \mathrm{t}_{\mathrm{s2}}\mathrm{d}_{\mathrm{c2}}(\mathrm{t}_4) + \delta \mathrm{t}_{\mathrm{c2I}}(t_4) \tag{3.2}$$

Where:

t_1 represents the moment when satellite S_1 receives and transmits the inquiring signal of the ground MCC.

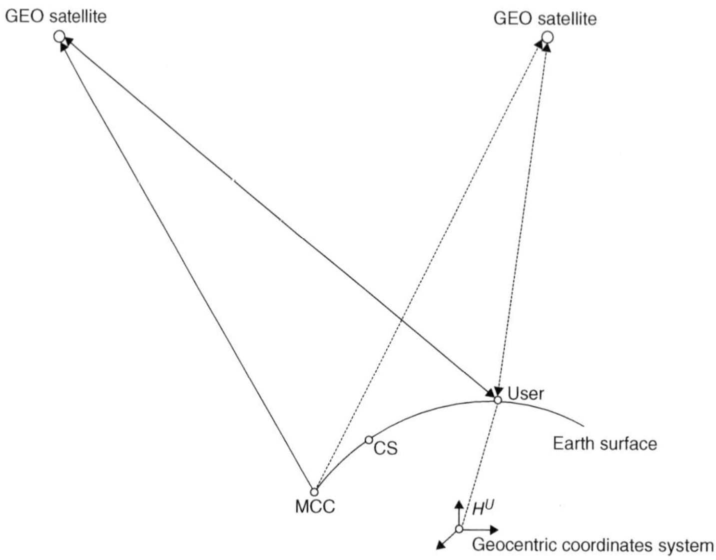

Figure 3.1 Principle of positioning.

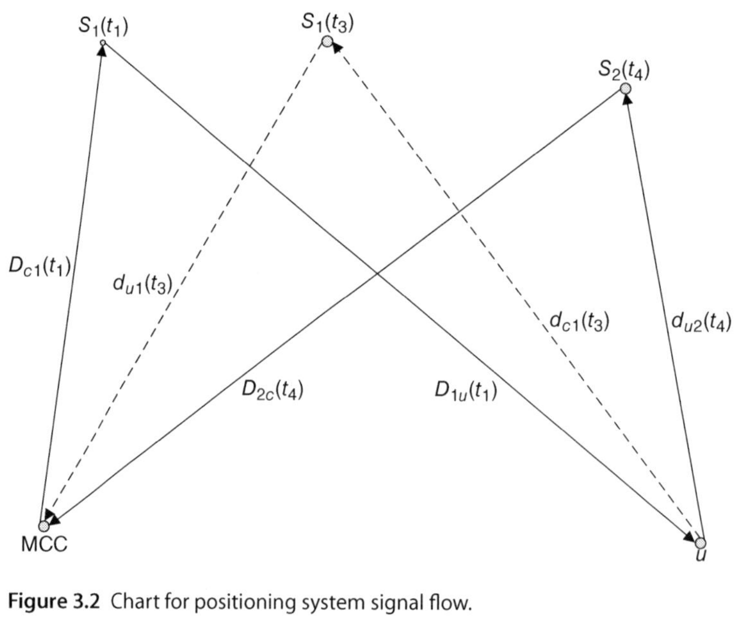

Figure 3.2 Chart for positioning system signal flow.

t_2 represents the moment when the user receiver receives the inquiring signal of satellite S_1.

t_3 represents the moment when satellite S_1 transmits the reply signal from the user.

t_4 represents the moment when satellite S_2 transmits the reply signal from the user.

$\delta t_{s1}(t_1)$ represents the equipment time delay of the outbound transponder of the satellite.

$\delta t_{s1}(t_3)$ represents the equipment time delay of the inbound transponder of satellite S_1.

$\delta t_{s2}(t_4)$ represents the equipment time delay of the inbound transponder of satellite S_2.

δt_u represents the time delay of the user receiver transmitting signal.
δt_{c1O} represents the outbound link equipment time delay from MCC to S_1.
δt_{c1I} represents the inbound link equipment time delay from MCC to S_1.
δt_{c2I} represents the inbound link equipment time delay from MCC to S_2.
c represents the light velocity.
D_{c1} represents the distance from the first satellite S_1 to MCC.
D_{1u} represents the distance from the first satellite S_1 to the user.
d_{u1} represents the distance returned from the user to the first satellite S_1.
d_{u2} represents the distance returned from the user to S_2.
d_{2c} represents the distance from S_2 to MCC when the user returns to MCC.

The time delay of signal in the equipment can be accurately measured, so the time difference between receiving and transmitting the signal is known. The transmitting moment of signal through satellite and users is at the hundreds-of-millisecond level considering the movement of satellite. In Figure 3.2, location of satellite S_1 has stretched relatively. Distances at all levels are represented in point coordinates as follows:

$$\begin{aligned}
d_{u1}(t_3) &= \sqrt{[X^{s1}(t_3)-X_u(t_2)]^2+[Y^{s1}(t_3)-Y_u(t_2)]^2+[Z^{s1}(t_3)-Z_u(t_2)]^2}\\
d_{1c}(t_3) &= \sqrt{[X^{s1}(t_3)-X_c]^2+[Y^{s1}(t_3)-Y_c]^2+[Z^{s1}(t_3)-Z_c]^2}\\
d_{u2}(t_4) &= \sqrt{[X^{s2}(t_4)-X_u(t_2)]^2+[Y^{s2}(t_4)-Y_u(t_2)]^2+[Z^{s2}(t_4)-Z_u(t_2)]^2}\\
d_{2c}(t_4) &= \sqrt{[X^{s2}(t_4)-X_c]^2+[Y^{s2}(t_4)-Y_c]^2+[Z^{s2}(t_4)-Z_c]^2}\\
D_{c1}(t_1) &= \sqrt{[X^{s1}(t_1)-X_c]^2+[Y^{s1}(t_1)-Y_c]^2+[Z^{s1}(t_1)-Z_c]^2}\\
D_{1u}(t_1) &= \sqrt{[X^{s1}(t_1)-X_u(t_2)]^2+[Y^{s1}(t_1)-Y_u(t_2)]^2+[Z^{s1}(t_1)-Z_u(t_2)]^2}
\end{aligned} \tag{3.3}$$

In the formula, the superscript integer represents the number of satellite, subscript $_c$ represents the MCC, and subscript $_u$ represents the user.

Time parameters for calculating the satellite location are isolated by the MCC according to outbound time scale and the sum of the measured distance.

The distance from the user receiver to the origin of the coordinates is as follows:

$$S_3 = \mathrm{r} + \mathrm{h}\cos\theta \tag{3.4}$$

In the formula, r represents the distance from the projection of the user receiver on the reference ellipsoid surface to the origin of coordinates; h represents the geodetic height of the point where the user is located; and θ represents the included angle between radius vector of the point where the user receiver is located and references ellipsoid normal. The relational expression is:

$$\mathrm{r} + \mathrm{h}\cos\theta = \sqrt{(X_\mathrm{u})^2+(Y_\mathrm{u})^2+(Z_\mathrm{u})^2} \tag{3.5}$$

Expression (3.5) shows the situation of the user responding to the inquiry signal of S_1 and transmitting the reply signal to two satellites.

Similarly, the expression of the user responding to the inquiring signal of S_2 and transmitting the reply signal to the two satellites is also given. It is only switching the mark number of S_1 with that of S_2.

In order to control the transmitting signal of users within a suitable level to meet the needs of MCC measurement and demodulation and to minimize the interference among users of CDMA system, the user can receive inquiring signals from the two satellites, conduct time difference measurement, and respond to inquiring signals from one of the satellites according to the lowest power. At this time, the ranging equation can only consist of one returned signal from one satellite. Similarly, the equation set for positioning can be restored: we will not go into step-by-step detail here. Expressions (3.1), (3.2), and (3.4) constitute the positioning solving equation.

After linearization based on (3.1), (3.2), and (3.4):

$$\begin{gathered} e_x^1(t_3)\delta_x + e_y^1(t_3)\delta_y + e_z^1(t_3)\delta_z + F[r^1(t_1), r^1(t_3), R_c, R_u^0, \delta_t^{s1}(t_1), t_u(t_2), \delta_t^{s1}(t_3)] - S_1 = 0 \\ e_x^2(t_4)\delta_x + e_y^2(t_4)\delta_y + e_z^2(t_4)\delta_z + F[r^2(t_1), r^2(t_4), R_c, R_u^0, \delta_t^{s2}(t_1), t_u(t_2), \delta_t^{s2}(t_4)] - S_2 = 0 \\ \cos L \cos B\delta_x + \sin L \sin B\delta_y + \sin B\delta_z + F[R_u^0] - S_3 = 0 \end{gathered} \tag{3.6}$$

Wherein:

$e_x^1(t_3)$ represents the direction cosine of S_1 satellite to the x-axis at moment t_3.
The rest e_y^1, e_z^1, e_y^2, e_y^2, e_y^2 may be deduced by analogy;
B represents the longitude of user receiver location;
L represents the latitude of user receiver location;
δ_t represents the propagation delay of equipment according to its superscript and subscript;
$F[C_1, C_2 \ldots \ldots C_n]$ represents the expression based on parameter C_i as the parametric variable.

According to (3.6), the user coordinates could be solved, which is simplified to:

$$\begin{gathered} e_x^1(t_3)\delta_x + e_y^1(t_3)\delta_y + e_z^1(t_3)\delta_z + 1 = 0 \\ e_x^2(t_4)\delta_x + e_y^2(t_4)\delta_y + e_z^2(t_4)\delta_z + 1 = 0 \end{gathered} \tag{3.7}$$

$$\mathrm{cosLcosB}\delta_x + \mathrm{sinLcosB} + \delta_y + \mathrm{sinB}\ \delta_z + 1 = 0$$

or

$$Ax + L = 0 \tag{3.8}$$

$$A = \begin{bmatrix} e_x^1 & e_y^1 & e_z^1 \\ e_x^2 & e_y^2 & e_z^2 \\ e_{ux} & e_{uy} & e_{uz} \end{bmatrix} \tag{3.9}$$

$$L = \begin{bmatrix} F_1 - S_1 \\ F_2 - S_2 \\ F_3 - S_3 \end{bmatrix} \tag{3.10}$$

$$X = \begin{bmatrix} \delta_x \\ \delta_y \\ \delta_z \end{bmatrix} \tag{3.11}$$

$$X = A^{-1}L \tag{3.12}$$

Coordinates of the user's location can be figured out through iterative calculation of formula (3.7).

3.2 Main Factors Affecting Positioning Accuracy

Considering the positioning process outlined in the previous section, and the geometrical relationship, the factors that affect positioning accuracy can be summarized as follows:

1. Accuracy of satellite location.
2. Accuracy of MCC time delay measurement.
3. Accuracy of system equipment time delay (including transmission delays of MCC, satellite transponder, and user equipment).
4. Accuracy of space propagation time delay correction (including time delay error of troposphere and ionosphere).
5. Relative location relationship between satellites and user (geometric figure).
6. User elevation accuracy that may be provided by system or user.
7. MCC location accuracy.

In this section, Points (2), (4), (5), and (6) will be discussed. The remaining error influence and the corresponding elimination method will be discussed specifically in Section 4.7.4 on dual-satellite wide area difference in Chapter 4.

3.3 Accuracy of MCC Time Delay Measurement

As is known from the principle of positioning, observation required for user positioning is accomplished by the MCC. Its observation qualities are the user-satellite distances of S_1, S_2 expressed in formulas (3.1) and (3.2). If time delay of the equipment on the transmission path is stable, its ranging random error only relates to the C/N_0 of the receiver input terminal, that is, design index of the measurement system, including the code rate of measurement signal, bandwidth of DLL delay phase-locked loop, signal accumulation time, or band width. Coherent despreading time delay measurement standard deviation (RMS) realized by a non-coherent delay phase-locked loop could approximate to:

$$\delta_{\mathrm{DLL}} = T_{\mathrm{ch}}\sqrt{\frac{B_{\mathrm{N}}}{2\left({}^{C}\!/_{N_0}\right)}\left[1+\frac{2}{T\left({}^{C}\!/_{N_0}\right)}\right]}\ (\mathrm{ns}) \tag{3.13}$$

Where:

T_{ch} represents the width of spreading code in ns, $T_{ch} = \frac{1}{F_{ch}}$, F_{ch} represents the spreading code rate.

B_N represents the loop bandwidth in Hz;

${}^{C}\!/_{N_0}$ represents the ratio of despreading signal carrier power and noise spectral density in dB/Hz;

T represents the coherent integration period in s.

${}^{C}\!/_{N_0}$ relates to satellite G/T value, transponder EIRP, system multi-user coherent interference, and noise performance and is an important index of RDSS system design. Under the condition that the RDSS system may possess a code rate of 8 Mbps, the minimum threshold of ${}^{C}\!/_{N_0}$ is best if higher than 48 dB/Hz. The system index should be designed based on this application target to control the distance measurement standard deviation of MCC within a 5~10 ns magnitude and match the elevation error provided by the system.

3.4 Space Propagation Time Delay Error

Space propagation time delay error includes the influence of additional propagation time delay caused by the troposphere and ionosphere. Hopfield and the improved model are usually adopted for propagation time delay correction of the troposphere.

$$\begin{cases} N = N_d + N_w \\ N_i(h) = N_{io}\left[\frac{H_i - h}{H_i}\right]^4 & \mathrm{h} \le H_i, \mathrm{i} = \mathrm{d, w} \\ N_i = 0 & \mathrm{h} > \mathrm{H_i}, i = d, w \end{cases} \tag{3.14}$$

$$\begin{aligned} H_{\mathrm{d}} &= 40136 + 148.72(T_k - 273.16)\ (m) \\ H_{\mathrm{w}} &= 11000(m) \end{aligned} \tag{3.15}$$

$N_i(h)$ represents the index of refraction at elevation of h in $N_a(h), N_w(h)$ dry and wet terms, respectively; N_0 represents the index of refraction of sea level.

Additional time delay due to atmospheric refraction adopts the following corrected formula 4 in the engineering application:

$$\begin{cases} \tau_T = \frac{1}{c}\left(\dfrac{K_d}{\sin{(E^2 + 6.25)}^{1/2}} + \dfrac{K_w}{\sin{(E^2 + 2.25)}^{1/2}}\right) \\ K_d = 155.2 \times 10^{-7}\dfrac{P_s}{T_s}(h_d - h_s) \\ K_w = 155.2 \times 10^{-7}\dfrac{4810}{T_s^2}e_s(h_w - h_s) \\ h_d = 40136 + 148.72(T_s - 273.16) \\ h_w = 11000(m) \end{cases} \tag{3.16}$$

In the formula, E, h_s, h_d, and h_w, respectively, represent elevating angle, height of observation station, and virtual height of dry and wet terms; T_s, P_s and e_s, respectively, represent air temperature, air pressure, and vapor pressure of the ground at the observation station.

Adopting formula (3.16) can generally correct at least 70% of the troposphere error; however, it is complicated, and the measurement equipment and parameters required can barely be realized in wide range and multiple users. Therefore, the two-satellite wide area difference method is often used for correction.

The influence of ionosphere is much stronger than troposphere. The height of ionosphere stretches from 60 to 1000 km. Typical ionosphere propagation time delay correction adopts the total electron content (TEC) on the electromagnetic wave propagation path for calculation, where TEC $= \int_s Nds$.

The additional time delay of radio wave propagation is

$$\delta_\tau = \pm 1.3436 \times 10^{-7}\frac{TEC}{f^2}(\mathrm{s}) \tag{3.17}$$

In the formula, f is the signal carrier frequency.

Apparently, TEC on the signal propagation path is complicated. Therefore, a new method of utilizing double frequency signal to correct its propagation time delay is created. As there have already been four frequencies in the RDSS system, which are S and L, the outbound frequency and inbound frequency, as well as upstream frequency and downstream frequency of satellite, adding a double frequency working system will inevitably cause additional costs to the system and user receivers. Refer to Section 4.7.4 in Chapter 4 for a more advanced method of space propagation time delay correction.

3.5 Geometric Figure and Positioning Accuracy

The geometric figure of the user and satellite has great impact on positioning accuracy, which is generally expressed in Geometric Dilution of Precision (GDOP). Besides GDOP, there are another four factors as follows.

1. PDOP: Position Dilution of Precision
2. HDOP: Horizontal Dilution of Precision
3. VDOP: Vertical Dilution of Precision
4. TDOP: Time Dilution of Precision

As for the RDSS system, the user's position coordinates can be calculated with iterative computations of formulas from (3.7) to (3.12). This geometric dilution of accuracy can be expressed, respectively, as:

$$A = \begin{pmatrix} \mathrm{e}_{\mathrm{x}}^{1} & \mathrm{e}_{\mathrm{y}}^{1} & \mathrm{e}_{\mathrm{z}}^{1} \\ \mathrm{e}_{\mathrm{x}}^{2} & \mathrm{e}_{\mathrm{y}}^{2} & \mathrm{e}_{\mathrm{z}}^{2} \\ \mathrm{cosLcosB} & \mathrm{sinLcosB} & \mathrm{sinB} \end{pmatrix}$$

Weight of the distance observation quantity S_1, S_2 through the satellite is defined as 1. Weight P_3 of the observation quantity through height determination S_3 is fixed according to the square of the inverse ratio of its distance observation error.

$$\mathrm{P} = \begin{pmatrix} 1 & 0 & 0 \\ 0 & 1 & 0 \\ 0 & 0 & P_3 \end{pmatrix}$$

$$\mathrm{Q} = (A^T PA)^{-1}$$

$$m_X = \sqrt{q_{11}}$$

$$m_Y = \sqrt{q_{22}}$$

$$m_Z = \sqrt{q_{33}}$$

$$\mathrm{PDOP} = \sqrt{q_{11} + q_{22} + q_{33}}$$

$$\mathrm{VDOP} = \sqrt{\vec{\mathrm{r}} \cdot \vec{\mathrm{q}} / |\vec{\mathrm{r}}\,|}$$

$$\vec{q} = \begin{pmatrix} q_{11} \\ q_{22} \\ q_{33} \end{pmatrix}$$

$$\mathrm{HDOP} = \sqrt{\mathrm{PDOP}^2 - \mathrm{VDOP}^2}$$

In the formula, q_{ii} is the diagonal element of matrix Q.

Thus, the relationship between positioning error and DOP can be expressed as:

$$\mathrm{m_P} = \mathrm{m_0 PDOP}$$
$$\mathrm{m_H} = \mathrm{m_0 HDOP}$$
$$\mathrm{m_V} = \mathrm{m_0 VDOP}$$

In the formula, m_0 is the standard error of unit weight, and its physical meaning is equivalent distance error, which is the equivalence of equipment distance measurement error and space propagation time delay error. Hypothetically, if the satellites are fixed at 80°E, 140° E above the Equator, the positioning accuracy covering the most area of the land and seas around China is better than 30 m, where it is better than 20 m to the north of northern latitude 28°. Nansha to the south of northern latitude 17° is around 40~100 m.

3.6 User Elevation and Positioning Accuracy

It can be deduced by solving formulas (3.1)–(3.3) through the user position that elevation accuracy directly influences positioning accuracy of the user. User elevation is provided via two means. One is the elevation produced by the MCC and stored in the ground elevation database. The other is the elevation provided by the user.

As for the elevation provided by the MCC elevation database, this is the data (DEM) generated through proper selection of calculations based on contour line data and a small amount of ground feature control point data. It is a lattice structure and the undulation shape of the ground surface is represented based on the elevation value of the cross points of the lattice within a certain area. The regular lattice of DEM adopts two structures based on a lattice of longitude and latitude as well as a Gauss planimetric rectangular coordinate system.

The digital elevation provided by the ground elevation database belongs to a normal height system. The elevation system is divided into an orthometric elevation system and normal height system according to physical characteristics. Orthometric elevation takes the geoidal surface as the elevation datum. The orthometric elevation of point B at any location on the ground is the distance from this point along a vertical line to the geoidal surface as shown in Figure 3.3.

Its expression is:

$$H_g^B = \frac{1}{g_m^B} \int g dh \tag{3.18}$$

In the formula, dh is the height difference measured from the zero point of elevation along the leveling line; g equals to the gravity at dh; g_m^B is the average gravity value from point B to the geoidal surface along the plumb direction of point B.

As g_m^B cannot be determined accurately, orthometric height cannot be obtained accurately, and only its approximate value is obtained.

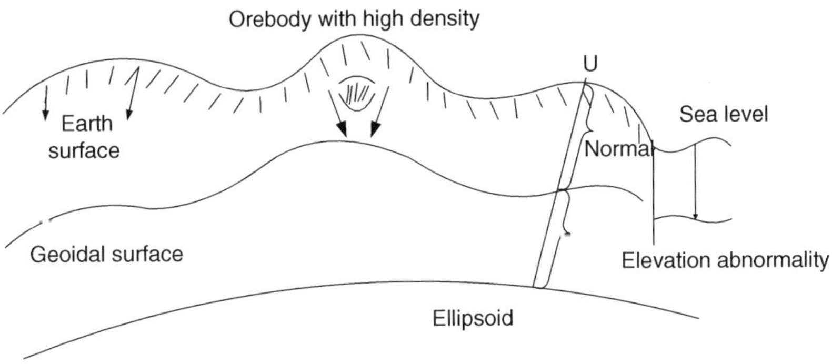

Figure 3.3 Schematic drawing of geodial surface.

Normal height system replaces g_m^B in formula (3.18) with normal gravity γ_m^B to obtain the elevation.

$$H_\gamma^B = \frac{1}{\gamma_m^B}\int gdh \tag{3.19}$$

As γ_m^B can be obtained through normal gravity, the elevation H_γ^B of the system can be obtained accurately. In China, the normal height system is used as the uniform system to calculate elevation. Normal height is not the distance from the point to the geoidal surface but the distance to a certain assumed curved surface, which is called the *quasigeoid*. The quasigeoid is identical to the geoidal surface at sea level, which has a slight difference to land. Therefore, the starting surfaces of the H_g and H_γ elevations are the same. The starting surface of elevation in China, namely the *elevation datum* – the *1956 National Elevation Datum* and *1985 National Elevation Datum*, both of which take Huanghai sea level obtained by the Qingdao Tide Station, China, as the uniform starting surface of national elevation. The difference between the two is:

$$H_{85} = H_{56} - 0.0286 \text{ m} \tag{3.20}$$

Quasigeoid is not superimposable over the ellipsoid of Earth. Therefore, in the actual application of satellite navigation and positioning, the concept of geodetic height is brought in, thus

$$H = H_\gamma + \xi \tag{3.21}$$

H is the geodetic height, which is the distance from the point to the ellipsoid.

ξ is the distance from quasigeoidal to ellipsoid, called the *elevation abnormality*. Elevation abnormality is often obtained via astronomical leveling or astronomical dynamic leveling.

The introduction of this concept used for navigation positioning solution through the elevation (normal height) provided by a digital topography base is the geodetic height, which is influenced by both normal height and elevation abnormality. Therefore, both digital elevation base and elevation abnormality base are stored in the MCC.

By taking the 1:250000 ground elevation database as an example and adopting longitude-latitude lattice structure, the latitude difference interval is 3 s, and longitude difference intervals are 3^{ε} and 6^{ε}. As for the 1:50000 ground elevation base, by adopting Gauss planimetric rectangular coordinates lattice, the north/south and east/west intervals are both 50 m to form a 50 × 50 m rectangular coordinate lattice. It can be seen that the elevation base of an RDSS coverage region is enormous. In order to realize engineering tracking and management, compression and decompression techniques must be adopted. Additionally, elevation accuracy provided by the ground elevation database is modeled and cannot fully describe the complicated topographic relief. Elevation accuracy is limited by the complexity of the ground surface. For the ground elevation digital base made according to a 1:50000 contour line with an interval of 10 m, the accuracy of a slow varying undulant plain and hilly area can reach 5 m. In the satellite position engineering, the equivalent distance error caused by elevation error is:

$$\Delta R_u = \Delta h \sin \beta \tag{3.22}$$

In the formula, Δh is the digital elevation diagram error; β is the elevation angle of observation satellite at point of user.

Near the Equator, ΔR_u is close to Δh. At mid-latitude, β is close to 60°; PDOP value is around 3; the positioning error caused by elevation data provided by the 1:50000 ground elevation base:

$$\begin{aligned}\Delta P &= PDOP \times \Delta h \sin \beta \\ &= 3 \times 5 \times \sin 60^{\circ} \\ &= 13\ (\mathrm{m})\end{aligned}$$

If the elevation is provided by the barometric altimeter carried by the user, the elevation of barometric altimeter and normal height belong to the same elevation system. The influence of its accuracy to the positioning is of the same nature as the influence of ground elevation base. Its exact accuracy varies according to correction accuracy of the barometric altimeter. Generally speaking, it is impossible to set up stations at known points of different heights. Therefore, it is difficult to calibrate the barometric altimeter.

If a radio (including laser) altimeter determines the elevation of the user, not only will the error of the radio altimeter be calculated for its accuracy, but the influence of the ground elevation database error will also be added.

4

Engineering Design of the Satellite Positioning Reporting System

A complete satellite positioning reporting system is designed according to the service type and positioning principles of the satellite positioning reporting system. This system is compatible with MSS, WAAS, and TDRSS. The satellite positioning reporting system with powerful functions also considers the relevant ITU service rules in terms of system level integration and service convergence to make it operate safely together with other satellite network systems.

4.1 System Composition

A satellite positioning reporting system consists of three subsystems: a space segment satellite system, a ground measurement and control system (MCC: also called the ground system for short), and a user system, as shown in Figure 4.1.

The space segment satellite system consists of two or more satellites operating in GEO. One basic RDSS system consists of two satellites. One to three satellites can be used as backup and to improve the coverage area as well as MSS and TDRSS service capabilities. The primary payload of satellite system includes a C/S (or Ku/S) transponder, L/C (or L/Ku) transponder, and corresponding frequency synthesis system.

The ground system includes at least one MCC and one CCS. The CCS consists of 20–30 correction stations (CS_1, CS_2, …… CS_n).

The user receiver system consists of various land, air, and sea user receivers and satellite-borne user receivers.

4.2 System Function Design

System function design fully considers user demand. An RDSS system can complete time service and communication function simultaneously based on compatibility of radio service. The more the integrated functions are, the higher the total cost performance ratio of the system. However, difficulty is caused in that the complexity of the satellite system increases. Optimization of businesses with the different natures mentioned previously is a key point of function design.

GNSS Systems and Engineering: The Chinese Beidou Navigation and Position Location Satellite, First Edition. Shusen Tan.

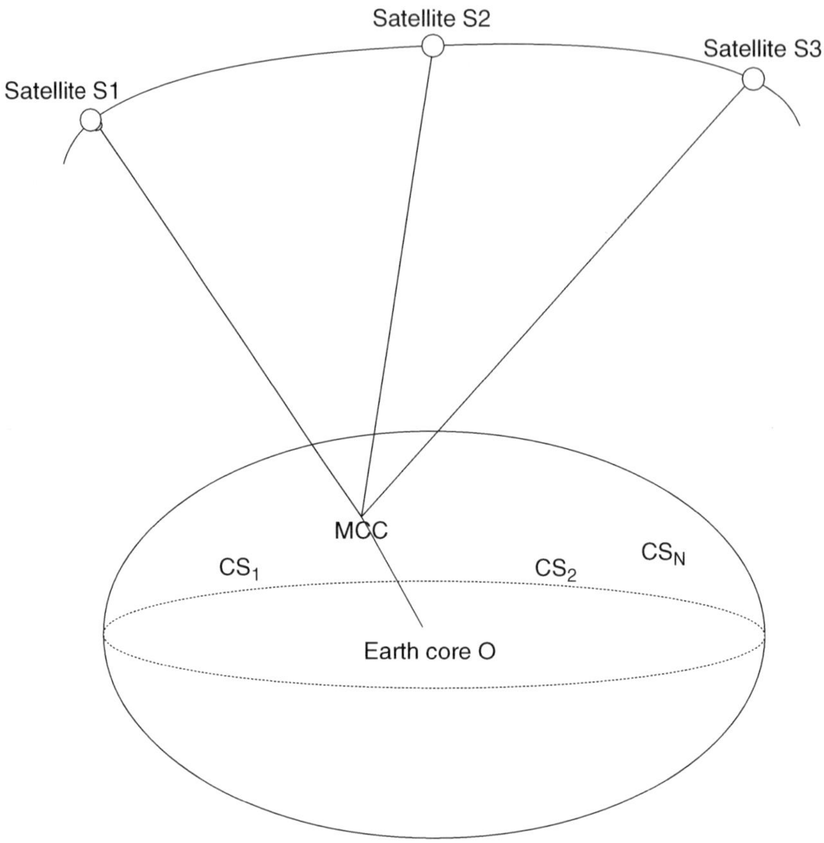

Figure 4.1 System composition.

4.2.1 Outbound Function Design

The so-called system outbound function refers to the capability of an outbound signal transmitted from the ground system MCC to the user through the satellite system. It includes the following functions:

1. *Continuous and Parallel Outbound Capability of Multi-Users*
   User quantity of an RDSS system is subject to the capacity of the outbound channel and working system of the system. Outbound capability is mainly manifested in:
   a. Sufficient effective transmitting power to make the user terminal easier to use and cheaper.
   b. Higher stability of transmitting signal frequency to facilitate the receipt and demodulation of the signal transmitted by a high dynamic user receiver.
2. *Continuously Tracking the Multi-User Signal System*
   Information transmission of many users, based on a time division system, is completed within one channel, which makes the signal tracking of one user difficult. Therefore, the outbound channel follows the QPSK modulation mode. The I sub-circuit provides public information and partial dedicated information transmission and the Q sub-circuit mainly completes the transmission of dedicated

information. It not only ensures that the user can continuously track the outbound signal of the MCC and improves ranging accuracy and success ratio of the MCC, but also completes information transmission of different users.

3. *Abundant Monitoring Function of System Integrity*
   The so-called system integrity refers to the warning capability that cannot be provided by the system according to design index operation. Such a warning distinguishes the degree of index reduction, response time of warning, false warning probability, and false dismissal probability. Integrity information in satellite positioning engineering includes positioning integrity, timing integrity, and communication integrity, from which orbit parameter integrity, ionosphere propagation delay correction integrity, and time information integrity are extracted. A user's demand for integrity aims to lower the application rick. The more accurate and prompt the integrity is, the lower the risk for the user is. Even if the real-time integrity warning cannot be accomplished, integrity monitoring is accomplished.
4. *Capability of Anti-Deception and Anti-Interference Provided for User*
   The outbound link provides the received signal for the user. However, according to stipulations of ITU, as for RDSS service, its power flux-density reaching the ground must not be greater than -139 dbw/m^2/4 kHz.

   The anti-interference capability of such a small satellite signal is very low and must be improved through system coding. RDSS adopts the CDMA system and the outbound signal adopts QPSK modulation. The coding and error correcting mode of the I sub-circuit fully considers its anti-interference capability. With proper code length, it can improve the spread spectrum augmentation to enhance the capability of anti-interference without influencing the first acquisition time of the signal and user capacity. Spread spectrum design of I sub-circuit has a code length of 255–1023 bit under the condition that code rate is not greater than 8.25 Mcps. The shorter the code length is, the weaker the capability of anti-interference is. However, the longer code length will influence the outbound capacity of the system as well as user's first acquisition time of the signal, which may reduce the responsiveness of the user. Practical application shows that the user's first acquisition time can be completed within one to two seconds by selecting from 255~1023 bit. In order to avoid the user from being deceived, it is necessary to implement information encryption.
5. *Anti-Attack Capability of Outbound Feed Link*
   The outbound feed link refers to the C(Ku) link between the ground system measurement control center to the satellites. The uplink feed signal from the ground, under the transponder system, will disturb the signal received by the user through transponder and further cause serious failure of the user's link S-frequency signal. The most effective method is that the satellite C (Ku) band MCC feed link adopts a spot beam that only covers the area of the MCC. Although the extended C (established at the WARCMoB-87 Conference in 1987) is the feed link frequency of RDSS, it is not suitable for this purpose due to the cost of extended C devices. In order to meet the anti-attack capability requirement of the feed link, the suitable measure is to implement space division of the wave beam. Then the Ku channel will have relativity big advantages. Feed receiving antenna of the same size have a ratio of wave velocity to width $\frac{Q_c}{Q_{ku}} = \frac{6.5GHz}{14GHz} = 0.46$.

6. *Service Area Covering Power*
   Satellite radio positioning reporting system service area is determined by the overlapping coverage between satellite outbound signal service area and inbound signal service area. If the outbound signal is of a spherical shape, its service area is the biggest. From the perspective of finishing the user position reporting function, this is the best option for selecting satellite RNSS navigation signals B1, B2, or B3. From the perspective of finishing users' short message communications, it is best to select the Bs navigation signal.

4.2.2 Inbound Function Design

The so-called system inbound function design refers to the capability of receiving the inbound signal from the user terminal to the ground system MCC through the satellite. It mainly includes the following functions:

1. *Receiving and Processing Capability of Multi-User Random Transient Burst Signal with a Low Signal to Noise Ratio*
   In order to reduce the volume, weight and power consumption of the user receiver, the responding transmitted signal of the user terminal possesses the features of low power consumption, transience, and burstiness. Carrier to noise ratio C/N_0 of the inbound signal processed by the user through satellite transmission can be reduced to 45~47 dBHz. The persistence length of signal is 30~100 ms. MCC is required to accomplish the acquisition, tracking, and demodulation of signal as well as distance measurement. The coding needs to be elaborately designed for the inbound signal to enable the whole channel to have excellent transmission characteristics, high quality factor G/T value of satellites, and strong capability of MCC signal processing.
2. *Capability of High Accuracy Distance Measurement*
   According to the requirements of RDSS positioning accuracy and two-way timing service accuracy, time delay measurement accuracy of MCC must be controlled within 5~10 ns.
3. *Anti-Attack Capability of Inbound Link*
   The satellite inbound link works at a high G/T value. The amplification coefficient of satellite channel is around 170 dB. It is relatively sensitive to the high-power attack of other systems. In a less serious case, the system cannot work normally. In a more serious case, the transponder may be damaged. Therefore, its destruction-resistant capability, namely, interference carrier to noise ratio is J/N_0 > 130 dBHz.
4. *Instant Large Data Volume Inbound Capability*
   In order to meet the TDRSS function, the system needs to possess the capability of transmitting and receiving large data volumes, which require the inbound link to work in a multi-signal system. The inbound signals with different power levels and different signal codes shall be processed properly.

4.2.3 System Processing Capability

The system possesses the capabilities of real-time processing positioning, communication, and timing service, mainly including:

1. *User Identity Authentication.*
   This includes the authentication of user's code and all registration information.

2. *Positioning Calculation and Positioning Accuracy Verification.*
3. *Providing the Elevation Calculation for the User.*
   This provides the user with elevation data searching for ground positioning without elevation. The height conversion for the user is completed with partial height data.
4. *Timing Calculation.*
   This provides the time synchronization calculation for the two-way timing service user.
5. *Information Exchange Processing.*
   Complete information exchange processing among users, including personal communications, group user information announcement and user positioning, and distribution of communication information according to administrative relationships.
6. *Generation and Broadcasting of System Integrity Information.*
   This includes warning and broadcast of user's information that cannot meet the index requirement of satellite system and MCC.
7. *User Billing*

4.3 System Technical Index Design

4.3.1 System Coverage Area

Design conditions of the RDSS coverage area accomplished by two satellites include:

1. *Orbital Inclusive Angle of Two GEO Satellites*
   In order to obtain excellent satellite-user geometry, the orbital includes that the angle of two working satellites at GEO needs to be 60° and is determined as per the coordination degree of the network of satellites in orbit.
2. *Observation Elevation Angle of the User to Satellite*
   In order to obtain excellent intervisibility and ranging accuracy, the observation elevation angle needs to be within 10~45°. An RDSS satellite does not just select a GEO satellite. It must also select an IGSO satellite. When necessary, an MEO satellite can be selected.
3. *Positioning Accuracy at Low Latitudes*
   When a GEO satellite is selected as an RDSS satellite, RDSS positioning accuracy is reduced when the latitude where the user is located decreases. The latitude of the user who obtains a positioning accuracy of 20 m is located north of 20°. The longitudinal range of the system's coverage area design is restrained by these conditions: 70~180°; latitude: northern latitude (southern latitude) 20~50°. The point position of the satellite is selected according to the service of the user domain.

4.3.2 System Capacity Design

System capacity design is subject to the following design conditions:

1. System outbound capacity design
2. System inbound capacity design

4.3.2.1 System Outbound Capacity Design

System outbound capacity refers to the communication broadcasting capability of the transmitting positioning result of every user calculated by the MCC to the user, which is proposed according to the total technical requirement of system and relates to complexity of system outbound link, the highest EIRP of each link of outbound signal, information rate, and length of information. It is expressed by number of users per hour.

The relational expression of outbound capacity with respect to C and outbound frame rate F, information number of every frame I_F, and outbound beam number N is as follows:

$$\mathrm{C} = 3600 \times F \times I_F \times N$$

Where F is the broadcasting number of positioning user completed in 1 s.

I_F is the information number of every frame, with code modulation to QPSK, $I_F = 2$

The number of outbound beam, N, is related to coverage mode of the satellite S frequency signal. Every satellite can commonly cover the whole area according to point beam. N is the total number of beams of the two satellites. In order to increase the system outbound capacity, a multi-point beam shall be adopted. N is subject to the relevant performance of system capacity and satellite. Currently, the RDSS of China is $N = 4$, which means there are two outbound beams for each satellite.

4.3.2.2 System Inbound Capacity Design

System inbound capacity refers to the capability of MCC to receive and process the inbound signal transmitted by the user in a unit period (1 h), which is determined by the following factors:

- $(\mathrm{EIRP})_u$ of the signal transmitted by user
- $(\mathrm{G/T})_L$ of the satellite receiving terminal
- Amplification coefficient of satellite inbound channel, A
- $(\mathrm{G/T})_M$ of MCC receiving terminal
- Allowable block rate of channel when MCC receives inbound signal, B
- MCC's reception channel configuration data demodulation terminal number, N

The inbound capacity of system shall be designed to meet the following conditions:

1. Inbound capacity should be greater than outbound capacity. It can be designed according to the inbound capacity to outbound capacity ratio of 1:1.3~1.5.
2. $\mathrm{C/N_0}$ value of MCC receiver data demodulation and ranging input port (or equivalent $\mathrm{C/N_0}$ of receiver input port) should meet requirements of ranging and data demodulation.
   a. *C/N_0 Required by Ranging*
   According to the analysis in Section 3.3, when ranging code rate $\mathrm{R_C} = 8$ Mbps and time delay measurement accuracy is 5~10 ns, the required $\mathrm{C/N_0} = 47$ dB·Hz.
   b. *C/N_0 Required by Data Demodulation*
   $\mathrm{C/N_0}$ required by data demodulation can be expressed in the following equation:

$$\mathrm{C/N_0 = E_b/N_0 + R_b + L - G\ (dB{\cdot}Hz)}$$

   In the formula: $\mathrm{E_b/N_0}$ is related to the modulation adopted by signal and error code $\mathrm{P_e}$ of data. When BPSK modulation is adopted and $\mathrm{P_e} = 1 \times 10^{-5}$, $\mathrm{E_b/N_0} =$ 9.6 dB

R_b: Data rate, it can select $R_b = 8$ kbps = 39 dBbps
L: Terminal despreading, modulation loss. Generally, L = 1~2 dB
G: Data coding processing augmentation. When a convolutional code with a beam length of 7 and efficiency of 0.5 is used for coding and demodulation terminal adopts Viterbi for decoding, G = 5 dB.
Therefore, as required by data demodulation, $C/N_0 = 44.6 \sim 45.6$ dB·Hz
Selection of $C/N_0 = 47$dB·Hz as the design parameter can meet the requirements of both.

In order to facilitate analysis and calculation, first assume there is no external interference for system and there is only one user working, at this time, the equivalent C/N_0 at the input port of the MCC receiver is expressed as follows:

$$\frac{1}{C/N_0} = \frac{1}{\left(C/N_0\right)_{up}} + \frac{1}{\left(C/N_0\right)_{down}} \tag{4.1}$$

Where:

$$\frac{1}{\left(C/N_0\right)_{up}} = (\mathrm{EIRP})_u - L_{C1} + (G/T)_L - K \quad (\mathrm{dB \cdot Hz})$$

$$\frac{1}{\left(C/N_0\right)_{down}} = (\mathrm{EIRP})_u - L_{C1} + A - L_{C2} + (G/T)_M - K \quad (\mathrm{dB \cdot Hz})$$

In the formula: L_{C1}, L_{C2} are the signal transmission losses from the user to the satellite and from the satellite to MCC, respectively.
K is Boltzmann's constant, K = −228.6 dBW/Hz
C of C/N_0 in the formula 4.1 is the user signal power of MCC receiver input port and determined according to the following formula:

$$C = (\mathrm{EIRP})_U - L_{C1} + A - L_{C2} + G_r \quad (\mathrm{dBW}) \tag{4.2}$$

In the formula, G_r is the MCC receiving antenna augmentation.
According to formulas (4.1) and (4.2), noise spectrum density N_0, equivalent to the MCC receiver input port, can be calculated.
Under the condition that there is no external interference for system and multiple users are working at the same time, MCC receiver receives and processes the inbound signal of the third user. Inbound signals of other users will be interfered with. At this time, signal to noise spectrum density ratio equivalent to the input port of the MCC receiver is $C/(N_0 + I_0)$ and determined according to the following formula:

$$\frac{1}{C/(N_0+I_0)} = \frac{1}{C/N_0} + \frac{1}{C/I_0} \tag{4.3}$$

In the formula, I_0 is the interference spectrum density of other user signals, which can be calculated using the following formula:

$$I_0 = \frac{4\Delta}{3T}\sum_{\substack{K=1\\K\neq i}}^{n} C_K/B_{IF} = \frac{4\Delta}{3T} C_{K\Sigma}/B_{IF} \tag{4.4}$$

In the formula, Δ is the spreading code element cycle; T is the data code element cycle; and $C_{K\Sigma}$ is the sum of power from other user's inbound signal to the input port of MCC receiver. When the signals transmitted by several users are almost the same, $C_{K\Sigma} = (n-1)C_i$. B_{IF} is the bandwidth of input narrow band filter of data demodulation element, which equals approximately twice as much data. For instance, if $R_b = 8$ Kbps, $B_{IF} \approx 16$ KHz.

Calculate formula (4.3) to determine value of I_0. Calculate formula (4.4) to determine n; namely, the number of user allowed by the system to work at the same time.

Ranging and number of data demodulation terminals, N, required to be configured can be calculated to meet the allowable blocking probability, B, when MCC receives the inbound signal.

As the user applies for inbounding at different time domains and space domains, assume the application is random and uniformly arranged on the territory according to a Poisson distribution state, the Erlang B-Equation of the system blocking rate described by communication theory can be used to calculate the blocking of a simultaneous inbound signal number:

$$B(N,A) = \frac{A^N/N!}{\sum_{i=0}^{N} A^i/i!} \tag{4.5}$$

In the formula, $B(N,A)$ is the blocking rate of channel. For example, $B(N,A) = 1\times 10^{-3}$.

N is the simultaneous inbound number

$A = \lambda t$, is the communication traffic, in which, t is the average length of inbound signal. For example, $t = 40$ ms. λ is the average signal number of input allowed per receiving channel in unit time (per hour), for example, average inbound capability.

According to formulas (4.4) and (4.5), the calculation of n and N shall meet $n \geq N$. Currently, the inbound receiving channel of RDSS in China is 2. Therefore, system inbound C is $C = 2\lambda$, and 2λ must meet the requirements of condition (1) mentioned previously.

4.3.3 System Positioning Accuracy Design

According to the concluding analysis in Chapter 3, the design conditions of system positioning accuracy are:

1. Distribution regularity of system PDOP along with the change of latitude: To the north of northern latitude of 20°, this is greater than 3; and the lower the latitude is, the more rapid the change is.
2. Accuracy that system ground digital elevation database can reach and the influence of this accuracy changing along with latitude reduction to the positioning accuracy. Using the 1:50000 contour line data elevation diagram, if the interval is 10 m, its model elevation accuracy can reach 5 m, generally, 10 m; the accuracy of using radio height measurement will match it. When its height measurement accuracy is 1–2 m, the error of the user's elevation is mainly subject to accuracy of digital elevation diagram; the final error can reach 5.4 m. Therefore, it is not economic in terms of engineering to improve the accuracy of the user's radio height measurement instrument without limit.

3. Accuracy of satellite position can be weakened by adopting differential technology.
4. After these design conditions are comprehensively considered, rational assignment of user's measurement accuracy is the key point in positioning accuracy design. According to the distribution principle of a user's positioning accuracy, comprehensively considering the precision of model of digital elevation diagram is 5 m, measurement error that matches this is 1.0~1.5 m, which is suitable. In other words, in formula (4.4), $^{\Delta T}/_{2} = 3 \sim 5$ ns. Because the distance sum from MCC to the user through the satellite and from the user to MCC through the satellite is adopted in positioning and two-way time services, the distribution of ranging accuracy adopts $^{\Delta T}/_{2}$. Therefore, EIRP, as well as total amplification coefficient of satellite, satellite inbound link EIRP, and MCC loop tracking parameters can be designed based on $^{C}/_{N_0}$ in formula (4.4).

Under RDSS service, satellite positioning engineering adopts a group of accuracy matching parameters under a wide area of differential technology:

- Ground digital elevation library elevation error 5 m.
- MCC measurement error 1.0 m.
- Satellite ephemeris difference residual 1.0 m.
- Propagation residual of ionosphere and so on 5.4 m.
- User's equivalent distance residual 7.5 m.
- When PDOP ≤ 3, the regional positioning error can be controlled within 22 m.

4.4 Signal System Design

There are two transmission signals in a system. One is the outbound signal from the MCC to the user through the satellite. The other one is an inbound signal from the user to the MCC through the satellite.

4.4.1 Outbound Signal Design

The outbound signal has two functions: one is providing the reference of time signal synchronous response for the user; the other is transmitting positioning information and messages from the user. System design of the outbound signal must follow the principles of both providing a continuous and stable tracking signal and ensuring sufficient communication capacity. In order to provide a continuous and stable outbound signal, there must be a public sub-circuit. In order to transmit dedicated information, there must be a dedicated channel. Therefore, QPSK modulation is suitable for the design of the outbound signal.

Considering compatibility with a general RNSS, the chip rate of the I sub-circuit and Q sub-circuit can be selected at 5.115 Mbps, and the information rate is the same. The I sub-circuit can select a Gold sequence with a code length of 1023 bits and the Q sub-circuit can select a longer PN sequence code, the initial phase of which can be provided by ID1 of the I sub-circuit so that the Q sub-circuit not only has stronger confidentiality but also fast acquisition performance. The format of the outbound signal adopting the QPSK modulation according to frame and super-frame is as shown in Figure 4.2.

The public segment can be used for broadcasting general information from other systems, such as differential correction information of the GPS wide area augmentation

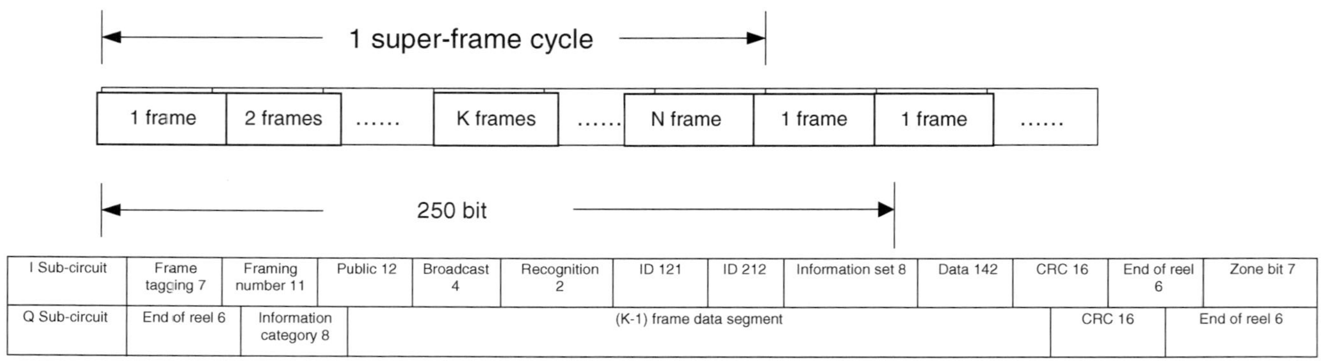

Figure 4.2 Format of the outbound signal.

	← Servicing segment →							
Synchronous head	Precision tracking	Frame tagging	User address	Framing number	Information category	Data	CRC	End of reel

Figure 4.3 Inbound signal format.

system. ID1, the address code of a Q sub-circuit user, shows that information of frame K in the Q sub-circuit is owned by an ID1 user. ID2, the address code of a I sub-circuit user, shows that information in the data segment of frame K in the I sub-circuit is owned by an ID2 user. The RDSS outbound signal working at the S frequency can be selected as $f_s = 10.23 \text{ MHz} \times 487 = 2491.005 \text{ MHz}$, which is 487 times the spreading code chip rate. Therefore, the outbound S band signal can be expressed as:

$$A_c\, C(t)\, D(t)\, \sin(2\pi f_s + \varphi_c) + A_P\, P(t)D(t)\cos(2\pi f_s + \varphi_p) \tag{4.7}$$

$C(t)$ is the fast acquisition code with a short code length.
A_c is the amplitude of a short code.
$P(t)$ is a long code with a long cycle and strong confidentiality.
A_p is the amplitude of a long code.
$D(t)$ is the data segment information, which can be encrypted according to one cipher per user or one cipher per time.

From formula (4.7), the outbound signal has the same confidentiality and security as the navigation signal of RNSS system

4.4.2 Inbound Signal Design

Inbound signal is in the formats of pseudo-code direct sequence spread spectrum, BPSK modulation and burst frame information, which is shown in Figure 4.3.

In the design, the synchronous head is special synchronous information designed for finalization of a fast acquisition burst signal, about 3 ms. The accuracy tracking segment is designed for realizing high accuracy ranging for MCC, approximately 6 ms. The user address is the recognition code of the user. The framing number is one of user response to the outbound signal, which is used for solving ambiguity of distance measurement. The data segment is communication information that is encrypted according to one encryption per user and one time per encryption. Synchronous head, servicing segment, and data segment adopt different coding spread spectra, respectively, and meet the requirement of signal acquisition, public servicing, and dedicated information transmission.

The inbound signal works in the L band with a center frequency of 1618.25 MHz and operating frequency of 1610–1626.5 MHz. Therefore, the highest bit rate is 8.25 Mbps.

4.5 System Frequency Design

According to the frequency assignment for aviation mobile service, the RDSS and MSS by all previous WARCs before the year 2000, frequency of 1610~1626.5 MHz was assigned as the uplink working frequency of ground-to-air. The S frequency of

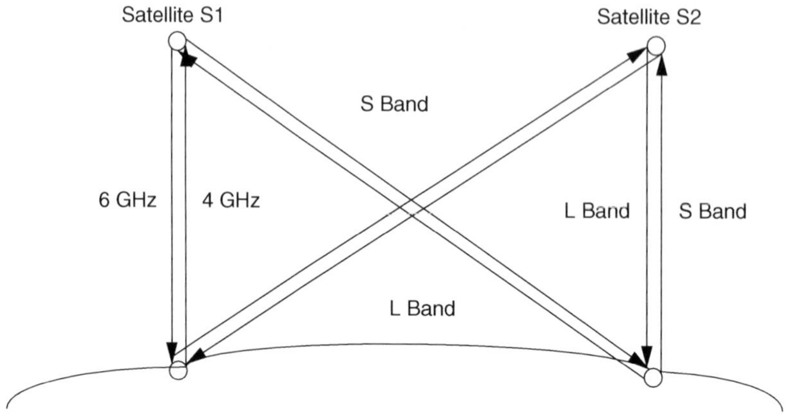

Figure 4.4 Link frequency assignment.

2483.5~2500 MHz was assigned as downlink working frequency of air-to-ground. However, uplink and downlink from MCC to satellite can select frequency C (6.0 GHz/4.0 GHz) or frequency Ku as the fixed satellite service. It is stipulated that the power density of total radiation of the users in frequency of 1610~1626.5 MHz at its synchronous orbit arc must not exceed −155 dBW$/m^2/$4 KHz in 0.01% of time and shall not exceed −158 dBW$/m^2/$4 KHz in 50% of that time. Outbound radiation of a satellite in frequency of 2483.5~2500 MHz must not allow the power regular density to exceed −154~ − 144 dBW$/m^2/$4 KHz. Typical frequency assignment is shown in Figure 4.4.

4.5.1 Influence of the Frequency Stability of a Transponder on System Performance

Inbound and outbound link frequency conversion is provided by the RDSS as a direct transmission type. If the local frequency (crystal oscillator) stability of responder on the satellite has an accuracy of 1×10^{-6} and the additional frequency deviation on the S band is ±3.5KHz, the tracking capability of the user receiver will be severely affected. Similarly, the L frequency signal transmitted by the user has the frequency deviation of ±2.4KHz through the L/C transponder, which severely affects receiving and tracking capabilities of the MCC. At the same time, the velocity measurement mechanism of the user is damaged. In order to overcome these defects, two ways can be adopted; one uses a high accuracy atomic clock as the standard of local frequency of the satellite; the other adopts satellite-ground frequency adjustment.

4.5.2 Satellite-Ground Frequency Adjustment

Adoption of satellite-ground frequency adjustment can not only overcome the problems brought by satellite local frequency scale stability, but also eliminate the Doppler frequency brought by movement of satellite relative to MCC at the same time. When the S-frequency signal of an RDSS system is used for the RNSS positioning service, only the Doppler frequency is generated by relative movement between user and satellite. When frequency adjustment accuracy of system on S band is above 1 Hz, its direct velocity

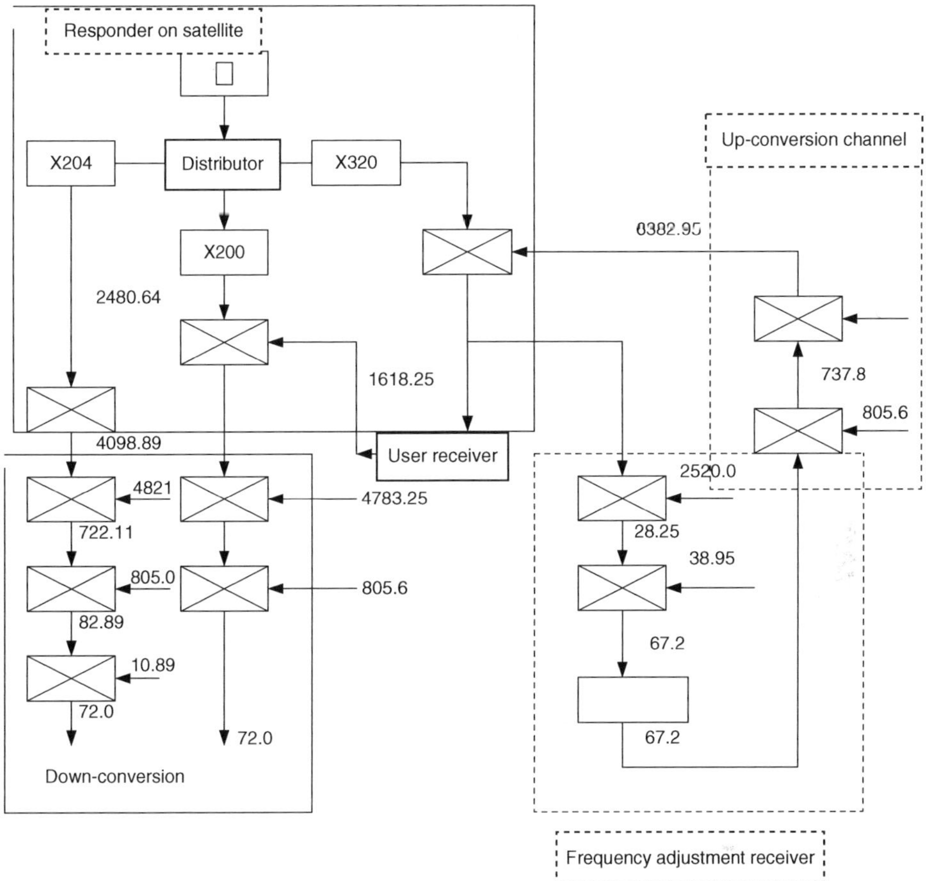

Figure 4.5 Satellite-ground frequency adjustment scheme flow diagram.

measurement accuracy is above 0.12 m/s. A typical scheme of frequency adjustment of two L band inbound signal beams is shown in Figure 4.5.

4.6 Engineering Design of Positioning Reporting Satellites

Positioning satellites are the multi-frequency transmission communication satellites operating in GEO. They include C/S (or Ku/S) transponders and L/C (or L/Ku) transponders. Compared to normal communication satellites, positioning satellites must have the following performance features.

4.6.1 Excellent Capability of Beam Coverage

Design of beam coverage must follow the following principles:

1. *Meet Autonomous Adjustment of Signal Power Transmitted by the User*
   Inbound link of the RDSS system allows multi-user random access. When the difference in user powers is great, the working performance of the system will be severely

influenced. A high-power user's signal restrains a low power user's signal, which sharply decreases the quantity of inbound users. The capability of distance measurement and demodulation of the system is reduced. Therefore, every user needs to have a common adjustment mechanism to transmitting signal to make the signal level received by the MCC the same critical level. The efficiency of system is highest. However, users are located at different positions. Antenna augmentation of user receivers is different along with the observation elevation angles. Coverage extension of vegetation to the user signals varies. The only basis for the user to adjust transmitting signal power is the satellite signal power received. The higher the signal power received, the stronger the restraint of transmitting signal power and vice versa. When a user can receive two satellite signals, a satellite signal path with weaker coverage can be selected according to the received different satellite signal power to meet the inbound requirement with a lower transmitting signal power. Based on this requirement, no matter whether it is S frequency outbound beam or L frequency inbound beam, both are required to cover the same region at the same time.

2. *Help to Improve System Outbound Capability*
   A user's outbound capacity is improved together with the increase in satellite outbound EIRP as well as the increase of outbound beam quantity. If the coverage area of the outbound beam is not controlled properly on a satellite, an excessive outbound beam will mean the user receives enhanced interference from adjacent beams. Therefore, during system design, the design must synthesize all kinds of factors.
3. *Help to Participate in Combination Operation of an RNSS System*
   In order to integrate joint positioning of RDSS and RNSS, the S beam signal of an RDSS satellite must participate in solution of RNSS. It is also expected that two S beams on GEO can realize the overlapping coverage of service area.

In conclusion, S and L beams of two operating satellites are allowed to form overlapping coverage on the area in order to reduce the interference from adjacent beam.

4.6.2 Design of EIRP and G/T Value

Positioning capability of RDSS is subject to communication theory. Link level and signal detection capability restrains user capacity of the system. The satellite payload design that meets the inbound and outbound capability is discussed here.

1. *Design of EIRP and User's Reception Level*
   - *Outbound Uplink*
     Links from MCC to satellite. One adopts the C frequency band satellite system to work with a frequency bandwidth of 16 MHz. The calculation of C/N_0, the quality of signal arriving at the satellite, is shown in Table 4.1.
   - *Outbound Downlink*
     EIRP of satellite shall be designed based on the minimum received power (sensitivity) that can be demodulated by the user. And then, according to EIRP and minimum received power of satellite, the outbound magnification times of satellite shall be designed.
     Satellite EIRP = minimum received power of user + path loss + atmospheric loss + system allowance
     A typical satellite downlink level is shown in Table 4.2:

Table 4.1 Channel Level with Bandwidth of 16 MHz in 6 GHz Frequency Band from MCC to Satellite.

Transmitting power of ground station	24.0 dBW (250 W)
Feed loss	−2.5 dB
Antenna augmentation of ground station	51.5 dB
EIRP of ground station	73.0 dB
Weather loss	−1.4 dB
Path loss	−200.2 dB
Received power of satellite (antenna of isotropy)	−128.6 dBW
Satellite G/T value	1.5 dB/K
Boltzmann constant, K	−(−228.6) dB
C/N_0	98.5 dB/Hz

Table 4.2 Typical 2491.75 MHz Frequency Signal Level from Satellite to User.

Satellite's EIRP	48 dBW
Free space path loss	−191.8 dB
Weather loss	0.7 dB
Received power of user receiver (antenna of isotropy)	−144.5 dBW
G/T value of user receiver	−24 dB/K
Boltzmann constant, K	−(−228.6) dB
Identical RDSS beam interference	−0.5 dB
C/N_0	59.6 dB/Hz
Data rate	16 Kb/s
E_b/N_0 received	17.6 dB
E_b/N_0 required by demodulation	6.6 dB
Total allowance.	11 dB

From Table 4.1 and Table 4.2, the outbound magnification times of satellite can be obtained:

$$\begin{aligned} \mathrm{K} &= \mathrm{EIRP} - \text{Satellite received power} \\ &= 42 - (-144.5) \\ &= 174\, dB \end{aligned}$$

2. *Design of Inbound Link Satellite G/T Value and MCC Received Level*
   According to the EIRP that can be transmitted by the user as well as the minimum received level MCC can demodulate, the G/T value of satellite inbound link is designed. It can be calculated via two steps.
   - Inbound Uplink
     According to the EIRP of selectable users, uplink level is calculated. A typical link level from user to satellite is shown in Table 4.3:

Table 4.3 Typical 1618.25 MHz Link Level with Bandwidth of 16.5 MHz from User to Satellite.

User's EIRP	18 dBW
Free space path loss	−188.35 dB
Weather loss	−0.5 dB
Single user's power received by satellite	−170.85 dBW
G/T value of satellite	−1.5 dB/K
Boltzmann constant, K	−(−228.6) dB
C/N_0	56.25 dB/Hz

Table 4.4 4 GHz Frequency Band 16.5 MHz Bandwidth Link Level from Satellite to MCC.

Single user's signal power received by satellite	−170.85 dBW
Total augmentation of satellite	174.85 dB
Single user's signal power on satellite.	4 dBW
Free space path loss	−196.42 dB
Weather loss	−1.5 dB
Received power of MCC antenna	−193.92 dBW
G/T value of MCC	32 dB/K
Received power of MCC	−161.92 dBW
Boltzmann constant, K	−(−228.6) dB
(C/N_0) down	66.68 dB/Hz
(C/N_0) total	56.64 dB/Hz

Ground MCC receives and detects the downlink signal transmitted by the satellite. The calculation of its level is shown in Table 4.4:

MCC design is able to detect C/N_0 of 47 dB/Hz. This system has an allowance of approximately 8.5 dB. Therefore, under general conditions when the user's EIRP 0 = 10 dBW, there is a satisfactory result for information receiving. It is feasible that satellite inbound uplink G/T value is −1.5 dB/K.

According to requirement of operation at the S and L frequency bands as well as the demand of EIRP and G/T value, one S/L shared antenna can be designed for a satellite with the extended aperture of 2.5~3.0 m.

4.7 MCC Engineering Design

Missions that are completed by MCC are as follows:

1. Signal transmission and signal modulation of system outbound link.
2. Signal reception and information demodulation of inbound link
3. Measurement of distance sum from MCC to user (calibrator) through satellite.

4. Satellite orbit determination and prediction.
5. Propagation time delay correction of ionosphere, and so on.
6. User's position solution, timing solution, and communication processing

4.7.1 MCC Outbound Link Design

The MCC outbound link subordinates to a fixed satellite service (FSS). The frequency bands in common use are C, X, and Ku. Uplink frequency bands are 5725~7075 MHz,7925~8425 MHz, and 14.0~14.5 GHz respectively. Downlink frequency bands are 3400~4200 MHz, 7250~7750 MHz, and 10.95~11.20 GHz respectively. When the Ku frequency band is selected, the influence of ionosphere propagation time delay between MCC and satellite can be omitted. Mature channel products can be selected according to the coordination difficulty of international frequency and economic feasibility. Its transmission terminal is synchronized with system time. Positioning or communication information as well as satellite orbit parameters are released according to the outbound signal format.

4.7.2 MCC Inbound Link Design

The inbound link completes the missions such as acquisition, tracking, demodulation, and high accuracy distance sum measurement for a burst user signal.

Fast signal acquisition is completed by the correlation accumulation of the synchronous head. The synchronous head consists of multiple segments of truncation *m* sequence codes. After correlation accumulation, many decisions are made by the false pseudo-correlation peak to the real signal that will need to be avoided. In order to increase the processing augmentation, the length of synchronous head can be properly increased. Certainly, this will be to the cost of increasing the inbound signal length. After weighing the advantages and disadvantages, the number of segments of code can be increased to 10 and above. The total cycle of synchronous head is approximately 3 ms so that the influence of Doppler Effect on accumulation of augmentation is not significant. At the same time, it is favorable to reduce the probability of acquiring false-warning so that acquisition probability is increased. The basic principle is shown in Figure 4.6. After segment correlation accumulation is implemented by N correlators to N segments of m sequence long code, N video accumulation is performed.

Accuracy tracking starts after fast acquisition finishes. At this time, pseudo-code accuracy tracking has been superior to half of a chip. It reaches the precision of $\leq 1/20$ chip through precise tracking at around 6 ms to complete a distance sum measurement that is superior to 10 ns.

Information demodulation is performed after accuracy tracking is met, the principle of which is the same as one of common continuous signal demodulation. In order to reduce the transmission signal power of the user and acquire low error rate, the constraint length of 7 and convolutional coding with coding efficiency of 0.5 are often adopted. The receiving terminal often adopts soft decision Viterbi decoding, which can acquire coding augmentation of 5 dB.

Measurement of distance sum adopts measurement of mantissa, short period start, and ambiguity judgment to shorten the measurement of propagation delay of radio

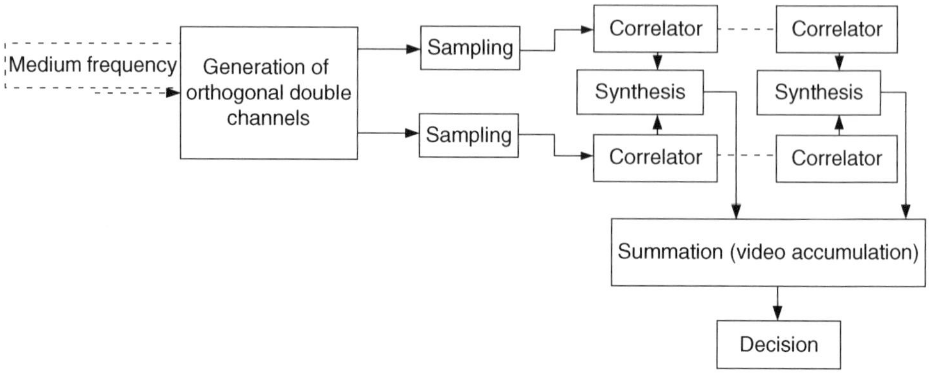

Figure 4.6 Fast acquisition principle.

waves by several tens of thousands km. The length of delay measurement can be controlled within 1 ms so the delay measurement capability of MCC is hastened, which is one of the measures to increase system user capacity.

4.7.3 Satellite Orbit Determination and Prediction

As RDSS adopts GEO satellites, its application features include frequent orbital maneuvers. East-west position holding must be performed every 15 days and south-north position holding every 30 days. During position holding, the satellite is influenced by force so determination of precise orbit is difficult. In order to achieve precise satellite position prediction, transient time after one orbit control can be as long as 12 h or more, which significantly reduces the availability of system. Therefore, RDSS satellite orbit determination will adopt a geometric method, least squares method, and Kalman filtering method to meet the requirements that orbit determination accuracy of orbit control, and so on at every phase is greater than 100 m to meet the demand of timing and positioning accuracy of the RDSS user.

RDSS satellite orbit determination can both realize single satellite orbit determination and double-satellite combined orbit determination. During orbit control, it is best at determining the orbit using a single satellite. The accuracy deterioration caused by orbit control of one satellite to another can be avoided. The so-called double-satellite (three satellites in total with one backup satellite) combination orbit determination is subject to the working system. The MCC measures the distance sum from one satellite to the orbit measuring station, which includes the distance to another satellite (see Figure 4.7). There are two observation paths from the MCC satellite to orbit determinator:

The first path constitutes single satellite orbit determination. The second path constitutes multiple-satellite orbit determination. Multiple-satellite orbit determination can utilize more measured data from orbit determinators. Single satellite orbit determination can only use the response data of an orbit measuring machine from the direct response beam. Therefore, in order to improve the accuracy of orbit determination, it would be best to use a multiple-satellite orbit determination during non-orbit control.

During orbit control, a geometric method can provide a high accuracy of orbit determination. In order to eliminate the influence of orbit error on positioning and timing, dual-satellite wide area differential technology is adopted.

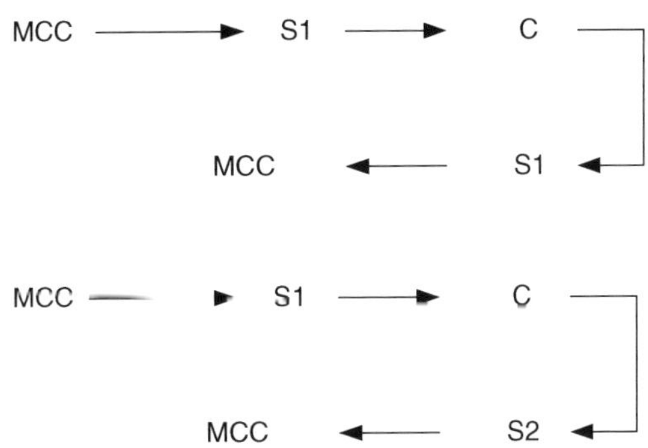

Figure 4.7 Combination of orbit determination distance and observation quantity.

In order to provide satellite position coordinates to the one-way timing user and tri-satellite autonomous positioning user, MCC broadcasts satellite position and velocity parameters X, Y, Z and $\dot{X}$, $\dot{Y}$, $\dot{Z}$ through the satellite.

When RDSS has an RNSS payload, its orbit determination technology becomes more abundant. Orbit determination and prediction can be performed according to the RNSS signal.

4.7.4 Dual-Satellite Wide Area Differential Processing

In the RDSS system, the concept of dual-satellite wide area differential technology is introduced, which is totally different to the concept of the RNSS satellite navigation system (such as GPS) wide area differential. It does not perform satellite clock correction or satellite orbit correction but implements correction for every positioning and timing service. The principle of the dual-satellite wide area differential is to perform distance correction of dual-satellite positioning and time through proper quantity of correction stations established in all service areas with the help of MCC, which obtains the total distance between user and satellite (user distance for short) through correction station to establish all areas based on longitude and latitude (L, B). Consequently, user positioning and timing correction is completed to achieve high accuracy positioning and timing.

The correction mechanism established according to the dual-satellite wide area differential helps to weaken the following error sources:

1. *Satellite Position Error*
   As the correction of user's distance improves the user's ranging accuracy, satellite position error is only reflected as the influence on DOP value. A slight change of satellite position has an extremely small influence on DOP value.
2. *Space Propagation Time Delay Error*
   The space propagation time delay error includes propagation time delay of troposphere and ionosphere. Space time delay error is corrected according to the known position of correction station, and the residual error is only related to the distribution and quantity of correction station. It is proved by the experiment that 20 correction stations can control the residual error of the user on land at a distance within 16 ns in the Chinese service area, for the user at sea, 26 ns. The tedious calculation of air humidity, temperature, and pressure on the propagation path of troposphere delay correction is omitted. The complicated and tedious test and correction of electron content of the ionosphere is also omitted.

3. *Equipment Time Delay Error*
   RDSS is the technology of actual distance positioning and timing, which is different from RNSS such as GPS, and so on that utilize pseudo-distance for positioning and timing. It measures precisely and deducts equipment time delays on paths of RDSS including MCC, satellite transponder, and user receiver. Since the system channel is too complicated to precisely measure, after the adoption of dual satellite wide area differential technology, all of the same parts of correcting path and positioning and timing path are totally offset. The equipment time delay functioning on the positioning and timing only includes the difference of time delays between calibrator and user receiver. Correction of both equipment time delays with one set of equipment can control system error within 1–2 ns so that the positioning and timing accuracies are greatly improved.
4. *Influence of Satellite Orbit Control and Attitude Control on Positioning and Timing*
   In two-way timing, the error of satellite orbit is corrected by additional time delay. In one-way timing and positioning, orbit error and space distance error are corrected together. Orbit control is generally completed within 1–2.5 min by controlling the impulse type air injection of the satellite motor. Once the external force stops, the satellite enters into a smooth and steady velocity transition. These corrected times can be shortened from 10 min to 1 min or even less so that the correction accuracy of space propagation time delay is even higher.

Additional time delay correction of orbit determination, positioning, and timing is simply elaborated as follows:

Additional time delay correction of two-way timing and orbit determination has nothing to do with orbit error. The model used is [4]:

$$\Delta\tau\ (\theta,\varphi,t) = \sum_{n}\left[A_n(L,B)\cos\frac{2n\pi(t-t_0)}{T} + B_n(L,B)\sin\frac{2n\pi(t-t_0)}{T}\right] \tag{4.8}$$

Wherein:

$$A_n(L,B) = \sum_{m_1,m_2}^{m_1+m_2\le M} a_{m_1,m_2}^{(n)}(L-L_0)^{m_1}(B-B_0)^{m_2}\cos^{m_1}(B-B_0)$$

$$B_n(L,B) = \sum_{m_1,m_2}^{m_1+m_2\le M} b_{m_1,m_2}^{(n)}(L-L_0)^{m_1}(B-B_0)^{m_2}\cos^{m_1}(B-B_0) \tag{4.9}$$

Being corrected together with additional time delay of ionosphere and troposphere. The model used is:

$$\Delta\tau(L,B,t) = C(L,B) + D(L,B)(t-t_0) \tag{4.10}$$

Where,

$$C(L,B) = \sum_{m_1,m_2}^{m_1+m_2\le M} c_{m_1,m_2}(L-L_0)^{m_1}(B-B_0)^{m_2}\cos^{m_1}(B-B_0) \tag{4.11}$$

$$D(L,B) = \sum_{m_1,m_2}^{m_1+m_2\le M} d_{m_1,m_2}(L-L_0)^{m_1}(B-B_0)^{m_2}\cos^{m_1}(B-B_0) \tag{4.12}$$

Expressions (4.8)–(4.12) constitute the time and space mathematical model of additional time delay, as for the user located at (z_0, B_0), if the correction of propagation is performed at t_0, then,

$$A_n(L_0, B_0) = a_{0,0}^{(n)} \tag{4.13}$$

Then the time delay correction of the user is:

$$\Delta\tau(L_0, B_0, t_0) = \sum_{n=0}^{N} a_{0,0}^{n} \tag{4.14}$$

Or

$$\Delta\tau(L_0, B_0, t_0) = c_{00} \tag{4.15}$$

In engineering practice, the MCC obtains the measurement data of all area calibrators first. Each calibrator takes sample of one point every 10 min, stores 24-h data and updates by new data overwriting old data. The additional time delay correction error from MCC to satellite is calculated through the calibrator on the MCC or through the MCC coarse positioning result of L, B for the positioning user. The corrected path value is used to perform positioning calculation soon afterwards. The two-way timing service is similar to the previous. The one-way timing service obtains the one way additional time delay on the path from satellite to every correction station through the MCC and extrapolates to the moment of time. Modeling is made with the geometric centers of all service areas to obtain the correction parameter of the radio wave. However, orbit determination correction takes the orbit measuring station as central to the model so the space time delay correction of orbit measuring station distance measurement within a certain time interval is obtained.

It is shown by system tests that positioning correction residual error of the whole service area is 12.63 ns. Residual error of one-way timing is 3.8 ns; that of two-way timing is 3.26 ns and that of orbit determination is 10 ns. Even during the orbit control, the requirement of positioning and timing accuracy can also be met.

After the RDSS satellite uses a higher frequency feeder link, such as spreading frequency C or Ku, the space distance between MCC and satellite has an extremely small ionosphere propagation error of signal frequency used. MCC distance sum measurement error is only related to the ionosphere residual error of the frequency used between satellite and the user, the correction method of which will be much easier. When positioning reporting outbound signal selects B1, B2, and B3, it is approximate to uplink inbound frequency. The simple arithmetic average frequency of both can be taken as equivalent frequency and corrected with ionosphere model.

4.7.5 MCC Service Processing

MCC service processing includes: positioning, timing (two-way and one-way), communication, and orbit determination.

The MCC receives the inbound application of calibrator and user receiver. Before these services are processed, identity authentication is performed to verify the validity of the user as well as the priority level and service type enjoyed.

The MCC adopts a multi-user parallel processing mechanism to complete the missions such as orbit determination, positioning, timing, and communication. The

MCC will still complete all the encryption and decryption of the outbound and inbound signals to improve confidentiality of the user. The MCC will also distribute all the communication and positioning results according to registration utilizing the ground transmission system to send to designated scheduling command system to complete position reporting, tracking, recognition, and warning.

Positioning is performed in two ways:

1. *Topography Library Positioning Method*
   The user does not carry any elevation aided information. The MCC will search the stored elevation database for positioning elevation required, which is suitable for fast positioning and position reporting of land users and sea/water level users, and also used for position reporting of bomb drop point to evaluate attack precision of a target.
2. *BYO Elevation Positioning Method*
   The user receiver carries land (water) elevation information in the inbound signal for positioning, which is suitable for positioning and position reporting of the user in the air. When applied with RNSS, in order to enhance the confidentiality of position reporting, RNSS autonomous positioning elevation information can be carried to complete position reporting.

4.8 RDSS Application Terminal Design

Since RDSS integrates positioning, timing service, and communication, its application system can combine the corresponding system according to different occasions, which completes the positioning, timing, and communication services of a single user and the position reporting, tracking, and monitoring of one another. For a user's convenience, the user receiver is divided into the single address and multi-address user receiver.

4.8.1 Single Address User Receiver

The single address user receiver is used for the commonly designated user. Basic functions of the single address user receiver include:

1. Capability of receiving outbound inquiry signals transmitted by multiple satellites and multiple beams. According to the frame tagging of inquiry signal, response signal from the user receiver is precisely transmitted to create the precondition for the MCC to perform a user's distance measurement.
2. Capability of demodulation of different designated address information to complete short message processing.
3. Capability of filling in the positioning (or communication) tag and information according to the format of the inbound signal.
4. Capability of combining the measurement of time difference and transmission power control of two satellites

Key technologies of common user receiver include:

1. Fast acquisition of satellite signal with an acquisition time ≤ 1 s.
2. Precisely tracking the outbound signal with the response accuracy of a transmitted signal superior to 10 ns.

3. Receiving and transmitting antenna low elevation angle augmentation to reduce the receiving difficulty and transmitting power of the user receiver.

The timing user receiver also has the timing processing capability.

4.8.2 Multi-Address User Receiver

The multi-address user receiver is a common user receiver with multiple receiving address codes that can be used for user carriers in scheduling command organization and flight formation. The number of its address code covers the dedicated address codes of the users within the jurisdiction or the address code of members of formation flight elements so that the MCC can notify the real-time position and coordinated and cooperative communication message of the user at one time to maintain a safe and well-organized flying pattern and coordinated actions.

4.8.3 RDSS Double-Model User Receiver

According to the satellite positioning reporting system, the Compass system has already integrated RNSS satellite payload into an RDSS link. The application terminal that implements RNSS continuous navigation positioning services and the single address user receiver that implements the PRSS service are integrated completely to constitute the RDSS double-model user receiver with an autonomous and continuous position and velocity services as well as the position reporting service to MCC as required. Under the cooperation of the multi-address user receiver, sharing individual user position information with a scheduling command organization is achieved.

5

Comprehensive Theory of RDSS and Engineering Design

5.1 Definition of CRDSS

CRDSS is short for Comprehensive Radio Determination Satellite Service. The CRDSS includes all the concepts of the International Telecommunication Union (ITU). The fundamental concept is to accomplish round-trip distance sum measurement from the Measurement Control Center (MCC) to the user through a transmitting RDSS satellite with a two-way round-trip ranging function. The difference in pseudorange between this satellite and other two navigation satellites is determined by the user. Through calculation of MCC, the position determination of the user and position reporting to MCC can be accomplished at the same time. Individual position service with a 70% coverage area on the land and offshore regions globally, as well as the position service with 85% coverage of near-earth satellites, can be accomplished through properly arranging two RDSS transmitting satellites on GEO. Compared with position reporting adopting GPS positioning and communication, its cost-efficiency ratio is increased remarkably. At least four satellites required by the traditional RNSS for positioning changes to three satellites or even two required for positioning. Under the same quantity of satellites of constellation, it is possible to select a lower DOP value to improve positioning accuracy. The greater advantage is reducing the cost to the user receiver and alleviating the burden on the user. Therefore, CRDSS is not a simple combination of satellite navigation and communication but an application mode that realizes higher precision, more flexible service, user's information sharing, and navigation system resource sharing. It highlights the application demand of the user and makes efforts to alleviate the burden on the user to avoid repeated setup of the system. It also enlarges the application scale and realizes the new idea and new scheme of satellite navigation industrialization. The spontaneous traditional service of satellite navigation is improved to an advanced stage of on-demand service, which is a significant technological effort not only for avoiding traditional low-level cut-throat competition and creating healthy competition based on service quality, but also for supporting independent innovation and developing a national industry of navigation. CRDSS is not abandoning the existing status of GNSS to establish a new system; instead, by taking China Compass as main body, it is exploring a new service mode and creating value-added services on the basis of existing several global navigation satellite systems.

GNSS Systems and Engineering: The Chinese Beidou Navigation and Position Location Satellite, First Edition. Shusen Tan.

5.2 Theory of CRDSS

5.2.1 Navigation Position Reporting Service

The typical RDSS position reporting service is introduced in Chapters 2, 3, and 4. In this chapter, the general CRDSS service, including navigation positioning, position reporting, and short text communication (navigation, position reporting service for short), will be introduced. The Compass satellite navigation system is a typical CRDSS system. The system composition is shown in Figure 5.1.

The system cannot only accomplish distance measurement, positioning processing, position reporting, and inter-user short message communication of users through MCC, but also accomplishes continuous navigation positioning of RNSS by the user. S^i_{CRDSS} is the CRDSS satellite that works in the orbit of GEO or IGSO. S^{i+1} is a RNSS service satellite that works in the orbit of GEO, IGSO, or MEO.

The CRDSS satellite possesses an uplink transparent transponder or a retexture transponder to MCC. At the same time, it also possesses a transparent transponder to the user u. The S^{i+n} satellite only has RNSS payloads. Regarding autonomous positioning principle of the user, this will be introduced into the RNSS service. In this chapter, only the position reporting service via tri- or multi-satellite is discussed.

The MCC obtains round-trip distance through S^i_{CRDSS}, S^{i+n} tri- or multi-satellite to the user u. Since positions of S^i_{CRDSS}, S^{i+n} are known, the distance from the tri-satellite to the user can be indicated, and Equation (5.1) can be used for position calculation of the user:

$$r^i_u = \left[\sum_{i=1}^{n} x^i_j - x_j\right]^{\frac{1}{2}} \tag{5.1}$$

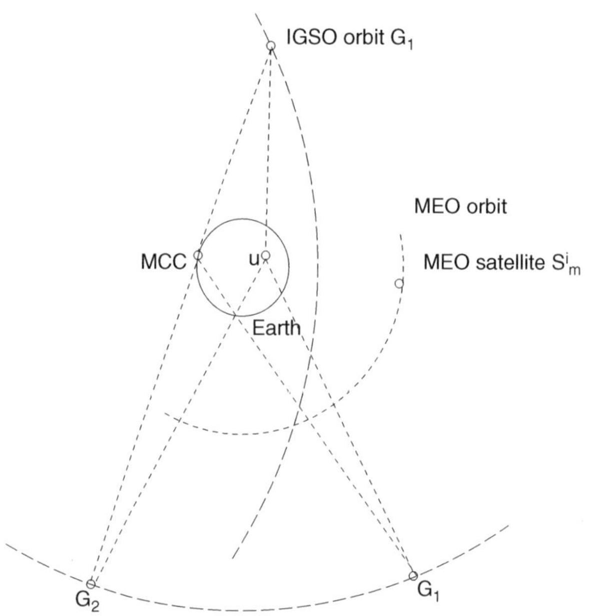

Figure 5.1 Selection of positioning satellite and principle of positioning of CRDSS.

r_u^i: Distance from Satellite i to the user, $i \geq 3$
x_j^i: Three-dimensional position of the satellite, $j = 3$
x_j: Three-dimensional coordinates of the satellite u, $j = 3$

$$r_u^1 = \frac{1}{2} D_u^1 - r_0^1 \tag{5.2}$$

$$r_u^2 = r_u^1 + r_u^{1,2} \tag{5.3}$$

$$r_u^n = r_u^1 + r_u^{1,n} \tag{5.4}$$

r_0^1: Distance from MCC to satellite S_{RDSS} can be obtained through the known position of MCC and $S_{RDSS.}$
D_u^1: Round-trip distance sum from satellite S_{RDSS} to the user measured by MCC.
$r_u^{1,2}$: The difference between the distance from satellite 1 to the user and the distance from satellite 2 to the user. It is obtained from the satellite clock error of satellite 1 and satellite 2 as well as the pseudorange data from the user to the satellite 1 and satellite 2.
$r_u^{1,n}$: The difference between the distance from satellite 1 to the user and the distance from satellite n to the user. It is obtained from the satellite clock error of satellite 1 and satellite n as well as the pseudorange data from the user to the satellite 1 and satellite n.
When $n = 3$, this is the position reporting service for three satellites.
When $n > 3$, this is a multi-satellite position reporting service.

5.2.2 RNSS Continuous Navigation Service

Starting from the principle of RDSS positioning, as shown in Figure 5.2, if the distance measurement and the positioning solution calculating must be accomplished by the

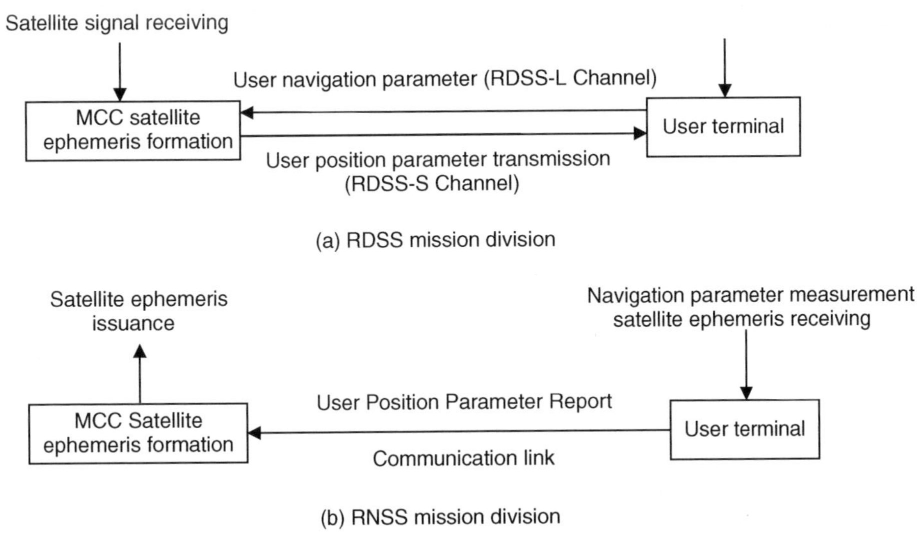

Figure 5.2 Service comparison between RDSS and RNSS.

user independently, the absolute synchronization of the time shall be realized between user and satellite to finalize the distance measurement from the user to the satellite. In engineering system, the synchronization of satellite and NAVT as well as the precise position measurement can be completed by the satellite through external operational control system. However, time synchronization of the user and NAVT is solved through the observed quantity of a navigation satellite signal. Hypothetically, if the difference between the local time of the user and NAVT is the clock error Δt_u of the user, when the distance is measured from the user who receives the navigation signal to the satellite, distance difference $C\Delta t_u$ caused by the clock error is introduced. At this time, the distance observed by the user is the sum of true distance $R_u^i(t)$ and $C\Delta t_u$, which is called the pseudorange ρ_u^i.

$$\rho_u^i(t) = R_u^i(t) + C\Delta t_u(t) \tag{5.5}$$

As for multi-satellite, the positioning equation for the pseudorange observation [5, 6] is:

$$\rho_u^i(t) = \left[\sum_{j=1}^{4} (X_j^i - X_j)^2\right]^{\frac{1}{2}} + C\Delta t_u(t) \tag{5.6}$$

In comparison with Equation (5.1), in Equation (5.6), the unknown number is added with the term $C\Delta t_u(t)$. The user solution includes four parameters that are coordinates of user $X_j(x, y, z)$ and clock error $\Delta t_u(t)$. Therefore, one more observed quantity must be added, and four unknown numbers can only be solved with the observed quantity consisting of four satellites at least. Hence, RNSS can be concluded as: the satellite transmits the radio navigation satellite service signal under the synchronization of the atomic clock; the user takes the local clock as reference to finalize the measurement of time difference and change rate of time difference to the satellite signal to determine the position and velocity of the user. RNSS service is a special condition under the non-synchronous observation of the RDSS service and is a subset where Δt_u is not equal to zero.

Therefore, RNSS is the radio navigation satellite system obtaining the autonomous navigation capability of the user by increasing the number of satellites. It is a subset of RDSS. On the contrary, with a fixed constellation number, better geometric conditions can be obtained by adopting the positioning principle of RDSS, equivalent to a 25% increase in the use ratio of the constellation so as to improve positioning accuracy of the user. Under different service conditions, different positioning equations are selected to obtain smaller DOP values, which can both improve positioning accuracy and obtain a position report of null information transmission. It is not that the positioning service of RDSS is inferior to positioning service of RNSS. In order to obtain the best application of RDSS and RNSS, the CRDSS system is generated.

5.2.3 Mission Comparison between CRDSS Service and RNSS Service

Both the RDSS and RNSS services include two basic services: One is the navigation parameter measurement and the other is the position parameter calculation. Navigation parameters (distance, change in rate of distance) measurements required by RDSS and position calculations are completed in the measurement and control center (MCC) rather than at the user. In RDSS service, the parameter exchange between the user and MCC is completed by the L/S channel, and no other communication link is required.

However, navigation parameter measurement required by the RNSS service and position calculations are both completed in the user's receiver. The position parameter of the user is transmitted to MCC through other communication methods. Its functional specialization is shown in Figure 5.2. If these two different mission divisions are integrated flexibly in MCC and user receiver, and selected flexibly based on different needs, the CRDSS system forms.

5.2.4 CRDSS System Position Reporting Capability Analysis

Position reporting capability includes: position reporting response time, position reporting precision and anti-shield angle, position reporting region coverage, and position reporting constellation satellite quantity. See Table 5.1 for performance comparison.

The performance comparison result is as follows:

1. Position reporting response time: time required from starting up to position reporting data received by MCC.
   RNSS + Iridium position reporting time includes time required for navigation signal captured by user, finalizing pseudorange measurement, ephemeris collection and communication access, and user positioning data receipt. It can be seen that CRDSS position reporting response time is extremely short.

Table 5.1 Position reporting performance comparison between CRDSS and GPS.

No.	Category Performance	CRDSS dual-satellite position reporting	CRDSS tri-satellite position reporting	CRDSS multi-satellite position reporting	RNSS + iridium position reporting
1	Report response time (s)	1	1	120	120
2	Positioning accuracy (m)	20~100	10	1.0	10
3	Anti-shield angle (degrees)	45	45	<15	<15
4	Region coverage	China and its surrounding areas	China and its surrounding areas	China and its surrounding areas	The whole world
5	Constellation satellite quantity (units)	35	35	Compass + GPS	GPS + Iridium ≈ 96
6	CRDSS satellite nos.	3GEO+3IGSO	3GEO+3IGSO	3GEO+3IGSO	3GEO+3IGSO
7	Application	Position reporting for rescue, etc.	Position reporting for rescue, etc.	Precise positioning and position reporting	Precise positioning for rescue
8	Cost-efficiency ratio	High	High	High	Low

2. Positioning accuracy and anti-shield angle: positioning accuracy is defined as position error with a reliability of 95%. Anti-shield angle is defined as the highest visible angle of elevation formed by the factors including terrain, ground features, civic buildings, mountainous regions, trees, and so on.
3. Positioning accuracy significantly relates to PDOP value of the satellite.
   For the RNSS + satellite communication scheme, PDOP value under an elevation angle of 5~10° is good. The bigger the elevation angle is, the worse the PDOP is, and positioning accuracy and success rate are both lowered. When the elevation angle is 15°, observation quantity of four satellites cannot be continuously obtained; therefore, the anti-shield angle ability is weak.
   CRDSS has higher positioning accuracy under high shield angle, therefore, it has a strong anti-shield angle ability.
4. CRDSS satellite: GEO and IGSO satellites of Compass System that possess the capability of position reporting. They are both positioning satellite and communication satellite.
5. Measures of improving the capability of position reporting: there are many measures of improving the capability of position reporting of CRDSS, including increase of CRDSS satellite number and satellite number of positioning satellite constellation.

Conclusion: the Achilles' heel of the realization for the position reporting scheme by adopting RNSS + satellite communication satellites is that it is difficult to improve the anti-shield angle. A great number of additional satellites must be added. When the shield angle is 45°, we can see over four satellites out of all the satellite of GPS/GLONASS [1] in only 96.3% of the time, and Iridium also loses visibility under the shield angle of 45°. However, Compass CRDSS still can finalize positioning and position reporting. Compared to GPS + Iridium, its position reporting time is increased by 46~50%.

5.2.5 CRDSS Global Coverage Analysis

On the basis of one RNSS global system, two GEO satellites with both RDSS payload and RNSS payload can be arranged to realize coverage of most regions of the globe through the support of the ground MCC center. This RNSS system can be Compass, GPS or GLONASS, Galileo. The missions of MCC are to make two payloads of GEO satellite synchronized on the same NAVT and to accomplish ephemeris collection and user navigation parameter measurement, position calculating, and so on. Namely, MCC possesses all of the functions of the RNSS operational control system and RDSS central control system. The so-called coverage of most regions of the globe means 79% of the Earth's surface area, 85% of the near-earth area for spaceflight users. Since RNSS has realized global coverage, RDSS service coverage only needs to cover most regions of the globe. RDSS of two GEO satellites is the biggest area of globe coverage beam as shown in Figure 5.3.

Height of satellite $H_S = 36000$ km, radius of Earth $R_e = 6300$ km, assume the satellite observation elevation to earth user $E_S = 10°$, coverage angle of one GEO satellite beam to the Earth's surface is [5]

$$\beta = 2\arcsin\theta\frac{R_e}{R_e + H_s}\cos E_s = 17° \tag{5.7}$$

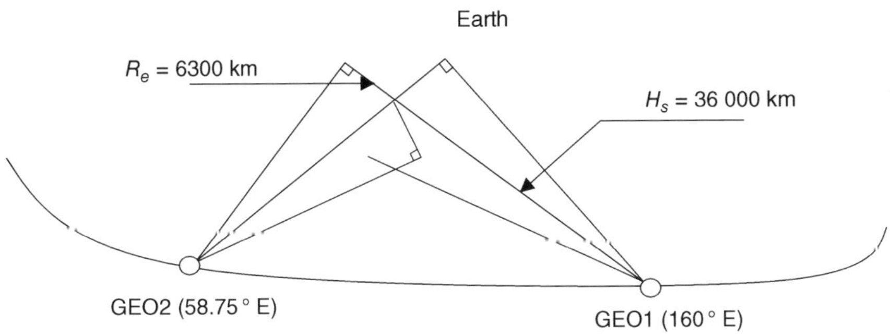

Figure 5.3 Coverage diagram of two GEO satellites.

The visibility angle for the user on the earth surface to see GEO satellite is

$$\alpha_1 = 180° - \beta - 2E_S = 143° \tag{5.8}$$

Visibility angle of the two GEO satellite is $\alpha = 143° \times 2 = 283°$, and the coverage rate is:

$$\eta = \frac{283°}{360°} = 79\%$$

As for the spaceflight user, observation elevation $E_S = 5°$, two satellites account for 85% of the global coverage. At this time, the included angle of the two satellites at the equator is precisely α. When the included angle of the two GEO satellites is 100°, there will be an overlap region of 43°. The coverage area is reduced to 67% of the user on the earth surface (10°W–133°E–130°W region) and 73% of coverage of the near-earth satellite user.

When GEO satellite is only 100° due to the limited distance of geostationary satellite orbit, a 43° overlapping area forms, which reduces the total coverage area, but in the overlapping area, two two-way distance sum measurement values from MCC through satellite to the user are formed, and a new observed quantity is formed. At this time, the user is only required to measure the distance difference between RNSS signal of one GEO satellite and another NGEO satellite signal to complete tri-satellite distance positioning. In this way, the RDSS position calculating method and its user's flexibility in use are enriched. In China, when the GEO satellites are arranged at 58.75°E and 160°E, the overlapping area happens to be east of Urumchi and Lhasa in China to land and ocean economy active zone of west of Guam in the Pacific, which has distinctly important strategic value. Starting from regional CRDSS principle, 2GEO+3IGSO satellites accomplish the navigation positioning and position reporting service of user's main economy active zone. At this time, coverage areas of two GEO satellites' regions start 10°W in the west to 130°W of the Pacific through the Indian Ocean and Eurasia. See Figure 3.1 in Chapter 3.

Compared with our realization of precise orbital determination and measurement and control of near earth satellite utilizing RNSS positioning and tracking together with TDRSS, this scheme is more economic and practical. Therefore, the CRDSS position reporting system can be deemed a future important infrastructure of the near-earth satellite system. The main great powers and regions such as USA, Japan, the European Union, and so on, in the world have accomplished relay coverage of near-earth satellite and offshore users in most regions of the globe through a relay satellite system consisting of two GEO orbit satellites with an interval greater than 100°. We use the RDSS service of the navigation satellite for compatible coverage, which has remarkable benefits.

5.2.6 Realization of the CRDSS Personalized Service

The so called personalized service is to realize the multilevel user service with different categories, different precision grades, and different real-time performances at the same MCC according to different demands of the user. According to previous conventional establishment mode, service of user with different levels can only be realized by establishing multiple service systems. Obviously, numerous service systems inevitably have similar functions. The total benefits of the navigation industry are low due to the repeated establishment of infrastructure and increase of total operating cost. In order to seek suitable service, the user makes selection with difficulty and echoes are based on different users' equipment, the result of which is carrying a heavy burden and has bad effects.

The principle of CRDSS is placing the resource of infrastructure uniformly in the MCC system including the central processing system and the reference station systems of different levels that are operated and managed by MCC uniformly. The user receiver mainly accomplishes the measurement of navigation parameter, including distance, change rate of distance, double difference pseudorange, and double difference phase position at the same observation time, as well as the triple differential positioning at different observation times, and so on. Then, it will be sent to MCC for processing through the RDSS link. As MCC masters a great amount of observation information of reference station, it forms a complete satellite precise time synchronization parameter, precise ephemeris, ionosphere correction parameter, differential correction parameter, and so on. According to the demands of the user, the different precision grade, different real-time performance position, and velocity can be calculated. The issuance and receipt as well as the re-processing of precise time synchronization parameter, precise ephemeris, ionosphere correction parameter, and differential correction parameters are prevented, which greatly relieves the stress on the system channels and reduces the complexity of the user terminals. Since the ephemeris, correction parameters, and suchlike transmitted by navigation signal are significantly reduced, information rate of transmitting message is greatly reduced to even maintain the format of time synchronization. For example, information rate is reduced from 500 bit/s and 1000 bit/s to 50 bit/s or even lower. Therefore, the spectrum of information processing is improved, which is not only favorable for improving measurement accuracy but also increasing the capability of anti-interference by 10–16 times. Indoor positioning, and so on are accomplished at low level.

In conclusion, the advantages of CRDSS personalized service can be summarized as follows:

1. Under the principle of RDSS, tri-satellite precise positioning is realized. Under the same quantity of constellation satellite, better geometric figures, and higher positioning accuracy can be acquired.
2. User positioning is accomplished at the same time with position reporting to central control system. The position reporting is potentially accomplished in the course of user transmitting navigation observation parameters to the central control system. Coverage area of the report system composed of RDSS link is big.

3. As all reference stations of satellite orbit determination, ionosphere parameter determination, and differential correction are uniformly managed and coordinated to work, the infrastructure and observation parameters are fully shared, which not only improves the positioning accuracy, reliability, and continuous operation capability of the system, but also reduces the total cost of the infrastructure of navigation.
4. Real-time high precision service is integrated with quasi-high precision service. Different real-time performance services are integrated to greatly reduce inconvenience and burden of the users who have to adopt different user receiver under different real-time performance. Personalized users only need to possess the observation receiver with certain level to obtain service of relevant demand through different service application. The service provider can realize the dream of providing service to the user for the sake of the user.

5.3 CRDSS system Engineering Design

The CRDSS global position reporting system design aims to accomplish index analysis and distribution of the space, ground, and user segments, as well as function design and application service mode. Its goal is to finalize precise navigation and positioning measurement, user's position information exchange, and communication services in most areas of the globe by utilizing the infrastructure of observation and control in the region including the precision time synchronization and orbit determination as well as information exchange to GEO, IGSO, and MEO satellites.

5.3.1 Application Object and Design Principle

5.3.1.1 Application Objects

CRDSS application objects include:

1. Land user: meeting the user's services requirements including high accuracy navigation, precision positioning, position reporting, short message communication, trip guidance, as well as high accuracy time service
2. Offshore user: meeting user's requirement of services including precision navigation, formation navigation position reporting, life-rescuing (short message and voice of distress), resource investigation, and environmental monitoring data collection.
3. Air user: meeting the aviation user's requirement of services including navigation above land and ocean, precision approach, air traffic control, and emergencies/lifesaving.
4. Near-earth satellite: meeting the requirement of services including satellite precise orbit determination, precision timing and telemetry, and telecontrol data relay service.

5.3.1.2 Design Principle

The basic design principle includes:

1. *Utilization of global GNSS public service resource*

   Fully utilizing civil or public service navigation signal of GPS, GLONASS, and Galileo satellites to form the best global service capability. Monitoring and reporting of the satellite above is completed in control segment. Compatible and sharing design will be implemented in the user segment. Application cost and operation complexity of the user will be reduced industriously through autonomous and non-autonomous positioning, as well as by position reporting to make satellite navigation meet all user requirements at different levels and become a tool that people cannot leave behind; even for a moment in their lives.

2. *User first, resource sharing*
   Starting from the demands of users, implementing uniform operation and resource sharing to basic service facilities of satellites, including wide and local area differential stations for real-time precision navigation on the basis of uniform time and space references. Continuous stable operation is maintained in a cost-effective manner to provide immediate service in urgent need so as to give both satisfactory service to the frequent user and immediate service to the low-level user for safety.
3. *Interactive on-demand service*
   Through the combination of CRDSS positioning and communication, the classic product service is transformed to an on-demand service in order to make the user select service and truly become the leader in the service market for navigation and communication life necessities for thousands upon thousands of families.

5.3.2 Constellation Selection and Coverage Area Design

5.3.2.1 Constellation Selection

The Compass constellation is taken as basic frame and GPS the main augmentation satellite. GLONASS and Galileo satellites are selectively utilized to form an RNSS service constellation. Through monitoring and differential correction of operational control and application systems, the integrity broadcasting mode is utilized to remind users of the best available choice of RNSS satellite. As the number of satellites on 2–3 systems is 50–70 or above, in the field of view there are 20–30 satellites that can be selected by the user. However, only four satellites are required to accomplish perfect positioning under optimum DOP value. Therefore, under the enormous visible constellation group, the optimization is not done depending only on the autonomous integrity of user's receiver. Otherwise, not only does the complexity of satellite selection by user receiver increase, but also the indexes such as precision and real-time performance might be lost. Therefore, the constellation is selected through the operational control and application system. Boundary conditions of satellite selection design include:

1. *Scope of service provided*
   As for the user demanding high performance that CAT-I is met and precision is superior to 1.0 m, the service scope only defines in China and its surrounding regions, which is easily satisfied. As for service precision around 8~10 m around the globe, the selection can be made voluntarily by the user according to its region condition.
2. *Optimum DOP value provided*
   The selection of observable satellites in 2–3 constellation systems considers that HDOP $\leq$ 1.5 and VDOP $\leq$ 1.5 can be achieved at any point in the service area. When a satellite constellation consisting of two systems with 24 satellites is utilized, the probability of meeting this DOP value is greater than 90%.

3. *Observation elevation of the selected satellite*
   The observation elevation of the user in every region relative to the optimum DOP satellite is greater than 15° in order to avoid multipath disturbance.
4. *Integrity of selected satellite navigation signal*
   The requirement of selected satellite integrity is fully satisfied without distortion and incontrollable clock error forecasting. After clock error and orbit correction, any errors meet the precision requirement.

The scheme is made after simulated analysis and actual monitoring of constellation integrity, orbit plane and phase characteristics, and real-time variation characteristics of DOP value in four satellite navigation systems to meet these design conditions, which is one of the important goals for wide area and local area augmentation services of multiple systems.

5.3.2.2 CAT-I Integrity Broadcasting Constellation Design

The mission of CAT-I integrity broadcasting constellation design refers to the design of appropriate GEO satellites in the service system of designated region, the integrity performance of which meets the requirement of precision approach of CAT-I. Its basic scheme is that: (1) every user of Class-I precision approach over airport must receive integrity information from two different satellites at the same time to perform consistency checking; (2) every observation elevation of integrity broadcasting satellite $\alpha \geq 25°$, and in order to realize the capability of CAT-I precision approach of user in China, at least three GEO satellites implement integrity broadcasting to meet the requirement of integrity checking. By taking tri-satellite GEO1, GEO2, GEO3 working at 140°E, 80°E, 110.5°E as an example, airports in low altitude areas basically meet this requirement. According to formulas (5.7) and (5.8) in Section 5.2.5, the design of E_s is as follows:

As known, $E_s = 25°$, thus,

$$\alpha = 180° - 16° - 2 \times 25° = 114°$$

$$\beta = 2 \arcsin \theta \frac{R_e}{R_e + H_s} \cos E_s = 16°$$

As for the GE03 satellite located at 110.5°E, the scope satisfied is:

$$110.5° - \frac{1}{2}\alpha \sim 110.5° + \frac{1}{2}\alpha$$

Namely, 110.5°-57°~110.5°+57°, and it obtains 53.5°E~167.5°E

Cooperating with GEO1 and GEO2, GEO3 not only meets the requirement of integrity observation elevation, but also meets the requirement of integrity availability. If only GEO1 and GEO2 are used to implement integrity broadcasting, the requirement of integrity observation elevation can be met but availability requirement cannot be met. Once one satellite fails, all of the CAT-I integrity broadcasting of airport will not satisfy the requirements.

Considering global rescuing and near-earth management, if two satellites GEO4 and GEO5 are arranged at 160°E and 58.75°E, the quantity of CRDSS satellite shall be 5, and they are arranged at GEO1, GEO2, GEO3, GEO4, and GEO5, respectively.

5.3.2.3 Air Traffic Control Coverage Area Design in the Region of China

For air traffic control in China, information can be exchanged between air and ground through the L channel, S beams of CRDSS to realize the situation awareness of the users and to avoid collision from happening. Every airport can smoothly finalize the mutual handover of aerospace target monitoring. According to the same calculation formula as Section 5.3.2.2, satellites GEO1, GEO2, and GEO3 are all suitable with respect to coverage area and observation elevation. Due to the number of the airplanes, design of communication capacity is the key.

It is assumed that there are 1000 airplanes flying in Chinese airspace at any given time, the position reporting is as follows:

User identification (9600 pieces) 13 bit
Flight height (1200 m) 11 bit
Position coordinates:

$$\text{Longitude} \times \times^{\circ}\ (\text{Degree}) \times \times\prime\ (\text{Minute}) \times \times.\times''\ (\text{Second})\ 8 + 6 + 10 = 24\ \text{bit}$$

$$\text{Latitude} \times \times^{\circ}\ (\text{Degree}) \times \times\prime\ (\text{Minute}) \times \times.\times''\ (\text{Second})\ 7 + 6 + 10 = 23\ \text{bit}$$

Total length of message of one time position reporting of each user: $13 + 11 + 24 + 23 = 71$ bit

If the position is reported every 10 s, the average information capacity of 1000 airplanes is $(1000 \div 10) \times 71 = 7100$ bit/s, which means the average capacity of outbound signal is 7.1 kbit/s. This capacity satisfies the requirement. GEO1, GEO2, and GEO3 possess sufficient communication capability to provide an air traffic management information sharing service in China.

5.3.2.4 CRDSS Coverage Area Design

Service of CRDSS meets the coverage requirement of 70% of land and offshore areas and 85% of near-earth satellites. They are mainly used for important services such as life-saving and ocean exploration; however, they have an extremely low frequency of user service. Such service is once in several minutes for a small number of users. Most services for users are provided only in emergencies. However, the response must be immediate and accurate with high reliability. Under the condition of sufficient RNSS payload satellites, the feature of the S and L beam of CRDSS is a big coverage area and low communication rate. Based on this design principle, the selection of satellites GEO4 (160°E) and GEO5 (58.75°E) of Compass GEO satellites is appropriate.

The calculation method for the design of coverage area is similar to previous two sections, the coverage angle of L and S beams of GEO4 and GEO5

$$\beta_{4S} = \beta_{4L} = \beta_{5S} = \beta_{5L} = 17^{\circ}$$

Namely the Earth coverage beam. Antenna gain $G = 23.5$ dB and user's angle of visibility $\alpha \geq 143^{\circ}$. The coverage of most of the land and ocean regions can be realized. It is easy to provide a short message with inbound low information rate ($D_i = 1000$ bit/s) and outbound $D_0 = 3000$ bit/s so as to complete the global life-rescue service. Its service area is shown in Figure 3.1. The link level calculation is shown in Table 5.2 and Table 5.3. When the service is provided to 10 users per second (1000 bit/user), the outbound capability is 144 000 per hour.

In order to improve the coverage capability for China's land and surrounding regions, one CRDSS working satellite with RDSS and RNSS payload is added at GEO3 (110.5°E).

Table 5.2 Typical signal level of 2483.5~2500 MHz from satellite to user.

Satellite EIRP [12.2 dB 120 W] (dBW)	43
Path loss of free space (dB)	−191.8
Received power of user receiver (dBW) (various isotropic antennas)	−148.8
G/T value of user receiver (dB/K)	−24
Bohr constant, K (dB)	−(−228.6)
C/No (dB/Hz)	55.8
Data transmission symbol rate loss [10 kbit/s] (dB)	40
Receipt Eb/N_0 (dB)	15.8
Demodulation required Eb/N_0 (dB)	9.6
Total allowance (dB)	6.2

Table 5.3 Typical signal level of 1610~1626.5 MHz from user to satellite.

User EIRP [3W] (dBW)	6
Path loss of free space (dB)	−188.35
Weather loss (dB)	−0.5
Satellite receiving single user power (dBW)	−182.85
G/T value of satellite [13 dB 560 K] (dB/K)	−4.5
Bohr constant, K (dB)	− (−228.6)
C/No (dB/Hz)	41.25
Total C/N_0 under adequate C/N_0 of earth central station (dB/Hz)	41.2
Data transmission symbol rate loss [1000 b/s] (dB)	30
Receipt Eb/N_0 (dB)	11.2
Demodulation required Eb/N_0 (dB/Hz)	9.6
Total allowance (dB)	1.6

When S band EIRP = 46 dBW and L frequency band satellite G/T = −2.5 dB/K, and when the user's transmitting power is P = 5 W, outbound information rate can reach 10 kbit/s and inbound information rate can reach 1 kbit/s. If the CRDSS capability for China's land and surrounding regions needs to be improved further, a CRDSS working satellite can also be added at GEO1 (140°E) and GEO2 (80°E). However, starting from the best economic benefits, CRDSS coverage area of China meets the requirement of service for China and surrounding regions as well as the Indian Ocean and West Pacific. The former is a dense area of users and the ocean is a sparse user area. A suitable GEO constellation is GEO1 (140°E), GEO2 (80°E), and GEO3 (110°E). GEO1 finalizes the coverage for the Pacific. GEO2 finalizes the coverage of the Indian Ocean. GEO3 finalizes the coverage for China and surrounding regions. Its beam coverage is shown as a pale line in Figure 5.4. In this scheme, this is the design reference for the CRDSS satellite.

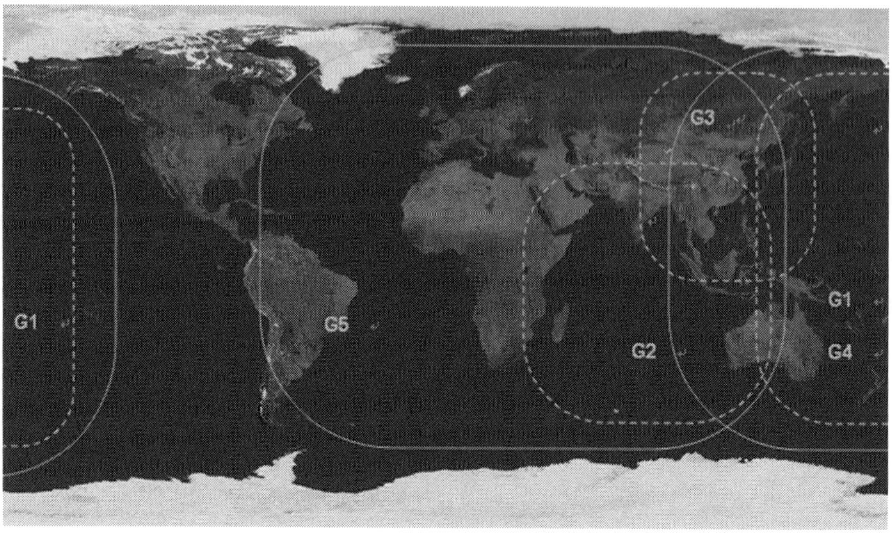

G1 region
G2 region
G3 region
G4 region
G5 region

Figure 5.4 Global rescue coverage based on GEO4 and GEO5 as well as Asia-Pacific coverage based on GEO1, GEO2, and GEO3.

5.3.3 Precision Analysis and Index Distribution

After the completion of constellation design, under the conditions of HDOP ≤ 1.5 and VDOP ≤ 1.5, the precision analysis and index distribution of other error resources are performed.

5.3.3.1 Total Precision Requirement

Index distribution of ground operation and control as well as application systems must meet the requirements of the CAT-I precision approach and ground precision positioning under the GNSS condition. Main indexes of precision are shown in Tables 5.4 and 5.5.

The service mode can be divided into two types, user active positioning service and passive positioning service. Users with high dynamic state and high real-time performance generally adopt the user autonomous and the continuous positioning mode, which is typical RNSS service mode. However, the RDSS passive positioning mode adopted by some of the users is not excluded. The position calculation is completed through operation and control and the application system. For reducing the cost to

Table 5.4 CAT-I requirement proposed by the ICAO.

Positioning accuracy (NSE 95%)	Horizontal 18.2 m; Vertical 7.7~4.4 m
SIS maximum non-continuity	8×10^{-6}/15 s
SIS maximum non-integrity	2×10^{-7}/approach
Warning time	6 s
SIS availability	0.9975
SIS as the space signal	

Table 5.5 Precision positioning service indexes.

Precision (95%)	1.0 m
Dynamic range	Static state, low dynamic state
Real-time performance	Several seconds to several minutes
Position reporting user	Land, offshore, aerospace
Guide map extraction time	Less than 120 s

the user, users of trajectory measurement and of precision positioning and position reporting generally adopt the RDSS non-autonomous positioning mode.

5.3.3.2 Observation Equipment Error Analysis

The first error source, no matter whether in time synchronization, precise orbit determination, or precise positioning of the user, comes from observation equipment and environment during observation, which is given the joint name of pseudorange error. This error consists of three parts: The first part is equipment time delay error, including channel time delay and processing time delay of the signal entering into the observation receiver; the second part is time accuracy of observed quantity, especially the time accuracy of pseudorange measurement – precise pseudorange observation time accuracy can reach the 1/100 of pseudorandom code period; however, during engineering design, nonstandard time synchronization extraction method can reach several millisecond or even bigger –and the third part is the observation error caused by the multipath effect; under the condition that only an anti-multipath antenna is adopted, even if it is a high elevation angle observation satellite, a multipath error of around 1.0 m exists. In the high real-time dynamic users, it can barely be restrained further. It is the main error source of dynamic precision improvement. In order to reach sub-meter level static state or RTK user positioning demand, carrier phase assistance, or multipath parallel narrow technologies are adopted for elimination. Further analysis will be conducted in later chapters.

5.3.3.3 Point Positioning and Single Reference Station Differential Positioning Error Analysis

Error sources and estimations of standard point positioning and single reference station pseudorange differential positioning are shown in Table 5.6. These estimations are suitable for conditions where the distance from user to reference station is 100 km. It can be concluded as follows:

Table 5.6 Standard point positioning and single reference station pseudorange differential positioning error table (in m).

Error Source	Error Name	Point Positioning	Single Reference Station Pseudorange Positioning
Satellite	Clock error	3.0	0.0
	Orbit error radial direction	1.0	0.0
	SA error (man-made)	30	0.0
Control system	Ephemeris forecast error	4.0	0.0
	Satellite clock forecast error	1.0	1.0
Propagation path	Ionosphere error	5.0	0.0
	Troposphere error	1.5	0.0
Error of user receiver	Noise	0.3	0.3
	Multipath	1.0	1.0
Total error of user distance (RMS)		7.4 (No SA) 30.9	1.45
User position error 2DRMS, PDOP≤1.5		92.7	4.33 (H) 4.33 (V)

1. Differential positioning method can eliminate most of the influence on the positioning accuracy due to satellite clock and orbit error.
2. Differential positioning method can significantly weaken the influence on positioning accuracy due to propagation error.
3. Satellite clock error and multipath error are the main error sources of precise positioning. It is very difficult to realize positioning accuracy from meter level to sub-meter level. The method of further improvement of positioning accuracy will be analyzed in the following chapters.

5.3.3.4 Base Network Pseudorange Differential Positioning Accuracy Analysis

It is demonstrated in practice that single reference station pseudorange differential positioning accuracy decreases as the distance between stations increases. When distance between stations is greater than 500 km, the propagation path between reference station and user has a poorer correlation. Propagation errors of ionosphere and troposphere cannot be eliminated completely. Propagation time delay error is increased by ±0.0216 m/km approximately. When distance between stations is 600 km, the propagation error is ±2.16 m. Position error is increased by ±6.48 m. In order to overcome the deficiency that the single station pseudorange differential precision changes along with the change of distance between stations, it is more efficient to adopt a reference station. This technology is widely used in wide area differential positioning, including American WASS, the European EGNOS, and the Chinese satellite navigation augmentation system. The difference between reference station and single station pseudorange differential is that the pseudorange correction value sent to the user is neither generated by observed value of single reference station nor just the pseudorange correction value.

Table 5.7 WASS pseudorange differential positioning error assignment (in m).

Satellite ephemeris error	0.1
Satellite clock error	0.1
Ionosphere time delay error	0.2
Troposphere delay error	0.1
Multipath error (after carrier phase processing)	0.2
Receiver noise error	0.1
Total error of user difference	0.35
User position error 2DRMS, DOP≤1.5	1.04

Its correction value consists of ionosphere correction parameter, troposphere correction parameter, correction value of satellite clock, and satellite orbit correction parameter isolated by observed value of multiple reference stations, which are represented in different components in the message format. Typical WAAS error assignment is shown in Table 5.7.

In reference station network (wide area differential) positioning, the reference station adopts a double-frequency monitoring receiver to independently calculate the ionosphere delay correction. However, three-dimensional ephemeris error, satellite clock error, and local clock error of reference station monitoring receiver are jointly processed utilizing system pseudorange in a big filter through operational control and application system. The two kinds of processing mentioned here are independent of each other. Therefore, there is the advantage of high accuracy. However, as there are many varieties of errors involving complicated elements and the correlation of errors is poor after isolation, precision is limited to meet a Class-I aviation precision approach with complicated processing. Its advantage is that a differential positioning of a region of 5000 × 5000 km can be completed with a small amount (20–30) reference stations. Its precision is at the 5–10 m level. By taking the ionosphere grid calculation as an example, first, the operational control and application system calculates the ionosphere correction value of zenith at hundreds of grid points in the entire region according to the observed quantity of double frequency pseudorange of all of the reference station double frequency monitoring receiver, and update with the interval from second to minute. After the ionosphere correction parameter broadcasted by the satellite is received by the user, complicated conversion needs to be done for use by single frequency user receiver. First, the longitude and latitude from user to ionosphere penetration point $\mathrm{T}_{\mathrm{u}}^{\mathrm{j}}$ is calculated, and its value [7] is:

$$\lambda_T^j = \lambda_u + \arcsin\left[\frac{\sin\alpha_u^j \sin A_S}{\cos\varphi_T^j}\right] \tag{5.9}$$

In the formula: $\varphi_{\mathrm{T}}^{\mathrm{j}} = \arcsin\varphi_{\mathrm{u}}\cos\alpha_{\mathrm{u}}^{\mathrm{j}} + \cos\varphi_{\mathrm{u}}\sin\alpha_{\mathrm{u}}^{\mathrm{j}}\cos A_{\mathrm{s}}$

$$\alpha_{\mathrm{u}}^{\mathrm{j}} = \frac{\pi}{2} - E_{\mathrm{u}}^{\mathrm{j}} - \arcsin\left(\frac{\mathrm{R_E}}{\mathrm{R_E} + H_{\mathrm{ion}}}\cos \mathrm{E}_{\mathrm{u}}^{\mathrm{j}}\right)$$

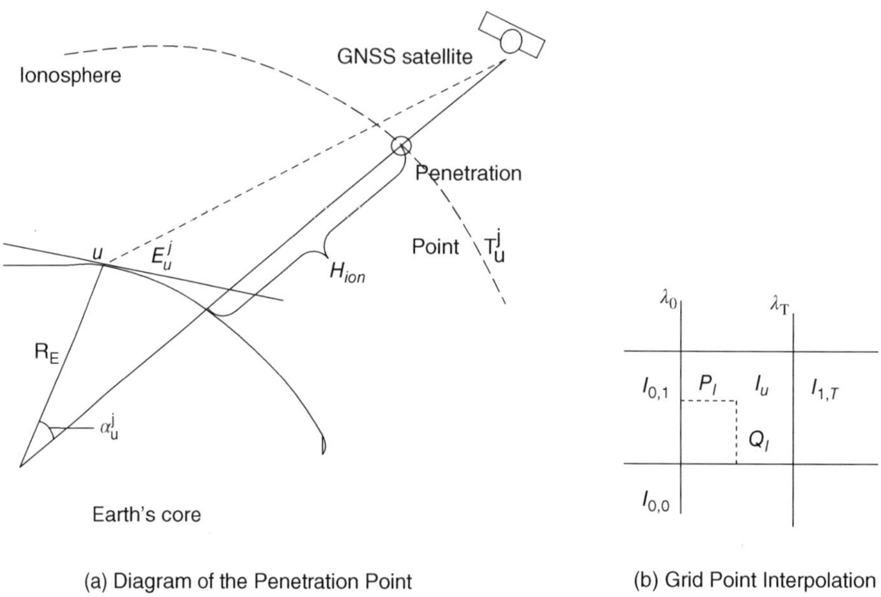

Figure 5.5 Ionosphere grid calculation diagram.

E^j_U is the satellite elevation angle; R_E is the radius of the earth; H_{ion} is the height of ionosphere; A_S is the satellite azimuth angle; λ_u, φ_u are the longitude and latitude of the user.

And then, based on Figure 5.5, the ionosphere time delay correction value at the penetration point is calculated; namely, interpolation is performed in the grid point at the perimeter to obtain:

$$I_u(T^j_u) = (1 - P_I)(1 - Q_I)I_{0,0} + P_I(1 - Q_I)I_{1,0} + Q_I(1 - P_I)I_{1,0} + P_IQ_II_{1,1} \qquad (5.10)$$

Where: $P_I = \frac{\lambda^j_u - \lambda_{0,0}}{\Delta\lambda}$, $Q_I = \frac{\varphi^j_u - \varphi_{0,0}}{\Delta\varphi}$

$\Delta\lambda$, $\Delta\varphi$ are intervals of longitude and latitude of ionosphere time delay correction grid points.

$\lambda_{0,0}$, $\varphi_{0,0}$ are the longitude and latitude of the grid point at the left bottom corner of penetration point.

5.3.3.5 Single Reference Station Carrier Phase Differential Positioning Error Analysis

The main difference between single reference station carrier phase differential positioning and the pseudorange differential positioning is that the observation data of the former is the carrier phase. Due to high accuracy of the observed quantity of carrier phase, which is close to a quarter of carrier wave length, when carrier frequency is 1200~1575 MHz, its observation accuracy is superior to 5–6 cm, which has the significant effect of restraining the multipath effect. Unlike pseudorange measurement, code related peak is distorted due to the influence of multipath that makes the measurement accuracy significantly lower. From the viewpoint of measurement, recovery of carrier phase adopts narrower processing band width and scale than those of recovery of pseudo-code phase. Random measurement error is two orders of magnitude higher than pseudo-code phase [8]. Therefore, the multipath effects can be weakened by ensuring continuous observation of the carrier phase. The pseudo-code distance measurement random error of a regular receiver and carrier phase measurement random error, respectively, are:

$$\sigma_{ch} = T_{ch}\left\{\frac{B_{ch}}{2(C/N_0)}\left[1+\frac{2}{T(C/N_0)}\right]\right\} \tag{5.11}$$

$$\sigma_{\phi} = \frac{\lambda}{2\pi}\left[\frac{B_{\phi}}{(C/N_0)}\right]^{1/2} \tag{5.12}$$

Where σ_{ch} is the pseudorange phase measurement error; σ_{ϕ} is the carrier phase measurement error; C/N_0 is the signal carrier-to-noise ratio; B_{ch} is code loop bandwidth from 1 Hz to several Hz. B_{ϕ} is the carrier recovery loop bandwidth, less than 1 Hz. T_{ch} is pseudorange chip width, approximately 100~1000 ns; λ is the carrier wave length, around 0.6 ns; and apparently, σ_{ϕ} is 1–2 orders of magnitude higher than σ_{ch}.

The observed quantity of carrier phase differential positioning is the phase difference at time of t_R, namely, the difference between local clock phase of receiver i at time of t_R and satellite j transmitting phase at time of t_T

$$P_i^j = P^j(t_T) - P_i(t_R) \tag{5.13}$$

Navigation solution includes the measurement value of the phase of four satellites at the time of t_R. As the distance from each satellite to the receiver is different, there are four different transmitting times of four satellites. With the function of receiving time, it can be expressed as

$$t_T = t_R - R_i^j(t_T, t_R)/C \tag{5.14}$$

Wherein, $R_i^j(t_T, t_R)$ is the travelling distance of signal from satellite j (at time of t_T) to receiver i (at time of t_R); C is the light speed. The equation is substituted into (5.13) to obtain

$$P_i^j(t_R) = P^j(t_R - \Delta t) - P_i(t_R) \tag{5.15}$$

Wherein, Δt is the propagation time of light; as Δt is small, it can obtain:

$$P^j(t_R - \Delta t) = P^j(t_R) - \Delta t dP^j/dt \tag{5.16}$$

And, $dP^j/dt = f^j$ is the transmitting frequency of satellite j. Thus, observation phase can be written as:

$$P^j(t_R) = P^j(t_R) - (f^j/C)R_i^j(t_T, t_R) - P_i(t_R) + N_i^j \tag{5.17}$$

N_i^j is an integer, the integer phase relative to transmitting time of satellite j. Formula (5.14) is the phase observed quantity expressed by receiver time. As there are fixed and random errors between receiver local time and satellite time, it is converted to navigation time for expression, thus:

$$t_R = t_G + T_i \tag{5.18}$$

T_i is the error between receiver and navigation time t_G, generally at the millisecond level. Substitute (5.15) into (5.16) for linear expansion, and convert to a navigation equation expressed by phase:

$$\begin{aligned} R_i^j(t_T,t_G) &= [(x^j(t_G) - \dot{x}^j(t_G)R_i^j(t_T,t_G)/C - x_i)]^2 \\ &+ [(y^j(t_G) - \dot{y}^j(t_G)R_i^j(t_T,t_G)/C - y_i)]^2 \\ &+ [(z^j(t_G) - \dot{z}^j(t_G)R_i^j(t_T,t_G)/C - z_i)]^2 \end{aligned} \tag{5.19}$$

In the formula, $(x^j, \dot{x}^j, y^j, \dot{y}^j, z^j, \dot{z}^j)$ are the coordinates and speed of satellite at the time of t_T in an Earth-fixed coordinate system. (x_i, y_i, z_i) are the coordinates of the receiver.

During the differential positioning, reference station i_0 and one reference satellite j_0 are brought in. Phase difference of satellite j and satellite j_0 at stations i and i_0 can be written as:

$$\begin{aligned} \Delta P_i^j - \Delta P_i^{j0} &= P^j - P^{j0} - (f^j - f^{j0})T_i - (f/C)(\Delta R_j^i - \Delta R_i^{j0}) \\ &- (f/C)T_i(\dot{R}_i^j - \dot{R}^{j0}) - N_i^j - N_i^{j0} \end{aligned} \tag{5.20}$$

$$\begin{aligned} \Delta P_{i0}^j - \Delta P_{i0}^{j0} &= P^j - P^{j0} + (f^j - f^{j0})T_{i0} - (f/C)(\Delta R_{i0}^j - \Delta R_{i0}^{j0}) \\ &- (f/C)T_{i0}(\dot{R}_{i0}^j - \dot{R}_{i0}^{j0}) + N_{i0}^j - N_{i0}^{j0} \end{aligned} \tag{5.21}$$

Formula (5.20) minus formula (5.21) can eliminate satellite phase terms shared in each receiver:

$$\begin{aligned} &(\Delta P_i^j - \Delta P_i^{j0}) - (\Delta P_{i0}^j - \Delta P_{i0}^{j0}) \\ &\quad = -(f/C)(\Delta R_j^i - \Delta R_i^{j0} - \Delta R_{i0}^j + \Delta R_{i0}^{j0}) - (f/C)T_i(\dot{R}_i^j - \dot{R}^{j0}) \\ &\quad + (f/C)T_{i0}(\dot{R}_{i0}^j - \dot{R}_{i0}^{j0}) - N_i^j - N_i^{j0} - N_{i0}^j + N_{i0}^{j0} \end{aligned} \tag{5.22}$$

As $(f^j - f^{j0})(T_i - T_{i0})$ is small and can be ignored, distance difference ΔR^j is the difference between actual distance estimated from the actual receiver to satellite and measurement value, namely,

$$\Delta R_i^j = (1/R_{icalc}^j)[(x^j - x_i)\Delta x_i + (y^j - y_i)\Delta y_i + (z^j - z_i)\Delta z_i]$$

When multiple correction values and carrier integer items are obtained, the carrier phase differential position calculation can be completed. Due to losing lock of tracking, cycle slip might happen, namely, N counting difference integer can be remedied through methods such as pseudorange smoothing, and so on.

Since the single reference station carrier phase differential positioning significantly weakens the satellite clock error and multipath effect, the positioning accuracy is quite high. However, propagation error is still increased along with the increase of distance between stations. Therefore, in order to obtain the distance observed quantity in centimeter level for positioning accuracy that is superior to meter level, density of the reference stations must be increased as the cost.

However, as the two-way data communication function of RDSS payload L/S channels of GEO satellite creates the conditions for transmitting differential correction parameters, precise positioning user can apply for differential correction parameter to operational control and application system at any time to obtain the precise position at user station. And work of the reference station as well as the formation of correction parameters is completed in the operational control and application systems. If precise position processing of some users is required to be accomplished by operational control and application system, then, it is only needed to return the observed quantity to operational control and application system for satisfactory result. During the two-way data exchanging, since the outbound S and L beam signals are synchronized with navigation time strictly, the round-trip distance observed quantity from user to operational control and application system processing center can be obtained, which is favorable for eliminating cycle slip and reducing quantity of satellites required for differential positioning. Therefore, involvement of RDSS has much more obvious functions and superiority by comparing to simple RNSS and communication service.

5.3.3.6 Narrow Correlation Pseudorange Differential Positioning Accuracy Analysis

Carrier measurement is able to weaken the influence of multipath on measurement accuracy to achieve the ranging accuracy of 6 cm. Another method of weakening the multipath observation error is still pseudorange measurement. However, such pseudorange measurement is not common wide correlation pseudorange measurement. It uses N ($N \geq 10$) narrow correlators and comprehensively measuring the navigation signal with influence of multipath to increase the pseudorange measurement accuracy by approximate one order of magnitude to reach around 6 cm of measurement accuracy similarly. The so-called wide correlator is the one with correlation space of $d = 1/2$. The narrow correlator often takes $d = 1/16$. Correlation measurement values (16) of different intervals are measured by 16 narrow correlators. Through linear combination or the least square method, the composite value of measurement can be obtained. As shown in Figure 5.6, assume that 16 narrow correlators on correlation peak of distortion are measured respectively, the observation value layout is shown as a triangle. Through identical multiple measurement values, the solution is obtained. The measurement error may be ΔT_w. Apparently, compared with ΔT_w, ΔT_σ is weakened significantly.

Figure 5.6 is the exaggerated situation. The pseudorange measurement accuracy can be increased to centimeter level by adopting multiple parallel narrow correlators. Pseudorange precision obtained by each narrow correlator can use variance expression of

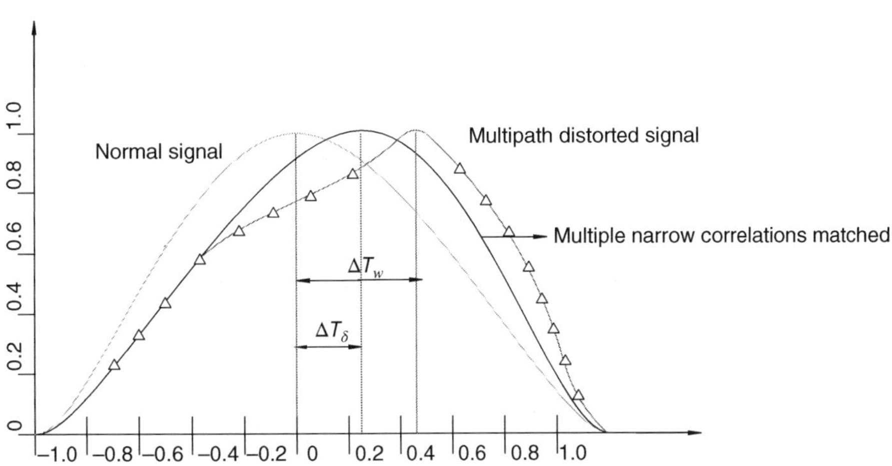

Figure 5.6 Weakening diagram of multiple narrow correlators on multipath error.

chip width normalization:

$$\delta_{DLC}^2 = \frac{B_L d}{C/N_0}\left[1 + \frac{1}{(C/N_0)T(1-d)}\right] \tag{5.23}$$

In the equation,

B_L: Loop Bandwidth (Hz)
d: Correlation space distance after normalization
T: Correlation processing accumulation time
C/N_0: Signal carrier-noise ratio
δ_{DLC} multiplying chip width is the pseudorange measurement standard difference expressed as time (s), and then multiplying light speed C to obtain the pseudorange measurement standard difference expressed as distance.

In some engineering measurements, a narrow correlation is usually adopted to perform the so-called high accuracy pseudorange measurement. It can be obtained by reasoning that a single narrow correlator can only improve the measurement accuracy under normal signal and cannot obtain the correct pseudorange value under multipath distorted signal. Therefore, single narrow correlation pseudorange measurement cannot overcome the multipath influence.

5.3.4 Selection of Precision Positioning Scheme

According to system design, the precision positioning scheme must meet the demands of the Class I precision approach and 1.0 m level high accuracy positioning. The former mission adopts suitable wide area differential positioning technology. In order to meet the requirement of a higher precision service, pseudorange differential technology and phase difference technology must be adopted depending on a single reference station. As well as setting up 30 real-time reference stations in China, there is a selectively reasonable quantity of quasi real-time reference stations set up in the service region. Besides pseudorange observation, carrier phase observation is also performed to provide pseudorange and carrier phase differential positioning service as per the demand of user so as to meet the requirement of a meter-high accuracy positioning service.

5.3.5 Guidance Service

Guidance service is designed for land users. The user makes a request of destination, vehicle, and transportation mode to the application system guidance center (called application guidance center for short). The application guidance center provides the most economic and efficient electronic map to the user to finalize the guidance service. This is an important part of the application system. The system composition is shown in Figure 5.7, consisting of application guidance center, navigation multimode user receiver, and abundant communication system. A navigation multimode user receiver possesses functions such as compatible reception of GNSS satellite signal, pseudorange measurement and positioning processing, as well as RDSS satellite communication and ground mobile communication. Besides finalization of navigation positioning, it is also involved in scheduling and commanding as well as time synchronization [9].

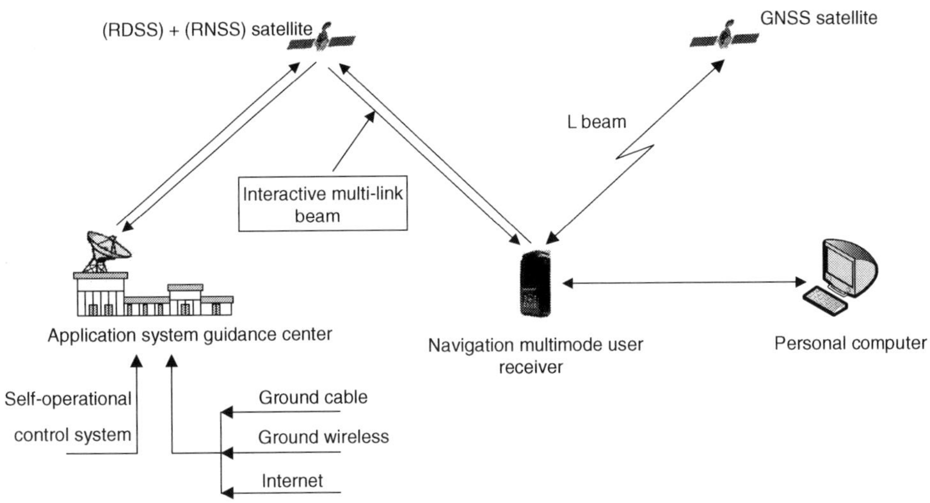

Figure 5.7 Guidance service system composition.

5.3.6 Two-way Data Transmission Link Design of the S/L- Frequency Band

5.3.6.1 Design Condition and Design Item Parameters

According to the preliminary argument and analysis, the design conditions of CRDSS system based on the Asia-Pacific regions are as follows:

Operational satellites are GEO1, GEO2, GEO3; three in total. Each satellite has three outbound and inbound S, L beams.

Satellite S frequency band transmission EIRP = 46 dBW;
Receiving sensibility of the user = 163 dBW;
Satellite receiving system G/T = −2.5 dB/K;
User transmitting EIRP = 6 dBW.

According to the design conditions mentioned here, a two-way data transmission link information rate can be acquired to make estimation for communication capability of the system.

5.3.6.2 Estimation of Outbound Link Level and Capacity

According to the design conditions in Section 5.3.6.1, the inbound level data rate of signal from 2483.5 to 2500 MHz can be calculated with reference to Table 5.1, the result of which is shown in Table 5.8. Outbound information symbol rate can reach 128 kbit/s; information rate is 64 kbit/s within the entire CRDSS region. The total outbound information amount per hour is:

$$C_{H0} = 64kb/s \times 3600 \times 9 = 2073.6 \times 10^6 bit$$

Outbound user number under each beam is:

64 kbit/s ÷ 71 bit/user × 3600 = 3,245,000/hour. It is 29,205,000 users/hour for the entire CRDSS region. According to 1000 bit/time of communication users, it can accommodate 2,070,000 users/hour.

Table 5.8 Calculation of outbound link level.

Satellite EIRP (dBW)	46
Free space path loss (dB)	−191.8
Isotropic antenna received power of user (dBW)	−145.8
G/T value of user receiver (dB/K)	−24
Bohr constant, K (dB)	−(−228.6)
C/N_0 (dB/Hz)	58.8
Data transmission symbol rate loss [128 kb/s] (dB)	51
Receipt Eb/N_o (dB)	7.8
Demodulation required Eb/N_o (dB)	4.6
Total allowance (dB)	3.2

5.3.6.3 Estimation of Inbound Link Level and Capacity

According to the design conditions in Section 5.3.6.1, data transmission rate can be calculated under an inbound link level of 1610~1626.5 MHz with reference to Table 5.2, the result of which is shown in Table 5.9. Inbound information rate is 2 kbit/s.

In the estimation of total inbound amount, multiple users in operation are considered. When a system receives and processes the inbound signal of the ith user, inbound signals from other user receivers will be deemed as interference. At this time, the equivalent signal/noise spectrum density of satellite receiving input terminal is C/(No+Io). The relational formula is:

$$\frac{1}{C/(N_0+I_0)}=\frac{1}{C/N_0}+\frac{1}{C/I_0} \tag{5.24}$$

Table 5.9 Inbound link level calculation.

User EIRP (dBW)	6
Free space path loss (dB)	−188.35
Weather loss (dB)	−0.5
Satellite receiving single user power (dBW)	−182.85
G/T value of satellite (dB/K)	−2.5
Bohr constant, K (dB)	−(−228.6)
C/N_0 (dB/Hz)	43.25
Total C/No under sufficient C/N_u of earth station (dB/Hz)	43
Data transmission symbol rate loss [Rb = 2 kb/s] (dB)	33
Multiple user deterioration (dB)	3
Receipt Eb/N_o	7
Eb/No required by VTB (dB/Hz)	4.6
Total allowance (dB)	2.4

I_0 is the density of signal power spectrum of the i_n user, which interferes the user i; in CDMA, there is

$$I_0 = \frac{4}{3}\frac{\Delta}{T}\sum_{k=1,k\neq i}^{h} C_k/B_{IF} = \frac{4}{3}\frac{\Delta}{T}C_{k\Sigma}/B_{IF} \tag{5.25}$$

When the power of every user is the same, $C_{k\Sigma} = (n-1)C_i$
Formula (5.21) can be written as:

$$I_0 = \frac{4}{3}\frac{\Delta}{T}(n-1)C_i/B_{IF} \tag{5.26}$$

Δ is the spreading code element period. When chip rate =10 Mbit/s, $\Delta = 0.1 \times 10^{-6}s$
T is the information code period. When information rate is $R_b = 1Kb/s$, $T = 1 \times 10^{-3}s$
B_{IF} is information demodulation unit input narrow filter bandwidth, which is around twice the information bandwidth. When $R_b = 2kb/s$, $B_{IF} = 2kHz$.

I_0 is calculated through formula (3–12), in which, C/N_0 is the $C/N_0 = 43dB/Hz$ obtained from Table 5.9. If a predomination of 3 dB is used for multi-signal deterioration, the total allowance of level is 2.4 db. If the total allowance needs to be further improved, the only way is to lower the information rate R_b.

After $C/(N_0 + I_0)$ is obtained, it can be calculated as follows according to (5.20):

$$\begin{aligned}\frac{1}{(C/I_0)} &= \frac{1}{C/(N_0+I_0)} - \frac{1}{C/N_0}\\ &= \frac{1}{[41.2]_{dB}} - \frac{1}{[44.2]_{dB}}\\ &= 44.2(dB/Hz)\end{aligned}$$

If it is represented by decibel, thus

$$[C]_{dB} = [C/N_0]_{dB} + [N_0]_{dB} \tag{5.27}$$

N_0 represents the noise, then $N_0 = KT$, K = −228.6dB/k, $T = 560k$. Represent the following with decibels $[T]_{dB} = 27.5dBk$, thus:

$$N_0 = -228.6 + 27.5 = -201.1dB/Hz$$

$$[C]_{dB}44.2 - 201.1 = -156.9dBw$$

$$[I_0]_{dB} = [C]_{dB} + \left[\frac{1}{C/I_0}\right]_{dB} = -156.90 - 44.20 = 201.1dBW$$

According to (5.22), n shall be obtained as follows:

$$I_0 = \frac{4}{3}\frac{\Delta}{T}(n-1)C_i/B_{IF}$$

$$\frac{4}{3}\frac{\Delta}{T}(n-1) = I_0 \times \frac{B_{IF}}{C}$$

When I_0, B_{IF} are represented with decibels, then

$$n = \left[10^{[(I_0)_{dB}+(B_{IF})_{dB}-(C)_{dB}]\div 10} \times \frac{3}{4} \times \frac{T}{\Delta}\right] - 1$$

$$= \left[10^{[-201.1+33-C-156.9]\div 10} \times \frac{3}{4} \times \frac{T}{\Delta}\right] - 1$$

$$= \left[10^{-1.12} \times \frac{3}{4} \times \frac{T}{\Delta}\right] - 1$$

$$= 58$$

The number of users who are operating at the same time is 58. Even if the spreading code chip rate is 5 Mbit/s, there are still 29. In the complete system, there are three satellites and nine inbound beams in total. Inbound information per hour C_{Hr} = 2000 bit/s × 3600 s × 58 × 9 = 3.7584 Gbit/h. Based on the position reporting service of inbound information per user of 71 bits, the number of users per hour N = C_{Hr}/71 = 57 925,000, which is twice the outbound capacity.

In order to ensure that the inbound information rate is maintained at 8 kbit/s, the measures that can be taken include: increasing the G/T value of the satellite to 0.5 dB/K and increasing the user transmitting EIRP to 12 dBW. High elevation angles of three observable satellites are fully utilized to improve the transmitting antenna augmentation of the user receiver (3dBi) to maintain the user transmitting power at 8 W. If it is required to reduce the transmitting power of the user receiver to the maximum extent, GEO3 satellite must be designed especially to make G/T ≥ 3.5 dB/K. First, the demand of the CRDSS service for Chinese users is satisfied.

5.4 CRDSS Navigation Positioning Satellite

5.4.1 Mission and Functional Parameter

The CRDSS Navigation Positioning Satellite includes a satellite with an RNSS payload and one with RNSS and RDSS payload (i.e., CRDSS payload). RNSS satellites include GPS satellite, Galileo, China Compass MEO, and IGSO satellite, which transmits the different navigation signals under different L frequency bands under synchronization of satellite clocks. Driven by MCC, RDSS payload transmits navigation signal of S frequency bands from MCC atomic clock and transmit inbound signal of L frequency band from the user. The satellite with RNSS and RDSS payloads together is called the CRDSS satellite, which generally operates at the GEO or IGSO orbit.

As its name implies, the RNSS satellite is a navigation positioning satellite that realizes the Radio Navigation Satellite Service. The CRDSS satellite, providing both RNSS and RDSS services, realizes global navigation positioning and position reporting and is key in realizing multilevel and personal services of satellite navigation. The RDSS service not only aims to provide position reporting and user information sharing but also is responsible for spreading positioning mode, improving positioning accuracy, relieving the burden of users, reducing the complexity of the user's receiver, and application processing.

The CRDSS Satellite has the following functions:

1. Realization of global RNSS service coverage;

2. Formation of global CRDSS position reporting coverage;
3. Assumption of interactive information exchange between MCC and user receiver.

The primary indexes include the following:

1. RNSS payload possesses the navigation signal of at least two frequencies Standard power reaching the ground is superior to −163 dBW;
2. RNSS payload is a 1500 km covering beam of the Earth;
3. CRDSS satellites of GEO1 (140°E), GEO2 (80°E), and GEO3 (110.5°E) finalize the coverage for West Pacific Ocean, Indian Ocean, and China and surrounding regions, respectively. The ocean is the area with the least users. China and the surrounding region is the area with a greater number of users;
4. The CRDSS satellite has an S frequency band EIRP = 43~52 dBW to realize a different coverage service area;
5. CRDSS satellite has an L frequency band G/T value = −4.5~3.5 dB/K to realize a different service area;
6. There are multiple measures of time synchronization payload and measurement including satellite-ground two-way time comparison payload, the C frequency band radar ranging time comparison payload, laser two-way ranging payload, and the conversion between the total two-way round trip distance measurement and user double difference distance measurement. Satellite borne pseudorange measurement accuracy is above 1.0 ns.

5.4.2 RNSS Satellite

The RNSS satellite is the one with the most CRDSS systems and maximum functions, consisting of GPS, GLONASS, Galileo, and Compass, which can also utilize the regional satellite systems of surrounding countries, such as the Japanese QZSS and Indian IRNSS. In order to meet the demand of stability and reliability of Chinese users, China Compass has the essential quantity required to meet the regional system service demand. Due to the utilization of the three-satellite distance measurement positioning principle, the RNSS regional constellation consisting of 5GEO+3IGSO+27MEO is able to provide satisfactory service for Chinese users. Therefore, CRDSS comprehensive radio determination service cannot specifically be designed for the RNSS satellite. However, it is included in the CRDSS constellation system, inspecting their satellite clock error, precise orbit, and making corresponding corrections as well as providing information such as monitoring and availability through GEO satellite RNSS payload messages. When necessary, it is able to broadcast correction parameters. When MCC is processing the CRDSS service, a more precise correction parameter is required to finalize a personalized service to specific users.

5.4.3 CRDSS Satellite

The CRDSS satellite can be divided into two schemes according to the global or regional coverage. One is the so-called regional service CRDSS satellite and the other one is the global service CRDSS.

5.4.3.1 Regional CRDSS Satellite

The regional CRDSS satellite is the CRDSS constellation that ensures the regional system meets the requirements of the CAT-I Aviation I category precise approach service and

1.0 m high accuracy positioning and position reporting services, by cooperating with the global or regional RNSS navigation satellite. The RNSS satellite will not be discussed, and only the constellations and satellites laid out on GEO orbit are discussed here.

1. *Regional CRDSS constellation*
   The regional CRDSS constellation is the GEO constellation designed for augmenting the navigation capability of regional RNSS and information exchange capability, with the missions as follows:
   a. Design the positioning accuracy of regional system RNSS to be superior to CAT-I precision, namely, plane precision of 18.2 m and vertical precision of 7.7–4.4 m;
   b. Meet the requirement of two-way parallel information verification of the aviation user. When GEO satellite is used to broadcast integrity, elevation of user observation in the mission area is greater than 25°;
   c. S/L frequency inbound and outbound links possess sufficient transmission capacity.
   According to this requirement, there must be at least three satellites for one regional CRDSS constellation. In order to meet requirement of wider area coverage, it is best to arrange the three satellites with interval of 30°. By taking Figure 3.1 as an example (see Section 5.3.2), the constellation consisting of three satellites, GEO1, GEO2, and GEO3 is suitable.
2. *Regional CRDSS satellite mission*
   Only the CRDSS satellite for regional service is discussed here. The RNSS payload is the same as regular RNSS satellite. However, difference corrections for other satellites in regional system are added, as well as the ephemeris correction parameter and integrity information. Partial S/L coverage capability of CRDSS is the key point of discussion, which meets the requirement of all-region RDSS positioning and report. This scheme is analyzed based in the region of China.
   It is assumed that constellation satellites are operating at GEO1 (140°E), GEO2 (80°E), and GEO3 (110.5°E), and the S/L beams of the three satellites cover the respective design areas. Every satellite to ground coverage area: east-west direction 120°, south-north direction 100~120°. Land area can be a little bit smaller to obtain higher antenna augmentation with greater information capacity. In order to improve the anti-shield angle capability of position reporting of the user, it is reasonable and feasible to arrange three IGSO satellites with CRDSS payloads.
3. *Satellite S/L beam design*
   If single beam meets the requirement of coverage, satellite S/L beam coverage is shown in Figure 5.8. The S represents the height from satellite to ground; Re represents the radius of the Earth; β is the included angle of satellite beams; and α is the included angle between satellite beam and the Earth. Here the latitudinal direction is taken as reference. According to Figure 4.1,

$$\frac{\beta}{2} = arctg\left(\frac{r_c}{H_c + r_0}\right) \tag{5.28}$$

$$r_c = R_e \sin\frac{\alpha}{2} \tag{5.29}$$

$$r_0 = R_e - R_e \cos\frac{\alpha}{2} \tag{5.30}$$

Substitute (5.29) and (5.30) into (5.28), thus

$$\frac{\beta}{2} = arctg\left[\frac{R_e \sin\frac{\alpha}{2}}{H_c + R_e\left(1 - \cos\frac{\alpha}{2}\right)}\right] \tag{5.31}$$

Substitute α = 120° in it, then

$$\frac{\beta}{2} = 7.9°$$

Included angle of beams β = 15.8°

Antenna augmentation $G_a = 10\lg\left[\left(\frac{360°}{15.8°}\right)^2 \eta\right]$

When antenna $\eta = 0.5$, $G_a = 24.1(dB)$

According to antenna augmentation, the relation expression of included angle of beams and antenna aperture is:

$$G_a = \left(\frac{\pi D}{\lambda}\right)^2 n \tag{5.32}$$

D is the diameter of antenna (m). λ is the operating wavelength (m). When λ is L frequency, the single wavelength $\lambda = \frac{3\times10^8}{1.6\times10^9} = 0.1875m$.

Expressed in decibels: $\eta = 0.5$, the diameter of the antenna (D) can be obtained.

$$D = \frac{360\lambda}{\pi\beta} = 1.35(m)$$

Obviously, a single beam antenna has very low augmentation due to small size. In order to realize the G/T value of ≥−0.5 dB/K in Table 5.9, satellite antenna augmentation Ga must be greater than 27 dBi. Multi-beam coverage must be designed with augmentation of each beam greater than 30 dBi. If Every satellite has three L, S beams respectively, the beams are arranged in east-west parallel sequence, antenna aperture $Dx = 3.0$ m along east-west direction and $D_y = 2.0$ m in a south-north direction, then the augmentation of L frequency band can reach:

$$\begin{aligned} G_a &= 10\log\left[\left(\frac{\pi}{\lambda}\right)^2 D_x D_y \eta\right] \\ &= 10\lg\left[\left(\frac{\pi}{0.1875}\right)^2 (3.0\times2.0)\times0.5\right] \\ &= 29.2(dBi) \end{aligned}$$

Corresponding beam angle $\beta_x = \frac{360\lambda}{\pi D_x} = 7.6^{\circ}$, $\beta_y = 10.74^{\circ}$.

Every satellite has coverage of 120° in east-west direction with some overlapping area. Three satellites finalizing the coverage areas in Section 5.3.2.4 also have overlapping areas. In the south-north direction, it can reach 40°S~60°N. As L/S antenna of every satellite shares one reflecting surface with 6 L, S feed networks and the augmentation of 3~4 dBi lower, then effective augmentation transmitted and received G ≥ 25 dBi to meet the requirement of EIRP ≥ 46 dBW in Table 5.8 and G/T ≥ −0.5 dB/K in Table 5.9. This is one of the schemes of realization of CRDSS. If a mesh antenna with aperture D ≥ 6 m can be adopted, the performance index can be better. Augmentation

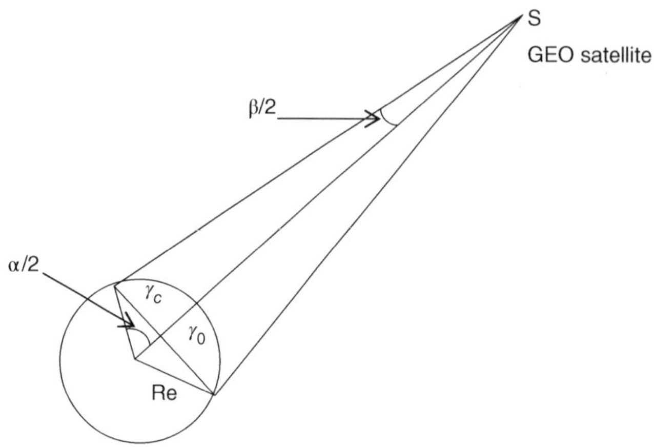

Figure 5.8 S/L beam coverage.

can be increased to 35 dB. Satellite G/T ≥ 6 dB/K meets the requirement of EIRP ≤ 6 dBW under the condition of inbound information rate of 8 kbit/s.

4. *S/L multi-beam transponder design*
In this scheme, take the C band as the uplink and downlink feed link between the MCC and satellite. Namely, the uplink is a 6 GHz frequency band and downlink is a 4 GHz frequency band. Refer to Figure 5.9 for the workflow of the C/S and L/C transponders. The three outbound S beams, which are the three frequency division CDMA signals, Fu1, Fu2, and Fu3 generated by MCC, convert to the space division beam at a frequency of 2483.5~2500 MHz through frequency conversion to cover the three designated areas, respectively. However, the three inbound L frequency

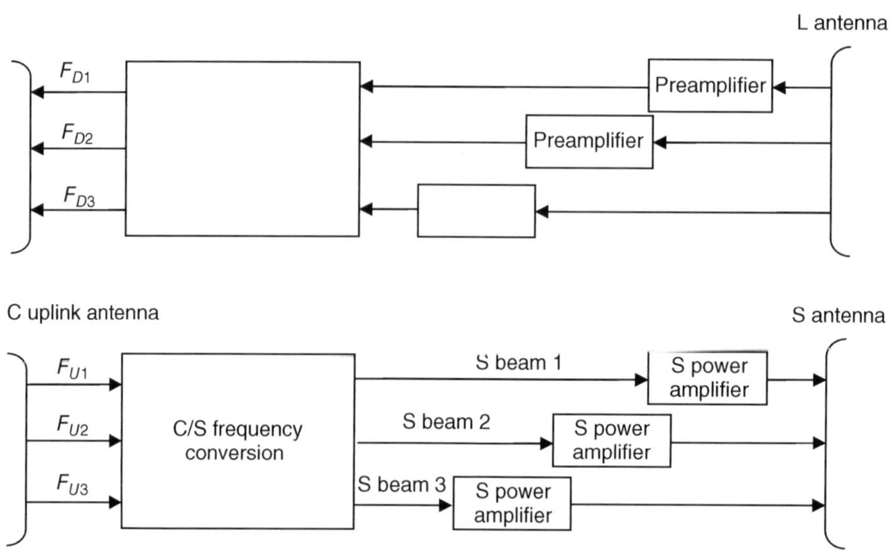

Figure 5.9 C/S, L/S transponder configuration.

bands (1610~1626.5 MHz) partition coverage beams convert to three inbound signals of the three frequencies F_{D1}, F_{D2}, F_{D3}, F_{D1}, F_{D2}, F_{D3}, and F_{u1}, F_{u2}, F_{u3}, through frequency conversion. The design is performed according to frequency coordination of the adjacent orbit.

5.4.3.2 Global CRDSS Satellite

The global CRDSS satellite is discussed through the GEO and IGSO satellite constellation and satellites focusing on global service. The RNSS mission and CRDSS mission aim to improve global service. RNSS constellation and payload are not required to meet the CAT-I precise positioning service.

1. Global CRDSS constellation
   The global CRDSS constellation consists of GEO and IGSO satellites designed to meet the requirement of the global RNSS service and users' information exchange capability. Its missions include:
2. Ephemeris design of GEO and IGSO satellites RNSS payload mainly aims to improve precision. Quick variable parameters in ephemeris can be reduced or eliminated.
3. In the constellation, S/L payload is designed to meet the requirement of a wide region with few users. Then, its coverage area is at its greatest extent. Design of distance between satellites aims to improve the maximum angle of visibility. Besides retention of the CRDSS satellites designed for Chinese service, the CRDSS payload is properly increased on some satellites of MEO.
4. Power flux-density of MEO satellites S/L payload can have satellite G/T lower than the regional CRDSS satellite. It is economic and feasible to realize service for wide area with few users with low rates of information.
5. MEO satellite CRDSS inbound and outbound payload beam design
   The S/L outbound and inbound beam is a globe coverage beam with an angle $\beta = 26°$ and satellite antenna aperture $D = 1.3m$ and satellite outbound beam antenna augmentation $G_s = 18.3$ dB.

When the S frequency power amplifier uses a TWTA of 120 W, actual outbound data rate can reach 10 kbit/s. When user EIRP = 6 dBW, the inbound data rate can reach 1 kbit/s. Refer to Tables 5.10 and 5.11 for level calculations.

Table 5.10 Link level calculation under antenna aperture D = 1.3 m.

Satellite EIRP (Gr = 23.5 dB, P = 120 W)	44
Free space path loss (dB)	−191.8
User's receiving power (isotropic antenna)	−147.8
User receiver G/T value (dB/K)	−24
Bohr constant, K (dB)	−(−228.6)
C/N_0 (dB/Hz)	56.8
Data transmission symbol rate loss [100 kb/s] (dB) (QPSK combiner)	50
Receipt Eb/N_0 (dB)	6.8
VTB demodulation required Eb/N_0 (dB)	4.6
Total allowance (dB)	2.2

Table 5.11 L link level calculation under antenna aperture D = 1.3 m.

User EIRP [P = 3 W] (dBW)	6
Free space path loss (dB)	−188.4
Weather loss (dB)	−0.5
Satellite receiving single user power (dBW)	−182.9
G/T value of satellite [G = 23.5 dB, T = 560 K] (dB/k)	−4.0
Bohr constant, K (dB)	−(−228.6)
C/N_0 (dB)	41.7
If C/N_0 of the earth station is big enough, total C/N_0 (dB/Hz)	41.2
Data transmission symbol rate loss [2 kbit/s] (dB)	−33
Receipt Eb/N_0 (dB/Hz)	8.2
VTB demodulation required Eb/N_0 (dB/Hz)	4.6
Total allowance (dB)	3.6

It is shown by the calculation result that the RDSS satellite S/C antenna size must be properly increased in order to further improve the outbound and inbound capacities. As configured in Figure 5.8, the complete coverage of the Earth is formed by utilizing an array antenna or multi-beam composition antenna.

CRDSS service under MEO satellite is different from the GEO/IGSO satellite as the outbound and inbound link cannot be formed continuously. Satellite processing and inter-satellite links must be utilized together to accomplish the missions.

5.5 CRDSS Ground System

The CRDSS ground system includes the system's operational control as well as the user's application service and terminal. The user terminal will be introduced in a later, specific chapter. In this chapter, mainly operational control of the CRDSS global position reporting system and the users' application service will be introduced.

5.5.1 Function and Composition of Ground System

Ground system is responsible for operational control and application service of CRDSS. Its functions can be summarized as follows:

1. Realizing time synchronization and precise orbit determination of a selectable GNSS satellite.
2. Ensuring time difference of the CRDSS system, namely maintaining the CRDSS time system, in which Compass Time BDT is the time reference for CRDSS.
3. Establishing a uniform ionosphere correction model and real-time correction parameter through the measurement of GNSS multiple frequency band navigation signals.
4. Finalizing the monitoring of integrity of the satellite used and space signal to provide integrity monitoring information for high dynamic and high reliable user.
5. Providing GNSS high accuracy real-time dynamic positioning capability.

6. Finalizing high accuracy quasi-real-time positioning service.
7. Providing measurement and control service for the near-earth satellite.
8. Implementing lifesaving and individual positioning services.
9. Providing guidance service for the user.

In order to ensure the completion of these missions, the ground system must have a powerful MCC and a proper quantity of reference station systems with reasonable layouts. This reference station system completes the collection of time synchronization, satellite precise orbit determination, and original observed quantity of system ionosphere correction parameters by properly dividing the work and configuring the corresponding GNSS monitoring receiver.

Maintaining the operational control of system basic function is not going to be elaborated on this chapter; here, we just analyze and elaborate on the functions of providing high accuracy and performance.

5.5.2 Measurement and Control Center (MCC)

The MCC is the measurement and information processing center for maintaining system operation and service and completing observation and calculation of user's distance in CRDSS, completing CRDSS satellite selection, orbit determination, time synchronization, navigation signal propagation time delay correction, navigation signal integrity detection and prediction, and realizing positioning services, including a high accuracy real-time dynamic positioning service, quasi real-time high accuracy positioning service, near earth satellite measurement and control services, rescue/lifesaving and individual position services, and guidance service.

5.5.2.1 CRDSS Service User Distance Measurement and Positioning Equation

CRDSS user distance measurement refers to the measurement of the total roundtrip distance from MCC to user through CRDSS, which is an important observation parameter for CRDSS positioning and has a strong influence on the tri-satellite positioning accuracy of the user. It is analyzed next according to the channel path of the system.

Figure 5.10 shows the relationships between the CRDSS satellite and MEO satellite operating in GEO orbit or IGSO orbit and MCC and the user. GEO1 and GEO2 are the CRDSS satellites operating in GEO orbit with the two-way round trip ranging function. GEOI is the CRDSS satellite operating in IGSO orbit with the two-way round trip ranging function. The round trip ranging observation means that MCC transmits the distance measurement signal to the user through CRDSS satellite outbound transponder C/S, and the user transmits the L-frequency response signal to MCC through CRDSS inbound transponder L/C after the inquiring signal is received. The total round-trip distance from MCC to the user through the satellite is measured by the MCC. Distances from three satellites to the user are obtained through the total round-trip distance of three CRDSS satellites with reference to CRDSS ephemeris. Tri-satellite user positioning can be accomplished by utilizing three spheres intersection principle and formula (5.1). If the observed quantity of three satellites is not present, dual-satellite positioning can be completed with the user's height data. If there is only one CRDSS satellite in GEO orbit, and the total distance is obtained and the user finalizes the measurement of pseudorange differential and carrier phase difference measurement of this CRDSS satellite RNSS L frequency band navigation signal and other satellite (GEO, IGSO, MEO), then

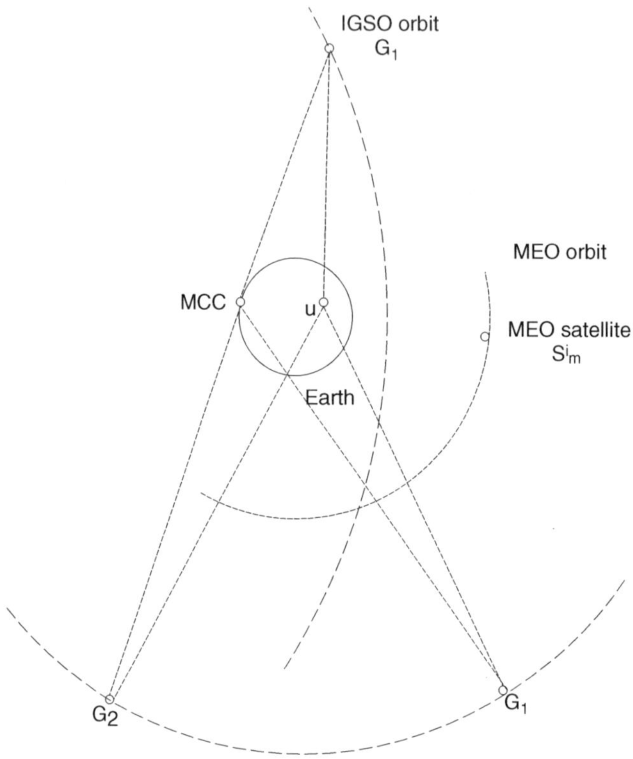

Figure 5.10 CRDSS positioning satellite selection and positioning principle.

the RNSS L-frequency band navigation signal, tri-satellite user positioning can be completed as well. This is the dominant mode of comprehensive RDSS position reporting. Obviously, obtaining the observed quantity of distance sum of one CRDSS satellite is the primary condition of determine the user's position. Its observation accuracy of distance sum is one of the critical factors that affects positioning accuracy. Equipment transmission delay on the observation path shall be deducted strictly. And actual distance in space is obtained as well to meet the positioning condition of formula (2.6). Based on this analysis, we can figure out observed quantity under different positioning equations:

- Dual-satellite positioning equation
  Dual-satellite positioning calculating is completed through formulas (3.1), (3.2), (3.12), (3.18), and (3.21) in Chapter 3.
- Tri-satellite positioning solving equation
  User position is calculated according to the rationale presented in Section 5.5.2.1.

5.5.2.2 Distance Measurement System Scheme

Distance measurement is completed by MCC, using the principle that follows:

MCC transmits the outbound inquiring signal through S_1, S_2, or S_I. Its identification moment is the enabling signal of a MCC distance measurement counter. After the identification moment signal is received by the user, the user will transmit a response signal with the user identification moment. The user identification moment is demodulated by

MCC as the closing signal to measure the total time delay sum of round trip from MCC to user u through S_1. It multiplies by light speed C to obtain the total distance sum. In order to achieve convenient engineering implementation, the transmitting signal and response signals possess the following features:

1. Transmitting signal is a continuous CDMA signal modulated by QPSK and repeatedly transmitted in the format of fixed length period frame.
2. Modulation code of the inbound signal transmitted for the user response is similar to the outbound signal.
3. As the distance signal is in a repeated frame format, the distance measurement can be finished within dozens of milliseconds. The distance sum measurement of hundreds of milliseconds can be completed within dozens of milliseconds.
4. In the answering signal, the user clearly fills the frame counting numbers of the receiving signal N and the transmitting response signal N + n (n = 1, 2, 3...32). As the processing time of frame format is sufficient, the simultaneous inbound possibility of multiple users is avoided and the mutual interference among multiple users is reduced.
5. MCC system possesses adequate capabilities of burst signal acquisition, tracking and high accuracy ranging. Incoherent delay phase-locked loop is adopted to realize the recovery of inbound signal. Time delay measurement standard error (RMS) can be approximately expressed as:

$$\delta_{DLL} = T_{ch}\sqrt{\frac{B_N}{2[C/N_0]}\left[1+\frac{2}{T[C/N_0]}\right]}\ (\text{ns}) \tag{5.33}$$

In the formula, T_{ch}: is the spreading code width (in ns), $T_{ch} = \frac{1}{F_{ch}}$, F_{ch} is the spreading code element rate;
B_N: Bandwidth of loop (Hz);
C/N_0: Despreading signal carrier power and noise spectrum density ratio (dB/Hz);
T: Coherent integration time (in s).
C/N_0, relating to performance of G/T of satellite, transponder EIRP, coherent interference of multiple users in the system and noise ratio, is the critical index of the RDSS system design.

In order to make the RDSS system reach 1.0 m level rapid positioning accuracy, the ranging accuracy of the incoherent delay locked loop is further analyzed as follows:

In RDSS specified service frequency band, the frequency bandwidth is only 16.5 MHz, no matter if it is outbound signal of frequency S or inbound signal of frequency L. Namely, S frequency signal bandwidth is 2483.5~2500 MHz and inbound L-frequency signal bandwidth is 1610~2626.5 MHz. Then, the maximum frequency of spreading code is F_{ch} = 8.25 Mbps. Satellite outbound S frequency band EIRP is 46 dBW limited by power. The potential precision of S frequency receiving signal recovered at user terminal is calculated as formula (5.13). The design condition is as follows (refer to Section 5.3.6):

$$C/N_0 = 58.8dB/Hz$$

Based on the data transmission rate of 128 kbps, it is T = $\frac{1}{1\times10^3}$ (S) design $B_N = 10$ Hz, then the response moment accuracy of the user receiver is

$$\delta^{s}_{Dll} = \frac{1}{8.25\times10^6}\sqrt{\frac{10}{2[58.8]_{db}}\left[1+\frac{2}{\frac{1}{1\times10^3}[58.8]_{db}}\right]} = 0.16(ns)$$

Due to the small random error of the influence from the user receiver, the influence can be ignored, and the ranging accuracy is completely depending on the ranging accuracy of MCC.

According to Table 5.9, when user EIRP = 5 dBW, there is $C/N_0 = 44.2$ dB/Hz, considering multiple user inbound interference, there is C/(N0+I0) = 41.2 dB/Hz. Inbound information rate Rb = 1 kbps, it is T = 1/Rb = 1×10^{-3} s, design BN = 2 Hz, substitute into formula (5.13), the total distance sum measurement accuracy is

$$\delta^{all}_{Dll} = \frac{1}{8.25\times10^6}\sqrt{\frac{2}{2[41.2]_{db}}\left[1+\frac{2}{1\times10^{-3}[41.2]_{db}}\right]} = 1.11(ns)$$

Ranging accuracy is 1.11 ns, equivalent to 0.33 m. Through difference correction of each item UERE = 2 ns, the available satellite with DOP ≤ 1.5 is selected. The real-time positioning accuracy can reach

$$\delta_H = 0.3\times2\times1.5 = 0.9(m)$$

According to this analysis, under the design conditions of this system option, error assignment, and environment parameters of obtaining 1.0 m real-time positioning accuracy are shown in the following Table 5.12:

5.5.3 GNSS Reference Station System

The GNSS reference station system aims to provide orbit determination, time synchronization, ionosphere correction, and differential correction observed quantity to the system. It is composed of a Compass RNSS reference station, RDSS reference station and GPS, Galileo, and a GLONASS reference station. Compass RNSS and RDSS reference stations and GPS reference stations are mainly studied in this system. Its design objective aims at two services of Class I precision approach and 1.0 m real-time positioning (RMC).

Table 5.12 Assignment of 1.0 m accuracy design parameter.

HDOP	1.5
UERE (m)	≤0.6
Outbound information symbol rate (kbps)	128
Inbound information symbol rate (kbps)	2
Loop bandwidth of user receiver (Hz)	10
MCC loop bandwidth (Hz)	2
Outbound spreading code rate (Mbps)	8.25

Refer to Table 5.1 for the assignment of parameter designs in the following chapters and sections.

5.5.3.1 Local Class I Precision Approach Reference Station System

In order to meet the mission requirement of the CAT-I precise approach for China and surrounding regions, the reference stations required have already had a mature conclusion. American GPS Wide Area Augmentation System (WAAS), European Geostationary Navigation Overlay Service (EGNOS), Chinese Satellite Navigation Augmentation System are all basically used for reference. A reference station system consisting of 20–30 reference stations can be designed to accomplish these missions, in which we will not go into unnecessary detail. After the GNSS satellite is used, are GPS, Galileo similarly added with 20–30 reference stations? This will be elaborated on in this section.

Besides meeting the requirement of the China Compass navigation system, as for GPS satellite, the GNSS reference station has the goal of supporting precise orbit correction, satellite clock correction, and integrity monitoring. The ionosphere shares the correction parameter of Compass GNSS. Thus, only 5–7 reference stations equipped with GPS or Galileo satellite receivers can realize the GNSS monitoring mission.

The compass monitoring receiver on a system monitoring station independently accomplishes observation parameter collection of Compass system satellite clock error, precise orbit, ionosphere correction, and satellite integrity determination. A few monitoring stations utilizing a single frequency receiver realize the observation of GPS, Galileo, and GLONASS satellites and provide pseudorange measurement, carrier phase measurement, and Doppler integration observation value required by the precise satellite clock correction parameter, precise orbit correction, and integrity assessment. The observation value is sent to MCC for calculation and assessment. Usage of corrected parameter and assessment mission are elaborated on in the following sections.

5.5.3.2 The 1.0 m Level Reference System

The accomplishment of arrangement of reference station system of 1.0 m level positioning accuracy is based on corrected precision of distance between reference stations relative to the position of satellite. Distance between reference stations is $D = 800$ km; satellite height is around $H = 20{,}000$ km; under the condition that position error of satellite is $\Delta\rho = 20$ m, the influence of differential stations on the position error of satellites is :

$$\Delta\rho_{\phi} = \frac{D}{H} \times \Delta\rho(m) = \frac{800}{20000} \times 20 = 0.8(m)$$

Stations are laid out as per 800 km, in China (5200 × 5600 km), there probably are 45 stations that need to be arranged. For the 20–30 reference stations, the number of reference stations is increased to 45 to meet the demand of 1.0 m positioning accuracy in China.

5.5.4 Multiple System Satellite Clock Error Determination

Multiple system satellite clock error determination includes Compass satellite clock error determination, GPS, and clock error determination of other satellites. The so-called satellite clock error refers to the error of the satellite clock relative to Compass time BDT. In this chapter, synchronization accuracy that is superior to 1.0 ns will be discussed to meet the requirement of 1.0 m positioning accuracy.

5.5.4.1 Compass Satellite Clock Error Determination

There are many ways of determining Compass satellite clock error, such as two-way satellite-ground pseudorange time synchronization [10], radar pseudorange time synchronization, laser pseudorange time synchronization, as well as clock error/orbit comprehensive determination, and so on.

- Two-way satellite-ground pseudorange time synchronization.
  The principle of satellite-ground pseudorange time synchronization, also called as satellite-ground radio two-way time synchronization, is that: ground station A transmits distance measurement signal with frequency of f_d at Compass time (BDT) t_{BDT}, and the signal is received by the satellite equipment at t_{sr}. The pseudorange p_s at corresponding time is measured and transmitted down to MCC. However, at Compass time t_{BDT}, the satellite transmits navigation signal with the similar frequency of f_d to the ground station, and the signal is received by the ground receiver. The pseudorange p_g at this time is measured. Due to the error Δt_{gs} between satellite time and Compass time, pseudoranges p_s, p_g both contain the clock error between ground time (BDT) and satellite clock. The relation expression is:

$$\Delta t_{gs} = -\frac{1}{2}(p_s - p_g) \tag{5.34}$$

  "–" shows that the Compass time (ground clock) is outstripping satellite clock.
  As the two-way distance measurement signal adopts extremely close frequencies, the same ionosphere propagation time delay can be offset in (5.34). Therefore, precision of Δt_{gs} depends on the precision of satellite-ground equipment pseudorange measurement. Its random error of pseudorange measurement can be weakened through multiple measurements. Time synchronization error is superior to 0.5 ns. In the time synchronization of multi-satellite, equipment time delay of two-way pseudorange measurement (zero) will become the main error source, in which, fluctuation term of the equipment time delay plays the leading role. The stabilization term of equipment time delay becomes the error of the system. In one 12-h period of satellite-ground time comparison, these errors can be corrected together in Δt_{gs} as the clock errors respectively. However, the fluctuation term of equipment time delay is extremely small, which is within around 0.1 ns.
- Radar pseudorange time synchronization
  Radar pseudorange time synchronization is accomplished by ground MCC. The pseudorange p_g of observed satellite is measured by MCC and at the same time, the space distance D from ground to the satellite is measured by radar. The satellite clock error is:

$$\Delta t_{gs} = p_g - \frac{1}{2}D \tag{5.35}$$

  As the inquiry response radar is used, time delay Δt_{tr} of transponder contained in D shall be deducted. Therefore, (5.34) is rewritten as:

$$\Delta t_{gs} = p_g - \frac{1}{2}D + \Delta t_{tr} \tag{5.36}$$

  This radar operates at C, Ku or an even higher frequency to eliminate the influence of propagation time delay of signal in ionosphere. In order to obtain higher accuracy, it is suitable to use laser radar.

- Laser pseudorange time synchronization
  Laser pseudorange time synchronization is the same as radar. It is nothing but the space distance from ground to the satellite D_0 is measured by laser. Formula (5.35) is suitable for laser pseudorange time synchronization. Merely, as for high accuracy of laser distance measurement, position correction between ground laser station and ground receiver, as well as the position correction between laser mirror on the satellite and signal reflecting antenna phase center, shall be taken into account.
  As CRDSS must consider the propagation path of RDSS signal as well as the clock error included in RNSS propagation path, namely GEO satellite clock error is mainly the usual error, the method elaborated on here must be used to its greatest extent to perform precise determination of clock error of GEO satellite and make it superior to 0.5 ns.
  When CRDSS adopts GPS or other satellites to get involved in 1.0 m level precise positioning, it becomes very important to realize synchronization of a GPS satellite and BDT. Most probably, the GPS clock error can only be weakened through differential correction. Otherwise, it is difficult to achieve 0.5 ns level. However, when ionosphere is corrected correctly, difference method is the effective measure and its relationship with difference station is slackened.

5.5.5 Multiple System Satellite Precise Orbit Determination and Application

As for real-time high accuracy positioning user, the pseudorange observed quantity of reference station is utilized to obtain the correction value of pseudorange. Although it significantly corrects the satellite clock error and orbit deflection, this pseudorange correction value is difficult to be provided in real-time for numerous users of regional system. For instance, for CAI-I precise approach user with strong real-time performance and high integrity, the user cannot select the satellite according to correction value of pseudorange of a few different reference stations and satellites as all of the satellite messages cannot finish broadcasting the pseudorange correction value of numerous satellites and reference stations within the minutes of approach. This method can only be used for high accuracy, such as 1.9 m level positioning accuracy in the CRDSS positioning method of calculation done by MCC. Then as for CAT-I users, it can only broadcast satellite clock error correction and orbit correction parameters. Therefore, as for multiple system satellite precise orbit measurement and application, generally, only the method of broadcasting clock error correction and ephemeris correction can only be used. The Compass Satellite can directly broadcast precise satellite clock error and orbit parameters, and the GPS satellite still adopts the method of broadcasting clock error and orbit correction.

For Compass GEO satellite precise orbit determination, Compass GEO satellites can use methods such as two-way pseudorange time synchronization, radar pseudorange time synchronization, and laser pseudorange time synchronization to determine the satellite clock error first, and then determine the satellite position through utilizing the observed pseudorange quantity obtained by the known reference stations and distance observation obtained by radar to acquire high accuracy orbit prediction. Observation equation of distance measured by the known reference station is expressed as:

$$\rho^i = [(X^i - X_j)^2 + (Y^i - Y_j)^2 + (Z^i - Z_j)^2]^{1/2} \tag{5.37}$$

ρ^i is the distance from the reference to satellite i obtained through observation pseudorange and observation distance of the known reference station;

X^i, Y^i, and Z^i are the parameters of the satellite three-dimensional position coordinates that need to be solved.;

X_j, Y_j, and Z_j represent the three-dimensional position coordinates of known reference station j.

After the time of every reference station is synchronized via two-way pseudorange time synchronization, the synchronized observation moment of every reference station $\rho^i(t)$ can be obtained to solve satellite position at every corresponding time t_k. Formula (5.37) can be written as

$$\rho^i(t_k) = [(X^i(t_k) - X_j)^2 + (Y^i(t_k) - Y_j)^2 + (Z^i(t_k) - Z_j)^2]^{1/2} \tag{5.38}$$

t_k is the observation moment of every observation pseudorange.

For Compass NGEO satellite precise orbit determination, the Compass NGEO satellite can only obtain satellite clock error through satellite-ground two-way pseudorange synchronization. After the satellite clock error Δt^i is known, distance from reference station to satellite i can be similarly obtained through pseudorange $\rho^i_p(t_k)$ of the reference station.

$$\rho^i(t_k) = \rho^i_p(t_k) - c\Delta t^i \tag{5.39}$$

Where c is the light speed.

For GPS satellite precise orbit determination, besides laser observation to obtain independent time synchronization through pseudorange observed quantity and laser observed quantity, the GPS satellite cannot independently calculate the GPS satellite clock correction parameter. It can only obtain the satellite clock error and satellite position together with pseudorange observed quantity obtained by known reference station. Its method is the same as Chinese satellite navigation augmentation system. Tests show that it can meet the requirement of the Class I precision approach.

5.5.6 Formation and Application of Ionosphere Correction Parameter

As there are many frequency bands applied in the CRDSS comprehensive radio determination, besides the pseudorange observed quantity of the L-frequency band, there is a total observed quantity of total distance consisting of the C-, S-, and L-frequency bands. If the precise correction is not made, it is hard to achieve high accuracy positioning. As for the users who only adopt the L-frequency band navigation signal pseudorange observation, the reliable ionosphere correction is to utilize dual-frequency receiver and two frequency observed quantities to obtain the space pseudorange from user to satellite. The relation expression is:

$$\rho_0 = \rho_{f1} - \delta\rho \frac{f_2^2}{f_2^2 - f_1^2} \tag{5.40}$$

ρ_0 is vacuum pseudorange

ρ_{f_1} is pseudorange observed with frequency of f_1.

$\delta\rho$ is pseudorange difference between frequencies f_1 and f_2.

f_1 and f_2 are the two navigation signal frequencies for receiver.

If a single frequency is used, ionosphere correction can be performed by utilizing grid ionosphere and model parameter obtained by system. These parameters are all in MCC system and can be obtained easily, which will not be given with more details here. Correction of user's observation can be performed by utilizing correction value obtained by dual-frequency pseudorange of adjacent reference stations. However, this correction requires a large amount of observation stations. Distance from user to reference station is generally 100 km. The relatively high accuracy can be obtained.

The method of pseudorange correction of total distance sum of multiple frequency bands (C, S, and L) will be discussed next.

CRDSS distance sum differential correction. This is a comprehensive correcting method of realizing multiple correction parameters. Look back at dual-satellite (tri-satellite) positioning equation in this chapter. In the total observed quantity of distance of S_1 and S_2, there are transmission distances of C1, C2, S, and L frequencies as well as transmission time delay residual error of signal in two transponders of C/S and L/C, which need to be corrected. Complicated utilization frequency and transmission time delay, expressed in one distance sum, cannot be corrected in segments. New propagation time delay must be set up to perform uniform total time delay correction. In the Compass No. 1 system, dual-satellite differential correction is performed, which can be used for CRDSS distance sum differential correction.

This is based on the comprehensive propagation time delay error correction of RDSS correction station established in the region. In the RDSS service area, several calibration stations are set up. The address of calibration station is known. One or two RDSS user receivers are placed on the station. Through the cooperation of MCC and RDSS user receivers, a number of total distance observed quantities S^i are formed. A distance sum observed quantity correction parameter is formed by these S^i observed quantities based on a grid arrangement. When user distance and observed quantity correction parameter are being calculated, the calculation can be performed according to the grid. Then, this observed quantity correction parameter ΔS_u^i includes both ionosphere correction and correction parameters of satellite transponders. Therefore, accurate corrected distance sum observed quantity S^i can be reached. This grid base is updated every 10 min and invoked by user positioning calculating module in MCC to meet the positioning accuracy requirement of RDSS and provide precision correction accuracy according to distance from the user to the correction station.

Therefore, in the CRDSS ground system, the quantity of calibration stations cannot be reduced. The quantity shall be increased according to requirement of 1.0 m positioning accuracy. In China, arrangement of 45 calibration stations is necessary.

5.5.7 GNSS High-Accuracy Real-Time Dynamic Positioning

The so-called GNSS high-accuracy real-time dynamic positioning is the real-time dynamic positioning that meets the CAT-I precision approach. Available satellites are Compass and GPS. Autonomous positioning is accomplished when MCC broadcasts ephemeris through CRDSS satellite and user receives ephemeris broadcasted by navigation satellite, namely, the autonomous navigation positioning implementing CAT-I approach, which will not be discussed here. Only the user's positioning accuracy (NSE95%), horizontal 18.2 m, vertical 7.7–4.4 m, and the positioning method solved by MCC will be discussed. In order to make argument convenient, the position error is specified as 5.0 m, which is equivalent of RMS's error of 2.5 m. In order to meet the requirement of this high dynamic users, MCC can select the observed constellation for the user according to approximate position and operation time of experimental area. The selection and replacement of the observed constellation of the user can be realized through RDSS link, and then, the original observed quantity of the user shall be received. Positioning calculating will be accomplished by MCC according to the possessed satellite ephemeris and all kinds of correction parameters. The procedure of high accuracy positioning of the user can be analyzed as follows:

1. Selection of satellite constellation
   According to the controllable capability of error, the observation satellites that are DOP $\leq$ 1.5 are selected to constitute the constellation of the user. Under this condition, the equivalent distance error is UERE $\leq$ 1.6 (m).
2. Measurement and receiving of the observation parameters
   After observation satellite is fixed, RNSS pseudorange observation or pseudorange distance observation is accomplished by RNSS and RDSS dual module user receiver. RDSS distance measurement response shall be made. In the response inbound message, observation parameters will be returned to MCC.
3. Selection and calculation of observation parameters
   When over four pseudorange observed quantities are obtained, RNSS pseudorange positioning equation can be used to obtain the user's position parameters. However, the correction of the observed pseudorange shall be completed first to make the end user's UERE $\leq$ 1.6 m.

Assessment of UERE is as follows:

1. Completing the assessment of user's observation error. Curve fitting can be performed to pseudorange observed quantity and pseudo-distance observation error can be assessed. If it is not qualified, the user shall be informed to improve observation conditions and operation states of receiver.
2. Assessment of satellite clock error correction parameter.
3. Assessment of satellite ephemeris correction parameters.
4. Assessment of ionosphere correction parameters.
5. Assessment of differential correction parameters. Under the condition that items (2), (3), and (4) are overproof, observation parameters of the reference station that is closest to the user shall be selected to form difference correction parameter and perform correction of user's observed quantity.
6. Positioning calculating. According to the user's demand, positioning accuracy is assessed and the result is reported.

5.5.8 CRDSS High Accuracy Quasi Real-Time Positioning

CRDSS quasi real-time positioning aims to meet the user's requirement of 1.0 m of precision (RMS). For instance, territorial resources monitoring user often needs to finalize the high accuracy positioning of the measured land within dozens of minutes. However, the user's receiver must be inexpensive, and the user is not required to set up a reference station for different observations or obtain the assessment result of high accuracy positioning without post-processing; then the user can shift the complicated post-processing and expensive observation equipment cost to service provider and service system to use high accuracy positioning. Obviously, the MCC will be responsible for selection of reference stations and extraction of data or forming virtual reference station correction parameters to replace the operation of reference station of the user. The post-processing performed by the user is done by the MCC in advance. As MCC has abundant precise ephemeris and ionosphere correction parameters and is capable of getting this mission done. It processing is basically the same as in Section 5.5.7. The critical point is selection of the virtual reference station and formation of virtual correction parameters.

5.6 Typical Application Scheme

5.6.1 High Accuracy Pseudorange Double Difference Application Scheme

High accuracy pseudorange double distance positioning is the main application mode of the CRDSS, the positioning accuracy of which can reach meter- or even tenth of a meter-level. The distance of the user and observation and position calculating are completed by the MCC other than the user receiver. The user completes the pseudorange double difference measurement among three satellites. CRDSS functions are shown in Figure 2.4 in Chapter 2. In the high-accuracy pseudorange double difference positioning scheme, simultaneous realization of user positioning, time service, and communications are fully reflected.

5.6.1.1 Pseudorange Double Difference RDSS Positioning Principle

In order to make the principle of pseudorange double difference RDSS positioning universal, Figure 5.1 can be replaced with Figure 5.11. User u can be the user in the air or

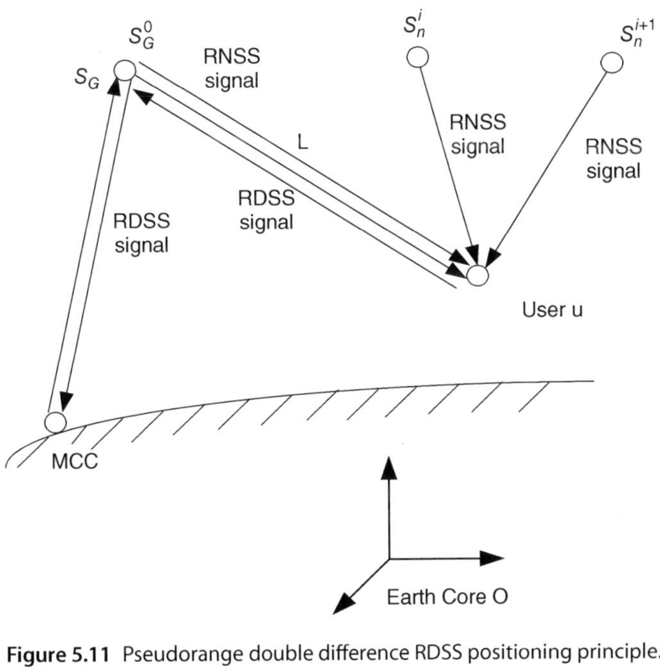

Figure 5.11 Pseudorange double difference RDSS positioning principle.

the user on the ground. Its positioning procedure is as follows:

MCC is one of the components of the system supporting RDSS and RNSS dual-module operational control and application. Its functions include:

1. Realizing time synchronization of the entire system.
   Including RDSS, RNSS satellite time synchronization, MCC system time, and satellite time synchronization.
2. Supporting satellite precise orbit determination of the entire system including precise orbit determination of Compass, GPS, and other navigation satellites. Obtaining satellite precise orbit parameters of Compass and GPS, and so on.
3. Supporting formation of precise ionosphere correction parameter of all service areas. Obtaining ionosphere correction parameters of all areas under different expressions, including ionosphere correction parameters of grid and parameter forms.
4. Supporting acquisition, correction, and application of the correction parameters of various differences. Correction parameters of wide area or local area or virtual difference is formed based on observation parameter obtained by the different reference station.
5. Providing service to the user.

Including user time synchronization service (timing service), positioning calculating service, position reporting service, and guidance service.

S_G is GEO satellite or IGSO satellite in the system that has integral RDSS and RNSS functions:

1. Transmitting an S-frequency signal through MCC uplink C-frequency band link to the user, being a continuous operation method.
2. Receiving an L frequency band signal transmitted by the user and form the user's inbound link by the L/C transponder.
3. Under the control of the satellite clock, transmitting RNSS navigation signals under every navigation frequency. Such as S_G^0 shown in the figure, which is also a component of RNSS navigation signal transmitted by CRDSS satellite.

S_n^i and S_n^{i+1} are the RNSS navigation satellites in the constellation including any navigation satellite of Compass and GPS with different system.

User u is the application terminal of the user, for responding the CRDSS's RDSS signal and receiving the RNSS signal to finalize double difference pseudorange measurement of S_G^0 and S_n^i, S_G^0 and S_n^{i+1}, or S_n^i and S_n^{i+1}.

The synchronization operation is as follows:

1. User u receives an S-frequency guidance signal of S_G transponder to finalize the acquisition, tracking and transmission of inbound signal.
2. When the synchronization inbound signal time is formed, double difference observation among S_G^0, S_n^i and S_n^{i+1} is completed to form double difference observed quantity of $\Delta R^{0,i}$, $\Delta R^{0,i+1}$, $\Delta R^{i,i+1}$.
3. In the data packet of inbound signal, observation data of $\Delta R^{0,i}$, $\Delta R^{0,i+1}$, $\Delta R^{i,i+1}$ are brought into MCC.
4. MCC receives inbound signal of user u and perform measurement of round trip distance sum *Rua* through satellite S_G to user.
5. MCC calculates the distance from satellites S_G, S_n^i, and S_n^{i+1} to user according to the observation of *Rua*, $\Delta R^{0,i}$, $\Delta R^{0,i+1}$, $\Delta R^{i,i+1}$.

6. MCC sets up the positioning equation according to the calculated positions of S_G, S_n^i, and S_n^{i+1} as well as the distance to the user to obtain the position coordinates of the user and determine the time of the coordinate position.
   Refer to formulas (2.6) and (2.11)–(2.14) in Chapter 2 for positioning calculation.
7. According to the application of user's inbound signal, the obtained user information will be transmitted to designated user to complete position reporting.

5.6.1.2 Pseudorange Double Difference RDSS Positioning Distance and Observation Parameters Correction and Precision Analysis

In the scheme of high accuracy pseudorange double difference positioning, there is one extremely complicated observed quantity, that is the total round-trip distance R_{ua} from MCC to the user through satellite S_G^0. As the distance consisting of two different C-frequency bands and one S-frequency band plus one L-frequency band is adopted, different ionosphere correction models are formed, respectively, and must be carefully analyzed. Considering the relative movement of satellite S_G^0 during the transmission of signal, propagation flow of the total round-trip distance can be rewritten as Figure 5.12.

Distance measurement signal is transmitted by MCC to the user u through satellite $S_G^0(t_1)$ and returns to satellite $S_G^0(t_2)$ and then MCC. Its total round-trip distance can be expressed as:

$$R_{ua} = D_{c1}(t_1) + c\delta t_{SG}(t_1) + D_{1u}(t_1) + c\delta t_u(t_2) + d_{c1}(t_2) + c\delta t_{SG}(t_3) + d_{u1}(t_4) + \delta t_{c1o} + \delta t_{c1I} \tag{5.41}$$

In the formula, t_1 is the moment when satellite S_G^0 receives and transmits the inquiring signal from ground MCC; t_2 is the moment when the user receives the inquiring signal of satellite S_G^0. This moment is especially important. It is also the synchronization moment of the time difference measurement of satellites S_G^0 and S^{i+1} as well as satellites S_G^0 and S^{i+1}. t_3 is the moment when satellite S_G^0 transmits the answering signal of the user. However, there is a fixed time delay after t_2, for example a complete frame signal delay T, then $t_3 = t_2 + T$. The moment T is kept for the user to perform packing flow of every observed quantity in the inbound information.

t_4 represents the moment when the user signal returns to the MCC.
$\delta t_{SG}(t_1)$ represents equipment time delay of the outbound transponder of satellite S_G^0 at t_1.
$\delta t_{SG}(t_3)$ represents the equipment time delay of the inbound transponder of satellite S_G^0 at t_3.

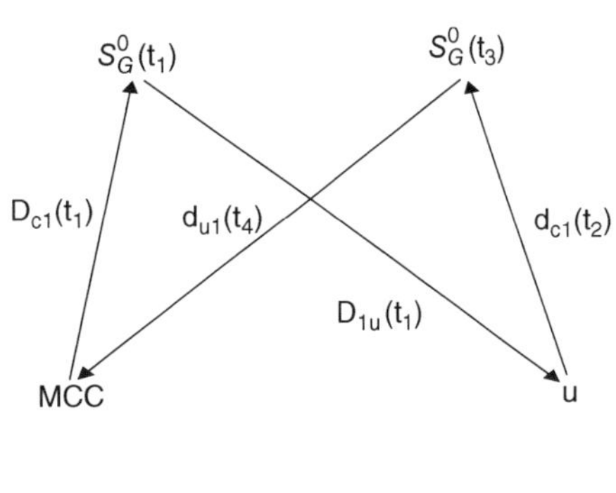

Figure 5.12 R_{ua} distance sum signal flow chart.

$\delta t_u(t_2)$ represents the total equipment time delay received and transmitted by the user at t_2.
δt_{C1O} represents the outbound equipment time delay of the MCC.
δt_{C1I} represents the inbound equipment time delay of the MCC.
c is the light speed.
$D_{c1}(t_1)$ represents the distance from MCC to satellite S_G^0 consisting of 6 GHz frequency band signal transmission time delay.
$D_{u1}(t_1)$ represents the distance from satellite S_G^0 to the user u consisting of S-frequency band signal transmission time delay.
$d_{C1}(t_2)$ represents the distance from the user u to satellite S_G^0 consisting of L-frequency band signal transmission time delay.
$d_{u1}(t_4)$ represents the distance from satellite S_G^0 to MCC consisting of 4 GHz frequency band signal transmission time delay.

Calculation of space distance r'_u from satellite to user through distance sum R_{ua} is not as simple as (2.7); namely,

$$r'_u = \frac{1}{2}R_{ua} - r_0 \tag{5.42}$$

In the formula, r_0 is the distance from MCC to satellite S_G^0.

In Formula (5.42), r'_u ignores equipment time delay in the signal transmission path and ionosphere time delay of distance measurement signal transmitted by 6 GHz, 4 GHz, S, and L frequency bands. If they are not corrected, the position error of the user calculated through r'_u obtained from Equation (5.42) reaches 100 m level, which can only be used for user tracking and rescuing service under ocean and ordinary topographic conditions.

More precise distance delay and correction must be performed for positioning accuracy of meter level.

The most efficient correction is still differential technology. In precise service area, RDSS reference station is set up. Correctors that are consistent with equipment time delay of user u to its greatest extent are arranged on the reference stations. According to influence of equipment time delay of the satellite and ionosphere change on the transmission time delay of radio wave, the operation frequency of the corrector is determined. As for precision of meter level, collection can be made every 10 min. As for higher precision of positioning, operation frequency shall be increased. Density of RDSS reference station shall be equivalent to density of GPS wide area difference reference station. Correction of user's distance sum R_{ua} shall be performed through distance sum R_{SA} measured by the RDSS reference station in the region. The correction value is:

$$\Delta R_{sa} = R_{sa} - R_{sa0} \tag{5.43}$$

R_{sa} is the total distance measured by the calibration station.
R_{sa0} is the total round trip distance from MCC to calibrator through satellite, obtained through known coordinates of MCC, satellite S_G^0, and the correction station.

i ΔR_{sa}^i correction values (i represents the quantity of reference stations) can be obtained through numerous correction stations. There are two ways of providing correction of user distance sum R_{ua}. One is a total grid distance correction for an adjacent user to calculate the r'_u correction. The other one is distance and correction of virtual corrector that is not far away from the user comprehensively calculated by MCC according to ΔS_{sa}^i for the user to measure the distance and correct.

The idea for the first grid distance correction is to integrate the distance sum measured by the calibrators on every reference station at every same moment in the whole region and compare to the distance sum calculated by the RNSS precise orbit determination to obtain the correction ΔR_{sa}^{i} of every reference station. Grid correction parameters at 10 km intervals are formed by ΔR_{sa}^{i}. In the positioning, the correction of the user ΔR_{ua} shall be searched according to the approximate position of the user.

Due to the relatively high operation frequency of reference stations, the radio signal has the same transmission path through satellite transmission equipment and ionosphere. Equipment time delay is the same as ionosphere transmission time delay, which can be offset and reach precision of ns levels. The main error comes from uncorrelated items of ionosphere generated due to the long distance between reference station and user. Therefore, in CRDSS precise positioning, calibrators on the reference station is indispensable. In order to ensure certain precision, the density of the reference stations with calibrators must be maintained at proper levels.

The calculation of the other correction parameter is that MCC selects measurement of calibrator with proper quantity and proper time span to obtain the correction that is the nearest to the user u and latest positioning time with reference of precise S_G^0 ephemeris.

5.6.1.3 L Frequency Band Double Difference Distance Correction

Another error in high accuracy pseudorange double differential positioning is the L-frequency band pseudorange correction in RNSS signal. This error mainly comes from refraction effect of ionosphere, which makes the propagation time delay deviate from the actual distance time delay. When the double difference measurement is performed by the user, its space relationship is shown in Figure 5.13.

The so-called pseudorange double difference $\Delta r^{o,i}$ is the distance difference from marking signal transmitted by RDSS at the same time (t_2 is explained in Figure 6.2, Chapter 6) after determination of bit synchronization of full frame format. $\Delta r^{o,i}$ does not include satellite clock error either. Its error source mainly includes satellite clock and ionosphere propagation time delay error. As for the user who adopts single frequency for operation, the correction value is still obtained through the receiver on the reference station. Certainly, as for the ionosphere correction, a double frequency receiver can also be used for direct correction. However, the user's burden and communication data volume will be inevitably increased by double frequency receiver.

Figure 5.13 Pseudorange double difference measurement.

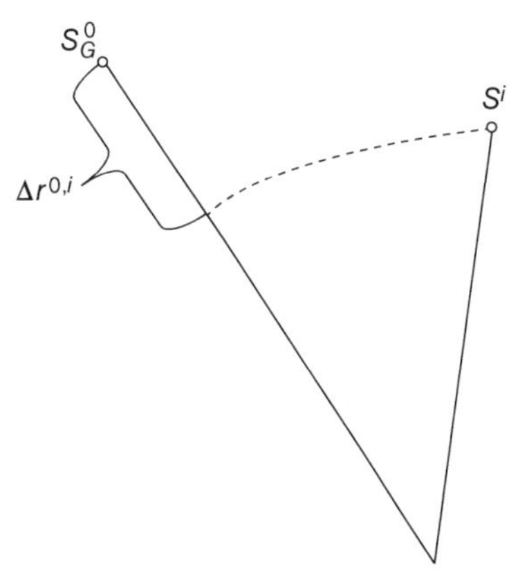

Correction parameters provided by the reference station are adopted as far as possible. This correction parameter can be obtained using the same method mentioned in the previous section. Since the position of the user is calculated in the MCC, this parameter is not required for broadcast in the satellite message and can be obtained by fully utilizing numerous abundant data acquired by the multi-frequency receiver on the reference stations in the region. Therefore, its correction parameter can reach precision of sub-meter levels.

5.6.1.4 Pseudorange Double Difference RDSS Positioning Accuracy Estimation

The pseudorange double difference RDSS positioning error is formed through the following influences:

1. Satellite ephemeris error
   In this scheme, precise ephemeris of MCC based on carrier phase can be adopted. Its satellite position error is superior to 0.5 m and radial direction error of 0.35 m.
2. Distance sum error.
   Utilization of grid or parameter correction. Under conditions where there is sufficient reference stations (calibrators) in the region, its distance sum residual error is superior to 0.5 m. Its error source mainly comes from ionosphere propagation error correction residual error, because in the measurement of distance sum, many various radio frequencies are used.
3. Single frequency pseudorange correction error.
   Correction residual error of reference stations (multi-frequency receivers) of sufficient quantities can be superior to 0.3 m.

These three errors are not incoherent with each other to constitute the main error source of positioning, 0.68 m. When HDOP = 1.5, VDOP = 2.0, horizontal error is 1.03 m and vertical error is 1.36 m.

In order to improve the positioning accuracy, efforts shall be made to eliminate the error of distance sum. The potential that can be exploited includes the following.

The precision of $D_{c1}(t_1), d_{u1}(t_3)$ in Figure 6.2 and formula (5.41) can be improved by utilizing precise ephemeris, which can reach 0.1 m. As the satellite equipment time delay error of $c\delta t_{SG}(t_1), c\delta t_{SG}(t_3)$ can be eliminated through difference, time delay difference between user's equipment and calibrators can be marked, the residual error is extremely small and can be ignored. The rest of the distance measurement error comes from $D_{1u}(t_1), d_{c1}(t_3)$.

By utilizing GNSS equipment on the reference station and the dual frequency observation of Compass or GPS satellite, the ionosphere residual error and operation frequency distribution shall be resolved. The relation expression is

$$\delta\rho_{f1} = \sigma_p \frac{f_2^2}{f_2^2 - f_1^2} \tag{5.44}$$

In the formula, σ_p represents ionosphere propagation time delay difference between operation frequencies of f_1,f_2.

$\delta\rho_{f1}$ is the ionosphere propagation delay error measured with operation frequency of f_1. f_1,f_2 are the signal frequencies used during pseudorange observation.

By utilizing the B1 and B2 of RNSS signals of satellite S_G^0, the ionosphere correction relative to S frequency (2483.5~2500 MHz) and uplink L frequency (1610~1626.5 MHz) shall be obtained respectively. The measurement value of $D_{1u}(t_1)$ and $d_{c1}(t_3)$ shall be further corrected. Finally, the precise space distance sum from MCC to the user through satellite S_G^0 can be obtained.

5.6.1.5 MCC Total Distance Measurement Scheme

In order to meet the target of positioning mentioned before, in MCC, there must be one high accuracy total distance measurement system, the measurement accuracy of which must be greater than 0.1 m. In this section, the realization of total distance measurement accuracy superior to 0.1 m will be discussed. Look back to Chapter 3, Table 5.9 lists 1610~1626.5 MHz inbound link level calculation from the user to the satellite. When a user's EIRP = 5 dBW, C/No = 44.0 dB/Hz, if the ranging accuracy of 0.1 m must be obtained, the distance sum measurement scheme for MCC is designed.

According to Chapter 3 (3.14), there is $\delta_{DLL}^2 = \frac{B_L d}{c/N_0}\left[1 + \frac{1}{(c/N_0)T(1-d)}\right]$

It can be seen that, when $d = \frac{1}{16}$, compared with $d = 1$ of regular wide correlation, under the same condition of C/No, δ_{DLL}^2 is reduced to one-sixteenth under the wide correlations. δ_{DLL} is reduced to one-quarter under wide correlation. Therefore, for the high accuracy distance measurement, narrow correlation technology shall be adopted

According to formula (3.14), assumed $B_L = 1$ Hz, d = 1/16, T = 20 ms, C/No = 44 dB/Hz,

$$\delta_{DLL}^2 = \frac{1 \times \frac{1}{16}}{[44]_{dB}}\left[1 + \frac{1}{[44]_{dB} 0.02\left(1 - \frac{1}{16}\right)}\right] = 0.000002493 \quad (\text{chip})^2$$

$$\delta_{DLL} = 0.01578969\ (\text{chip})$$

When the chip rate is 5 mbps, $\delta_{DLL} = 0.316$ (ns) $= 0.09$ (m), which meets the requirement of distance sum measurement accuracy of 0.1 m.

As the inbound signal is the burst random accessing signal, in order to make the distance measurement system possess excellent capability of acquisition, tracking and measurement, the inbound signal design must have synchronous head and tracking distance measuring segment with sufficient length. In order to meet the requirement of 0.1 m precision, tracking distance measuring segment of inbound signal shall not be less than 20 ms and its inbound signal format is shown as Figure 5.14.

As MCC adopts a large antenna with an aperture greater than 13 m, its multipath influence is extremely small. Signal transmitted by user will not be distorted in the satellite transponder. Therefore, it is feasible to adopt a narrow correlator for obtaining high accuracy total distance measurement in MCC.

Synchronization segment	Tracking and measuring segment (>20 ms)	Data segment...

Figure 5.14 Basic design requirement of inbound signal.

5.6.2 Design Scheme of the Double Module User Terminal

The user terminal is a standard CRDSS double module compatible user receiver, the basic functions of which are proposed as follows:

1. Receiving the S frequency signal and demodulating outbound information.
2. Transmitting the L frequency band signal according to received signal marking response and transmitting the packet of measurement information and communication information of the user to MCC along with the inbound signal.
3. Receiving navigation signal of Compass and GPS and pseudorange measurement.
4. Differencing pseudorange among satellites S_G^0, S^i, S^{i+1}.

Basic technical indexes are as follows:

1. Receiving signal
   S Frequency band: 2483.5~2500 MHz
   L Frequency band: B1 1561.098 ± 2.046 MHz; L1 1575.42 ± 1.023 MHz
2. Transmitting signal
   L Frequency band: 1610~1626.5 MHz
   Modulation mode: BPSK or BOC
3. Synchronization accuracy 0.3 (ns) transmitted according to receiving signal marking response.
4. Receiving sensitivity is superior to −160 dBW
5. Transmitting signal EIRP = 6 dBW
6. Number of channels received by S frequency band $n_{ch} \geq 10$
7. Number of channels received by RNSS navigation $n_{now} \geq 10$
8. Time synchronization accuracy of pseudorange differential and response signal ≤ 0.3 ns
9. Processing time of signal transmission ≤ 1 frame (outbound signal) length.

The simplified block diagram of user terminal is shown in Figure 5.15.

Antenna must be at least a three-frequency band antenna.

Frequency band ranges are 2480~2500 MHz, 1610~1626.5 MHz, and 1260~1559 MHz. Due to the wide range of frequency bands, it can be a combined antenna with multiple frequency bands.

In order to ensure the receptivity of RNSS, RNSS frequency band (1260~1559 MHz) antenna must have better low-elevation augmentation.

Pre-amplifier of S frequency band must possess the features of low noise and high augmentation. Tracking and extracting baseband processing module is a multi-channel signal processor with capability of processing outbound beam of GEO satellites. However, only one stable beam with stronger signal is tracked at a time. It shall possess the function of automatic discrimination and switchover of beam. Information demodulator processes the RDSS public information and dedicated information of the user with the error ratio superior to 1×10^{-6}. Information modulator adopts modulation modes of BPSK or BOC. According to requirement of MCC receiving and processing system, determination is made uniformly. Transmitting power of power amplifier of L frequency band is 3~4 W. In order to reduce the total power consumption, transmitting efficiency shall be improved to the greatest extent. As the transmitting duration is from dozens to hundreds of milliseconds, heat radiation is not the main technical obstacle.

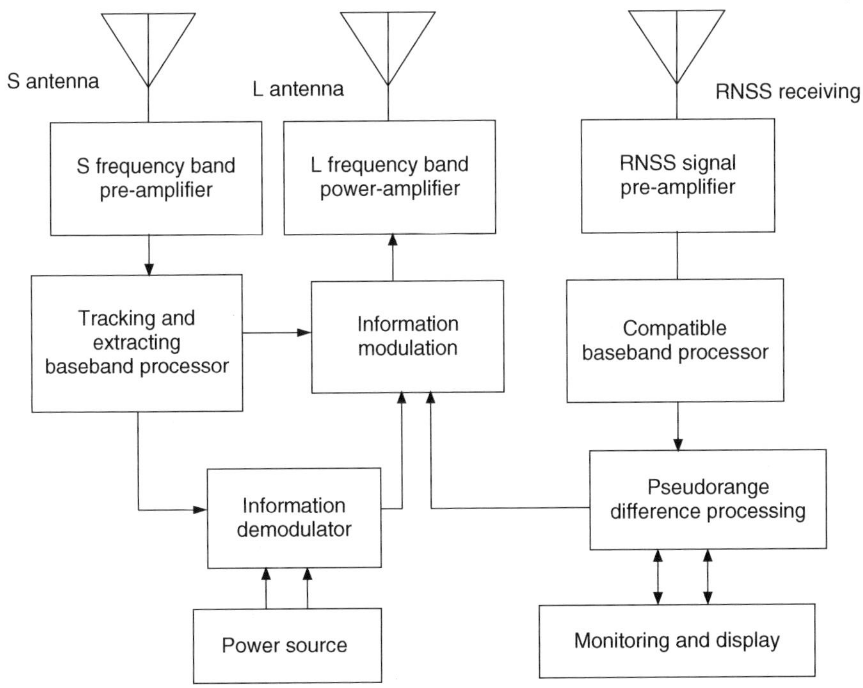

Figure 5.15 Block diagram of a double module user terminal.

Pre-amplifier of RNSS is a radio frequency part with a relatively wide frequency band, which shall possess flat augmentation to ensure the receptivity of different navigation signals. Compatible baseband processor has the mission of finalizing the tracking, despreading, demodulation, and pseudorange measurement for Compass and GPS signal. Its number of channel $n_{naw} \geq 3$, namely, signals of at least three satellites are tracked. Therefore, there shall be one redundant channel as the satellite signal management. Pseudorange differential processing cannot be done by an independent processor. Double difference measurement and normalization processing are completed by compatible baseband processing.

In order to maximally relief the burden on the user, this user receiver cannot receive tedious satellite message and only discriminate demodulation of frame and characters to obtain integral system time to ensure the time accuracy of double difference observation. However, it shall possess the capability of monitoring and management on some level.

In order for precise positioning, epoch time of the pseudorange differential shall be synchronized with the marking time of satellite S_G^0 signal received by S frequency user receiver, namely t_2 in equation (5.41). Its synchronization epoch error shall be superior to 0.1 μs. The synchronization error can be accurately analyzed through code phase counting in pseudorange measurement.

Basic requirements of every module of the user terminal are as follows:

S-frequency pre-amplifier

1. Frequency bandwidth 2483.5~2500 MHz
2. Pass-band augmentation 28 dB ± 1 dB

3. In-band augmentation variation $\leq \pm 0.5$ dB
4. Noise factor: ≤ 0.8 dB
5. Output amplitude (1 dB compressing point): $\geq$0 dBm
6. Suppression degree of three order intermodulation $\geq$60 dB_C
7. Input SWR $\leq$1.5
8. Output SWR $\leq$2
9. Impedance 50 Ω

RNSS pre-amplifier:

1. Frequency point includes GPS L1, Compass B1
2. Pass-band augmentation 30 dB $\pm$ 1 dB
3. In-band augmentation variation$\leq \pm 0.5$ dB
4. Noise factor ≤ 0.8 dB
5. Suppression degree of three order intermodulation $\geq$ 60 dB_C
6. Input SWR $\leq$ 1.5
7. Output SWR $\leq$ 2

Compatible baseband processor:

In order to meet high accuracy pseudorange measurement, the compatible baseband processor is the processor based on the narrow correlation and incoherent DLL with the function indexes as follows:

1. Capability of processing Compass and GPS L_1 frequency point signal at the same time.
2. Being capable of setting up Compass, GPS satellite constellation, satellite numbers, frequency point, acquisition and tracking, Doppler frequency separation, ranging code phase searching region, threshold of acquisition, and so on.
3. Multiple independent programmable tracking channel, which are compatible with satellite tracking despreading, pseudorange measurement, carrier phase measurement, carrier Doppler measurement, and information demodulation used for Compass or GPS.
4. Tracking sensitivity: At $C/N_0 \geq 39dB/Hz$, the error rate $\leq 1 \times 10^{-6}$
5. Pseudorange measurement accuracy 0.1 m
6. Pseudorange epoch accuracy 0.1 μs

S frequency signal tracking and extracting baseband processor:

The mission of this baseband processor is to extract inquiring signal marking time. After a length of time delay of a complete frame signal, it accomplishes the measurement of distance sum R_{ua} as the response time of the user, entering the MCC through a satellite inbound transponder as the time mark of distance sum R_{ua} measurement. Therefore, accuracy of marking time extracted makes a great contribution to the measurement accuracy of R_{ua}. It is the one of the critical technologies of the user receiver and needs to be designed as meticulously as a compatible baseband processor. Precision extracted by marking time is superior to 0.3 ns. In order to rapidly track multiple S frequency outbound beams, numerous processing channels must be designed to realize the initial contact between the MCC and user adopting any available channel. Therefore, this baseband processor possesses similar capabilities such as multiple channel signal acquisition, tracking, Doppler frequency searching region, ranging code phase searching region, acquisition threshold, and so on. At 39 dB/Hz, error rate of information demodulation is superior to 1×10^{-6}.

L Frequency pre-amplifier:

1. Transmitting signal frequency 1610~1626.5 MHz
2. Modulation code rate: $R_{ch} \leq 8.25Mbps$
3. Harmonic and clutter power at band: $\leq -130dBW/4KHz$ (within RNSS receiving frequency)
4. Modulation phase noise error: $\leq 3^0$
5. Transmitting carrier phase noise:

$$\leq -65dB_c/100Hz$$

$$\leq -75dB_c/1kHz$$

$$\leq -85dB_c/10kHz$$

$$\leq -100dB_c/100kHz$$

6. Carrier frequency stability $\leq 5 \times 10^{-7}$
7. Suppression degree of carrier ≥3.0 dB

The user terminal is a CRDSS double module compatible user receiver with the functions of positioning, searching, and communications. Under the support of the standard communications format, it can transmit the positioning demand of the user within the communication data segment according to user-defined protocol. Various difference corrections processed by MCC can finalize the precise positioning of the user by MCC and then transmit to command and dispatching organization. Or, the ionosphere correction parameters, precise clock error, precise ephemeris, precise differential correction parameters required by precise positioning of the user can be broadcasted through S frequency outbound information. Therefore, user receivers possess multi-functional modules with multiple communication interfaces to tailor according to different users and different spectrum receivers. In this section, only the basic function has been introduced.

6

Anti-Interference and Low Exposure Technology of the Satellite Positioning User

6.1 Self-Adaptive Spatial Filtering Principle

The anti-interference principle of the satellite navigation receiver is to improve navigation satellite signal strength, suppress interference signal strength, or both. Narrow-band interference can be suppressed through the frequency domain and time domain filtering technology. Manmade malicious interference has a wider frequency band than full bandwidth of a satellite signal. Filtering technology is not able to achieve the ideal effects. Self-adaptive spatial filtering is an effective measure against man-made malicious interference. So-called self-adaptive spatial filtering is a design of antenna and signal processing system. Ideal value of signal to interference plus noise ratio (SINR for short) after being processed is achieved to make the demodulation of the satellite signal smooth.

The self-adaptive spatial filtering antenna is generally divided into two categories: the first is to improve antenna augmentation of the receiving signal and to use a low side lobe to align with interference signal to achieve the improvement of SINR as shown in Figure 6.1.

The second method is to align the main lobe with interference and adopt suppression or offset to weaken the augmentation of main lobe so as to achieve the improvement of SINR as shown in Figure 6.2.

However, low exposure means that the beam transmitted by the user is narrowed and aligns with the receiving satellite. As the transmission beam of the user is narrower, EIRP is controllable so that the demand of MCC multi-user receiving processing can be met. The probability of being detected by a hostile force is reduced.

As the number of satellite signals expected to be received by the satellite navigation receiver is greater than four and the interference signals generally coming from one or two directions will be effective. Therefore, the signal number that is suppressed is generally less than the number of signal that is expected to be received. Therefore, generally, a widespread method is to align the antenna augmentation nulling point with interference, which can reduce the number of elements significantly.

As for the RDSS user, the satellite signals that are expected to be received are the outbound signals of two operation satellites. Through determining the time difference of the two satellites, the response signal of the user returned from only one satellite is able to constitute the distance sums S_1 and S_2 in the solution of satellite navigation (see Chapter 3). This creates many selections of schemes for self-adaptive nulling antenna design.

GNSS Systems and Engineering: The Chinese Beidou Navigation and Position Location Satellite, First Edition. Shusen Tan.

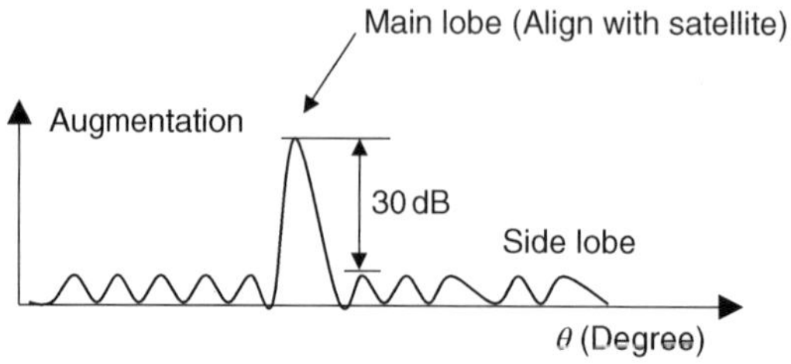

Figure 6.1 Improvement of the main lobe.

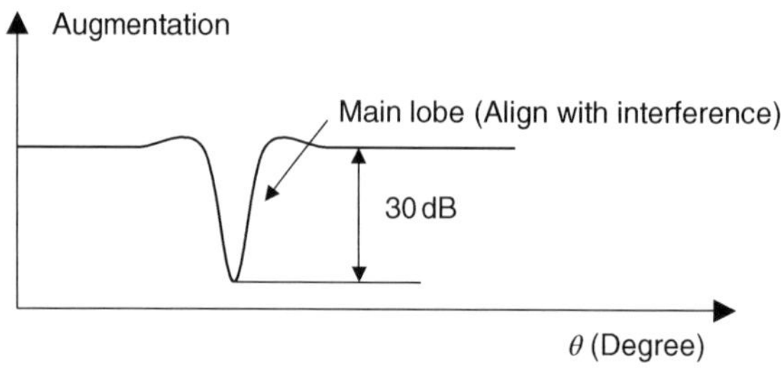

Figure 6.2 Suppression of the main lobe.

The first iteration of the RDSS user receiving nulling antenna uses the augmentation of main beam received. It is favorable to improve the quality of received signal of ship and vehicular command under complicated environmental conditions. There are two main beams, augmentation $\geq 3dB$, side lobe augmentation $\leq -2dB$, and SINR of antenna $\geq 30dB$. The cost is a large number of elements in the antenna making it complicated and expensive. However, it has good adaptability to the environment.

The second design index of nulling antenna is: comprehensive augmentation $\geq 2dB$; amount of anti-interference source ≥ 2. SINR of antenna (or anti-interfering tolerance) $\geq 30dB$. Starting from the feasibility of engineering, it has a small amount of antenna elements, simple structure, and is inexpensive with obvious results. In order to improve adaptability to the electromagnetic environment, time domain filtering can be performed accordingly.

6.2 Basic Algorithm of Self-Adaptive Filtering

Figure 6.3 is the basic block diagram of an N antenna array digital beam formation consisting of an N antenna array, digital receiving channel, digital sampling (N route),

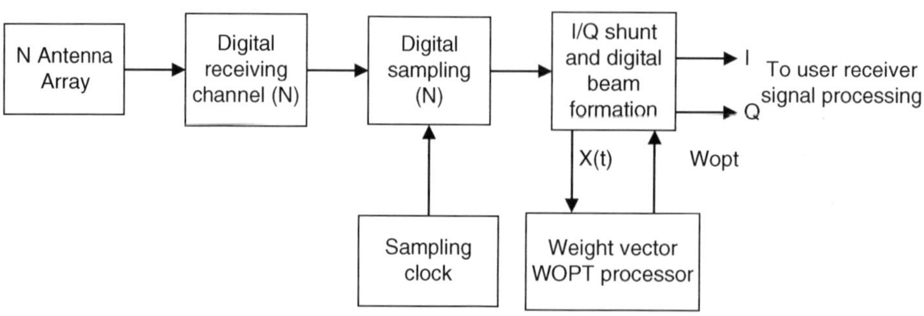

Figure 6.3 N antenna array digital beam formation block diagram.

I/Q shunt and digital beam formation, and weight vector processor. The N antenna array consists of N microstrip antennas. N digital receiving channels enlarge the signal received by antenna and conduct down-conversion to ensure the necessary bandwidth of signal to perform I/Q decomposition in digital domain. And under the weighted control of weight vector processor, the signal consisting of I, Q two route beams are formed to be input to RDSS signal processing section for relevant processing.

A critical part of digital beam formation is the weighted processing of signal received by the antenna of each array element to both improve the satellite receiving signal and suppress interference.

In the RDSS user receiver, orientation of satellite signal beam is determined approximately to fix the best weight vector for fast solution through *a priori* value of the satellite signal and preliminary position of the user. In the 1960s, research emphasis of array signal was placed on the formation of phased-array antenna. Up until the 1970s, processing emphasis of array signal turned to self-adaptive nulling control algorithm and self-adaptive side lobe set-off. Control of nulling alignment interference of antenna directivity diagram was studied under the unknown environment. The main methods included LMS algorithm, power inversion algorithm, and main beam constrained nulling algorithm proposed by Berhie Widrow in 1966. These algorithms all required the receiving channel of the same antenna array elements. In 1976, B. Widrow and J.M. McCool jointly proposed the weight vector perturbed algorithm processed with one receiving channel (radio frequency repeated weight). After that, the most influential algorithms included orthogonal sequence perturbed algorithm, sequential perturbed algorithm. From 1976 to 1986, an estimation algorithm for space signal arrival direction (DOA) was proposed. Typical bending form algorithms include MUSIC, ESPRIT, and the minimum inner product method.

The basic principle is that array antennas have different sensitivity to the incident signal, which causes different journey time delay. When there is only one incident signal, the signal received by antenna n(n = 1 ~ N) is [5],

$$X_n(t) = f_n(\theta_k, \varphi_k)J_k(t - \tau_{kn})e^{jw_k(t-\tau_{kn})} + n_n(t) \tag{6.1}$$

as shown in Figure 6.4.

Where xoy is the reference plane of antenna array.

the x-axial direction is the reference direction

O is the antenna array phase center

θ_k is the elevation angle of incident signal, which is also the included angle between incident plane and xoy.

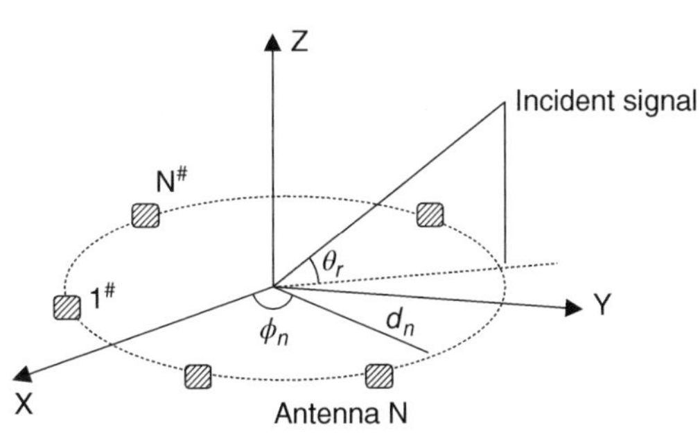

Figure 6.4 Schematic diagram of the uniform circular path delay.

φ_k is the included angle between wave surface and xoz.
d_n is the distance from antenna n to antenna array phase center.
φ_n is the azimuth angle of antenna n.
$f_n(\cdot)$ is the directivity function of antenna n.

Thus:

$$\tau_{kn} = \frac{d_w}{c} \cos\theta_k \cos(\varphi_k - \varphi_n) \leq \frac{D}{c} \tag{6.2}$$

Where:

D is array caliber
c is light speed.

Record $X_t = [x_1(t), x_2(t), \cdots, x_n(t)]^T$ as received signal of the N array antenna. $N(t) = [n_1(t), n_2(t), \cdots, n_N(t)]^T$ as the channel noise.

RDSS received signal tracking algorithm highlights the two satellite signals that are expected to be received for searching, called double beam tracking arithmetic (DBTA) [5]. The satellite position and approximate position of user are utilized to control the weighting system and align the two received beams with RDSS satellite. The processing method includes DBTA based on enlarged SNR, weighted DBTA, and equal-weighted DBRA. These three algorithms can form double beams. However, weighted DBTA is favorable to control the weak signal in the two satellites to accommodate the inconsistent situations of satellite signal at different places. It is shown by the physical simulation that array augmentation of the two satellites can, respectively, reach $G_1 = 7.49$ dB and $G_2 = 10.49$ dB.

So, how to realize the offset of interference signal? As for the satellite navigation system using CDMA signal modulation mode, satellite signal is total ignored prior to interference offset calculation when the signal of user terminal is completely overwhelmed in the noise of receiver below 20 dB; it is only required to align the nulling point of the array antenna with interference direction. The goal can be achieved by using a power inversion algorithm. In order to express this easily, assume the antenna array as N + 1 elements of antenna array, X_{om} is the reference output of antenna. In the iterative algorithm, take

$$W = [W_1, W_2, \cdots, W_N]^T$$
$$X(t) = [X_1(t), X_2(t), \cdots, X_N(t)]^T$$
$$e_m = W^H X_m - X_{om}$$
$$d_m = X_{om}$$

Record $V_d = E\{X_0^m(t)X(t)\}$

$$R_{xx} = E\{X(t)X^H(t)\}$$

The power inversion algorithm can be obtained based on the LMS, BLMS, LRS, and Newton algorithms. It is shown by the analog simulation that the suppression capability to interference signal ≥ 30 dB. These five algorithms have difference nulling depths with orders from big to small as

$$RLS \geq LMS > BLMS > LRS > Newton$$

Order based on convergence time from short to long

$$RLS < BLMS < LMS < Newton < LRS$$

Order based on the complexity of calculation from big to small:

$$Newton > RLS > BLMS > LRS > LMS$$

The advantage of power inversion algorithm is that structure and direction of the signal are not required to be known in advance and is able to obtain effect of nulling offset. However, the augmentation of satellite signal cannot be ensured.

6.3 Self-Adaptive Nulling Antenna Engineering Design

Self-adaptive nulling antenna engineering design of a satellite navigation receiver must consider the following factors:

1. Users' requirements for antenna size, carrier, and installation mode
   As for a mobile user, vehicle-, warship- (ship), and airplane-mounted users require a smaller antenna array. The acceptable diameter of array antenna must be less than 0.3 m. Therefore, there will not be many array elements in the antenna.
2. Quantity of interference source that can be offset.
   There must not be many interference sources that can be offset. Generally, 1–2 interference sources are acceptable.
3. Performance of interference suppression.

Considering these factors, as for the self-adaptive nulling antenna of RDSS receiving system, generally the main beam of interference offset of four array elements is adopted as a design option. The spatial filtering combining time-domain filtering technology is adopted, namely, space-time 2D signal processing or space-time signal joint processing. If improving main beam augmentation is adopted to improve the signal to interference ratio, as for the array antenna augmentation of 30 dB as the technical requirement, assuming unit antenna augmentation of 3 dB, its antenna array elements will be ≥ 500. Adopt total augmentation of four array elements of 9 dB. Therefore, as for RDSS receiving nulling antenna, four array element double beam tracking algorithm is advisable.

6.4 Low Exposure Transmission Array Antenna Design

Regular RDSS user receivers all utilize all-directional L, S double frequency transmission, and receiving sharing antenna to meet the requirement of receiving signal of operation satellite in different location and different carrier condition. This can also radiate a response signal to the two satellites to meet the demand of S_1, S_2 distance measurement. As for the common coverage area of the two S beam satellites, only when the user receiver measures the difference of the arrival times of them will the user receiver only returns to the MCC from one satellite to constitute the distance sum S_1, S_2 similar to equations (3.1) and (3.2). Then, all directional L beam transmitting antenna can be designed as orientation adjustable beam antenna. The composition of its beam is similar to control and receiving nulling antenna, namely, the adjusting direction of orientation is the same. It merely aligns the main beam with the satellite that has minimum path loss to improve the augmentation with the goal of power composition. As for the four unit beam array antenna, the augmentation can reach 9~11 dB. Therefore, this narrows the all-directional beam down to less than 18°. Meanwhile, through the energy test of S_1, S_2 receiving beams, the transmitted power of user receiver is controlled to make the EIRP to the minimum nominal rating so that the exposure of the signal transmitted by the user can be greatly reduced, which increases difficulty of detection. It is difficult for the normal ground detection station to detect the signal transmitted by the user receiver and the detection range of air detection station is narrowed down significantly. The detection becomes more difficult. Even if the air detection station can detect the signal, it is also difficult to form the condition of coherent positioning. Therefore, in order to improve the viability of the RDSS user, it is economic and feasible to adopt the technology of beam forming antenna.

Based on these reasons, anti-interference and low-exposure beam forming antenna of the RDSS user receiver are proposed to enable the receiving system to resist two malicious interferences and the transmitting system be less exposed by four units receiving and transmitting shared L, S double frequency antenna through the respective formation of the receiving and transmitting beams. Simulation system tests show that when special level allowance of communication space is ignored, the user receiver, EIRP = 4~6 dBW, has optimal system performance. Adopting four-unit transmitting beam forming antenna, the transmitting power for each unit is only 0.6–1.0 W. The requirement for the user's power device is greatly reduced, which is favorable for improving the reliability of user receivers. Cost will finally become much lower by replacing the superpowered device with signal processing technology.

To sum up, after the anti-interference performance of RNSS is analyzed, we can see that the anti-interference and low exposure of the RDSS user is much easier to realize, which builds the important foundation for the vitality of RDSS.

7

Concept of Satellite Navigation and the Principle of Positioning and Velocity Measurement

7.1 Concept of Satellite Navigation

As stated in Chapter 1, the system that realizes the satellite radio navigation target is called the radio navigation satellite service, or radio navigation system for short. Its basic concept is that it is a radio positioning system determined by non-inquiry response continuous high-accuracy timing, spatial (three-dimensional) positioning, and a movement velocity vector accomplished by the user by utilizing the radio signal transmitted by the satellite. Such positioning and velocity measurement performance is continuous and suitable for dynamic performance of the user in motion. The user scope includes aerospace, aviation, navigation, and ground users at any point on the globe or near Earth space. In addition, according to the designated destination, the user can continuously obtain navigation parameters such as course, navigational speed, off-course difference, and so on, and the path and time to arrive the destination can be estimated according to the current position and navigational speed. Based on this integrated concept, the integrated concept of satellite radio navigation was born in late 1970s. WARC-79 approved the satellite radio navigation frequency of GPS proposed by the United States Federal Government. Previously, the US Navy's radio navigation system or Transit that could only determine the plane position of the static user on the ocean was only a sprout of satellite radio navigation. Therefore, satellite radio navigation is a youthful engineering application science, the concept of which is still being developed and evolved. Due to the application of GPS as well as people's expectations of satellite radio navigation, ICAO proposed the concept of GNSS, the global navigation satellite system, which is a global service system of satellite radio navigation consisting of radio navigation satellites. It is not just a certain unique global satellite navigation system. ICAO also proposed the demands and establishment plan of GNSS. Refer to Table 7.1 for the demand of civil aviation. From the items and indexes listed in Table 7.1, based on the demand of satellite radio navigation system proposed by civil aviation, established GPS and GLONASS currently do not meet all of the performance requirements of civil aviation. Therefore, different designs of systems are adopted under different flight conditions, including the global navigation satellite system, regional satellite navigation augmentation system, and local satellite navigation augmentation system, all of which fall under the joint name of radio navigation satellite service. Besides the established GPS, GLONASS currently and Galileo, Compass under construction, the so-called RNSS also includes regional augmentation systems such as EGNOS, WAAS, MSAS, and a local augmentation system that meets the requirement of CAT1, CAT2, and CAT3. According to ICAO's proposal

GNSS Systems and Engineering: The Chinese Beidou Navigation and Position Location Satellite, First Edition. Shusen Tan.

Table 7.1 Civil aviation requirement proposed by the ICAO [6].

RNP Type (Total Error Probability 95%)	Flight Condition	NSE 95% Horizontal/ Vertical	SIS maximum Non-Continuity	SIS maximum Integrity	Warning Time	SIS Availability
RNP-20	Over ocean airline fright	19.9 nm	10^{-6}/h	10^{-7}/h	5 min	0.999
RNP-12.6	Over land airline, low density	12.44 nm	10^{-6}/h	10^{-7}/h	3 min	0.999
RNP-4	Over land airline, high density	3.87 nm	10^{-6}/h	10^{-7}/h	1 min	0.9999
RNP-1	Over land airline, very high density	0.44 nm	10^{-6}/h	10^{-7}/h	15 s	0.9999
RNP-0.3~0.5 0.3 nm/125 ft	Preliminary approach, non-precision approach, departure	100 m	10^{-6}/h	10^{-7}/h	10 s	0.9975
RNP0.03/50 ~0.02/40	CAT1	18.2 m /7.7~4.4 m	8×10^{-6}/15s	2×10^{-7}/Approach	6 s	0.9975
RNP-0.01/15	CAT2	6.5 m/1.7 m	4×10^{-6}/15s	1×10^{-9}/ Approach	1 s	0.9985
RNP-0.003	CAT3	3.9 m/0.8 m	2×10^{-6}/15s	5×10^{-10}/ Approach	1 s	0.9990

(SIS is space signal)

and concept, people draft the GNSS system to suit themselves. GNSS is generally divided into two stages, GNSS1 and GNSS2.

The so-called GNSS1 is the first step of establishing global GNSS. The GNSS1 plan is accomplished by utilizing the established GPS and GLONASS through regional augmentation. Then the GNSS1 of European navigation plan is the European Geostationary Navigation Overlapping Service (EGNOS). GNSS2 is divided into Preliminary GNSS2 and GNSS2. Preliminary GNSS 2 accomplishes the primary and intermediate service of GNSS 2. GNSS 2 accomplishes the premium service. GNSS 2 is a system without GPS or GLONASS. Galileo plan under establishment is GNSS2 of Europe, which provides services under different flight conditions through Galileo global system, Galileo global system + regional augmentation, and Galileo global system + regional augmentation + local augmentation respectively. Schemes of establishment are shown in Table 7.2 and Table 7.3. The so-called regional augmentation utilizes GEO satellite as observable navigation satellite of the user, which, at the same time, improves positioning accuracy, availability, and integrity, provides an integrity warning service, and so on. Since the system is augmented through GEO satellite, its coverage is close to one-sixth to one-third of the Earth's surface; that is the reason why it is called regional augmentation. Local augmentation uses pseudo satellites on the ground or in the air to get the system performance augmented so that its augmented coverage meets the demands of CAT1, CAT2, and CAT3; therefore, its application scope is extremely small. However, it has high requirements of integrity and continuity. Due to a small application area, it can be satisfied easily. In GNSS, LEO satellite is suggested being used for augmentation. For instance, Jean Cnenebault and Jean-Pierre Provenjano made a rationalization proposal of INES European LEO relay navigation system. INES augmentation system with orbit altitude of 1500 km and inclination angle of 62.75°, having eight or nine orbit planes with seven satellites on each plane, 56–63 satellites in total, was designed. However, no response has been received by the engineering sector.

Based on this practical engineering plan, a regional satellite navigation system consists of an NGEO satellite such as IGSO or MEO and GEO with integrity warning time response capability. This system can meet the requirement of CAT-I precision approach en route to the greatest extent. Chinese satellite navigation system takes regional system as preliminary system. Its best positioning aligns with a higher level of CAT-1 precision approach. If the number of NGEOs is not sufficient, the index of non-precision approach should also be satisfied.

From the index advocated by ICAO, the referred SIS maximum non-continuity is the continuity risk. SIS maximum non-integrity is integrity risk advocated by Galileo. Definition of indexes will be introduced in specific chapters.

7.2 Satellite Navigation Principle

The basic principle of satellite navigation is that the navigation satellite operating on the designated orbit transmits radio navigation signals continuously under the control of the satellite clock, and the user receiver receives navigation signals from at least four satellites. The navigation ranging code is recovered and compared with the ranging code imposed by the local clock to complete the pseudorange measurement of each satellite. Satellite ephemeris is demodulated from the satellite navigation signal to obtain the

Table 7.2 Galileo A-level public and professional application service performance requirement [7].

A-Level service (Public and professional application)			
Parameter	**Galileo global segment**	**Global segment + Regional segment**	**Global + Regional + Local**
Typical * Application	Relaxation A_0	Traffic service (Vehicle and ship management) A_1	Land survey (Time transfer) A_2
Coverage	Global	Regional	Local
Access level[a)]	Free Charge (OAS)	Controlled CAS-1	Controlled CAS-1
Accuracy (95% Probability) Horizontal Vertical	10 m 10 m	10 m 20 m	0.001~0.1 m 0.0001~0.2 m
Integrity Risk Warning limit Horizontal Vertical Warning time	No support of space signal	Guaranteed, to be confirmed 10^{-3}/h 25 m 50 m 60 s	Guaranteed, to be confirmed
Continuity risk Max. interruption rate	To be confirmed	To be confirmed 2 min	To be confirmed
Availability	>0.9/Day	>0.7/Day	>0.7/Day
Shield angle	5°	25°	15°
Time to first fix (TTFF)[b)]	1~2 s	1~2 s	1~2 s
License	No	Yes	Yes

a) CAS, CAS1, CAS2 are, respectively, open service, commerce controlled, and life safety and encryption controlled;
b) Recaptured positioning without losing effective ephemeris data.

Table 7.3 Galileo B-level service performance requirement [7].

B-level service (Life safety application)			
Parameter	**Global segment**	**Global segment + Regional segment**	**Global + Area + Local**
Typical Application	En route B_0	CAT-1 precision approach B_1	CAT-III precision approach B_2
Coverage	Global	Regional (EGNOS)	Local
Access Level	CAS2	CAS2	CAS2
Accuracy (95% Probability) Horizontal Vertical	2 nm Not available	16 m 4 m	0.8 m 0.8 m
Integrity Risk Warning limit Horizontal Vertical Warning time	Only receiver autonomous integrity Cannot meet requirement	2x10^{-7}/150s 20 m (to be confirmed) 12 m (to be confirmed) 6 s	2 m 2 m 1 s
Continuity risk Max. interruption rate	10^{-4}~10^{-8}/h	8x10–^{6}/15 s To be confirmed	2x10^{-6}/30 s
Availability	0.99	0.99	0.99
Shield angle	5°	5°	15° (avoid low elevation angle)
Time to first fix (TTFF)	To be confirmed	15 s	To be confirmed
License	Yes	Yes	Yes

position. The user receiver sets up the positioning equation based on the known satellite position and at least four measured pseudoranges to calculate the position of the user through the autonomous calculation.

According to this positioning principle, the early Transit and цикаda used Doppler frequency measurement and ephemeris parameters of the radio signal transmitted by only one near-Earth satellite to determine the plane position of static user (the positioning principle is as shown in Figure 7.1), which did not belong to the integrated radio navigation satellite service. The velocity of movement needs to be known for the positioning of user in motion.

7.2.1 Solutions of the Navigation Mission

In radio navigation satellite service, the basic mission of navigation is to determine the user's coordinates of space and time as well as the velocity component. The extended state vector of the user in the inertial coordinate system is $R_u = (x.y.z.\Delta t.\dot{x}.\dot{y}.\dot{z})$, namely that, the user's space coordinate is $(x.y.z)$; the user's time correction is Δt; the user's velocity component is $(\dot{x}.\dot{y}.\dot{z})$.

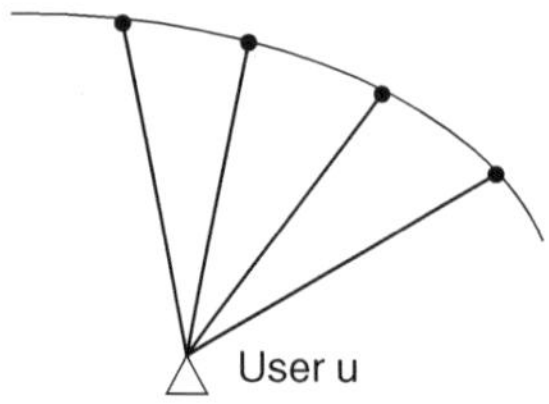

Figure 7.1 Positioning principle of transit.

Utilization of the radio method cannot directly measure the vector elements of the user. From the received radio signal, some parameters can be measured, for instance, transmission time delay of the signal in the space or Doppler frequency shift. Corresponding geometric parameters of the radio signal parameters measured for navigation are called navigation parameters. Therefore, propagation time delay of the signal in the space, τ, and corresponding distance to destination of Doppler frequency Y as well as the radial velocity approaching the destination V_u are the navigation parameters. The basic relationship is:

$$r = c\tau \cdot V_u = f_d \lambda$$

In the formula, is the light speed; is the wave length of the satellite transmitting signal.

Locus of points in the space with same navigation parameters is called position surface. The intersection line of the two position surfaces determines the position line, which is the locus in space with two fixed values of navigation parameters. The intersection point of three position surfaces or two position lines determines its position coordinates. Under many circumstances, due to the nonlinearity, two position lines can intersect at two points. At this time, additional position surfaces or other information related to target position are used to obtain the single solution of the user.

In order to get the user's vector R_u solved, the functional relationship between navigation parameters and user's vector element is used. The corresponding functional relationship is usually called the navigation function. Specific form of navigation function depends on many factors: form of navigation parameter, kinetic characteristics of the satellite and user, selected coordinates, and others. The navigation function of the user's coordinates in space can be determined by utilizing various distance measuring methods, distance difference measuring method, angle measuring method as well as the combination of these methods. In order to obtain the navigation function containing the velocity components of the user, radial velocity is adopted. The satellite navigation user adopts pseudorange measurement to determine the state vector of the user $R_u = (x.y.z.\Delta t.\dot{x}.\dot{y}.\dot{z})$.

7.2.2 Concept and Definition of the Pseudorange

Under precision time control of a high accuracy atomic clock, the satellite transmits the radio navigation signal. User receiver processes the received signal and recovers satellite time signal to complete the pseudorange measurement by comparing it with the time of local clock. The basic relationship is shown in Figure 7.2.

Time expression of pseudorange R_p is

$$R_p(t) = (n^l(t) - n^r(t)) \times C = R_u(t) + \Delta T(t) \times C = [t_R + \Delta T(t)] \cdot c \tag{7.1}$$

In the formula, $n^l(t)$ is the clockface time recovered by receiver and marked by satellite clock time; $n^r(t)$is the clockface time recovered by receiver at local time of t; $R_u(t)$is the

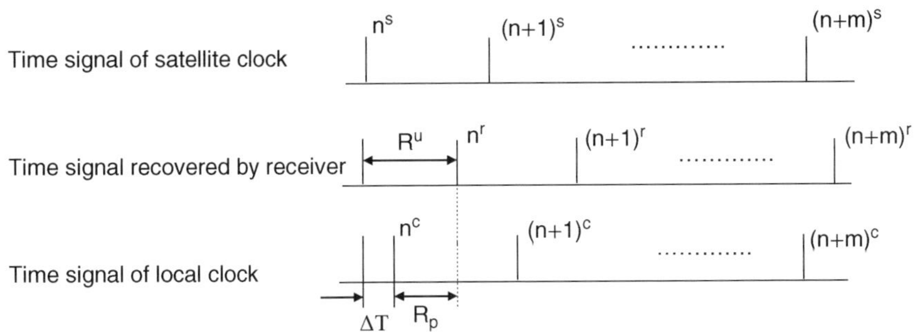

Figure 7.2 Schematic diagram of pseudorange and clock error.

spatial distance from the local user to the satellite at moment t; $\Delta T(t)$ is the difference between local satellite clock and local clock at moment t; t_R is the propagation time delay of the signal from satellite to the user; and C is the light speed.

The pseudorange represents the product of multiplying the sum of relative time error between satellite and user at designated moment and the propagation time delay of signal from satellite to the user by light speed. Therefore, pseudorange contains definitions of two values, one is moment t generated by pseudorange, which can be expressed as satellite time or local time. The other one is size of pseudorange. Pseudorange measurement is to accomplish the extraction of $R_P(t)$. It can be seen from the expression of pseudorange that radio navigation parameter pseudorange contains the distance between satellite clock and user clock. This difference emerges as an unknown value. Pseudorange emerges as navigation parameter. Position surface is a spherical surface taking satellite centroid as center of sphere and pseudorange as radius. Radius contains unknown clock difference $\Delta T \cdot c$. Therefore, measurement of pseudorange to three satellites can produce a system of ternary linear equation with four unknown numbers of $(x.y.z.\Delta t)$. An indefinite parameter occurs in the solution of the equation. In order to eliminate the uncertainty, the measured value must be added to measure the pseudorange to the fourth satellite so that the solution of quaternion equations is obtained. According to these measured values of pseudorange, the user's position determined is the intersection point of the four position surfaces.

7.2.3 Navigation Positioning Equation

From formula 7.1, the observed quantity of pseudorange $R_p(t)$ is obtained, after conversion [3]:

$$\left(\rho^j\right)^p = \rho^j + c\Delta t_k \tag{7.2}$$

In the formula, $\rho^j$is the spatial distance from satellite j to the user's receiver.

$(\rho^j)^p$ is the measured value of pseudorange from satellite j to the user's receiver.

Δt is the clock difference between receiver clock and satellite clock.

Formula (7.2) can be rewritten as

$$\rho^p = \rho + c\Delta t_k - c\Delta t^j \tag{7.3}$$

Wherein, Δt^j is the clock difference between satellite j's clock and the navigation system clock.

Δt_k is the clock difference between the receiver clock and navigation system clock.

When multiple satellites are observed:

$$(\rho^j)^p = \left[\sum_{i=1}^{3}\left(x_i^j - x_i\right)^2\right]^{1/2} + b - c\Delta t^j \tag{7.4}$$

In the formula, x_i^j, (in x_i, $i = 1, 2, 3$) represents, respectively, the components of the coordinate values in ground coordinate system used by satellite j and the receiver; b is the equivalent distance error of the receiver's clock error.

Since the satellite clock error Δt^j and position of satellite are known, four pseudorange equations can be written according to the positions of the four satellites and the observed quantity of pseudorange measured so that the position and clock error of the user can be solved.

Adopt tailor progression to extend formula (7.4) and omit the high-order item so as to obtain the system of linear equations:

$$\sum_{i=1}^{3}\frac{\partial F^j}{\partial X_i}\Delta X_i + b = (p^i)^p - F^j \tag{7.5}$$

Function in the formula

$$F^j = \left[\sum_{i=1}^{3}\left(X_i^j - X_i^0\right)^2\right]^{1/2} - c\Delta t^j \tag{7.6}$$

X_i^0 is the initial position of the receiver and ΔX_i is the corresponding correction number.

After the partial derivative in the equation is obtained, formula (7.6) can be written as:

$$e_1^j\Delta x_1 + e_2^j\Delta x_2 + e_3^j\Delta x_3 - b = F^j\left(p_i^j\right)^p$$

In the formula, $e_i^j (i = 1, 2, 3)$ is the direction cosine of satellite j's observation direction to a three-coordinate axis.

When four satellites ($j = 1, 2, 3, 4$) are observed, the equation can be written as

$$\mathbf{AX} = \mathbf{L} \tag{7.7}$$

$$A = \begin{bmatrix} e_1^1 & e_2^1 & e_3^1 & -1 \\ e_1^2 & e_2^2 & e_3^2 & -1 \\ e_1^3 & e_2^3 & e_3^3 & -1 \\ e_1^4 & e_2^4 & e_3^4 & -1 \end{bmatrix}$$

$$X = \begin{bmatrix} \Delta x_1 \\ \Delta x_2 \\ \Delta x_3 \\ b \end{bmatrix}$$

$$\mathbf{L} = \begin{bmatrix} F^1 - (\rho^1)^p \\ F^2 - (\rho^2)^p \\ F^3 - (\rho^3)^p \\ F^4 - (\rho^4)^p \end{bmatrix}$$

Solution of unknown parameter X is

$$X = A^{-1}L \tag{7.8}$$

After the first solution is obtained by utilizing the iterative method, updating the initial value to obtain the solution again so as to obtain a big difference between the initial value of the receiver position and initial value of the clock error solved.

When more than four satellites are being observed and integrities are qualified, the solution is obtained through balancing. The error equation can be obtained when the observed pseudorange quantity is deemed as distance measurement value with error, namely, replace $(\rho^j)^p$ in formula (7.5) with $(\rho^j)^p + v^j$.

$$V^j = \sum_{i=1}^{3} \frac{\partial F^j}{\partial x_i} \Delta x + b - \left(\rho^j\right)^p + F^j \tag{7.9}$$

Or

$$V = AX + L \tag{7.10}$$

Solve with the least-squares method

$$\left(A^T A\right)^{-1} X = A^T L \tag{7.11}$$

$$X = \left(A^T A\right)^{-1} A^T L \tag{7.12}$$

Accuracy of the solved parameter can be estimated simply according to the least-squares method:

$$m_{x_i} = \sqrt{q_{ii}} \cdot m \tag{7.13}$$

Wherein:

m represents pseudorange measuring error.
q_{ii} is the element at row i and column i in the Matrix Q.

$$Q_L = \left(A^T A\right)^{-1} \tag{7.14}$$

When only four satellites are being observed, the accuracy can be estimated according to (7.13).

Usually, when over four satellites are being observed, combine the observed quantities of different four satellites to perform autonomous integrity operation of the receiver to determine if there is abnormal satellite observed quantity.

When six satellites are being observed, which observed quantity of satellite does not meet the requirement of integrity can be determined.

7.3 Geometric Precision Factor

From the analysis in last section, precision can be estimated using (7.11)–(7.13).

$$X = \left(A^T A\right)^{-1} A^T L \tag{7.15}$$

$$m_{x_i} = \sqrt{q_{ii}} \cdot m$$

$$Q_L = \left(A^T A\right)^{-1}$$

$$A = \begin{bmatrix} a_1^1 & a_2^1 & a_3^1 & -1 \\ a_1^2 & a_2^2 & a_3^2 & -1 \\ \cdots & \cdots & \cdots & \cdots \\ a_1^j & a_2^j & a_3^j & -1 \end{bmatrix} = \begin{bmatrix} a_1^1 & \cdots & a_i^1 & -1 \\ \vdots & \ddots & \vdots & \vdots \\ a_1^j & \cdots & a_i^j & -1 \end{bmatrix}$$

In the formula, a_i^j is the direction cosine of observed direction of the satellite j to i coordinate axis.

It can be seen from the arranged formula (7.15), precision of positioning parameters to be solved depends on precision m of the pseudorange and coefficient matrix A. However, A has nothing to do with precision of pseudorange and only depends on the geometric relationship between user receiver and observed satellite. In order to describe these geometric relationships, geometric dilution of precision (GDOP) can be used for description, coming from the Loran-C Navigation System

However, in the satellite navigation system, this has been extended to three-dimensional positioning and timing.

Besides GDOP, there are still the following parameters:

PDOP: Position Dilution of Precision
HDOP: Horizontal Dilution of Precision
VDOP: Vertical Dilution of Precision
TDOP: Time Dilution of Precision

$$m_T = \sqrt{q_{44}} \tag{7.16}$$

$$m_T = TDOP \cdot m \tag{7.17}$$

In the formula, m is the error of pseudorange. When m is represented by time delay, m_T is the timing error.

Similarly, as for the three-dimensional position error,

$$m_p = \sqrt{q_x^2 + q_y^2 + q_z^2}$$

Define

$$PDOP = \sqrt{q_{33}^2 + q_{22}^2 + q_{11}^2} \tag{7.18}$$

Thus:

$$m_p = PDOP \cdot m \tag{7.19}$$

HDOP, *VDOP* can be defined using a similar method. Because that orientation of the three-coordinate axes are different from the vertical height direction of receiver's position, VDOP is required to be represented on the projection of three component errors along the coordinate axis on the vertical line of the receiver.

$$VDOP = \sum_{i=1}^{3} (\vec{x_i^0} \cdot \vec{q}) \quad (7.20)$$

In the formula, $\vec{x_i^0}$ represents the unit vector of three-coordinate axes.

$$\vec{q} = \begin{bmatrix} q_1^1 \\ q_2^2 \\ q_3^3 \end{bmatrix} \quad (7.21)$$

$$m_v = VDOP \cdot m$$

From this derivation, geocentric radius vector of the point location where the user is located replaces the ellipsoidal normal of this point.

As for the horizontal position, there are

$$HDOP = \sqrt{PDOP^2 - VDOP^2} \quad (7.22)$$

$$m_H = HDOP \cdot m \quad (7.23)$$

GDOP is the comprehensive geometric dilution of precision, which represents the influence of the geometric relationship of observed satellite on the calculation of user's position and comprehensive precision of the clock error.

$$GDOP = \sqrt{PDOP^2 + TDOP^2} \quad (7.24)$$

From the definitions here, it can be seen that the bigger the GDOP, the lower the precision. Therefore, GDOP value means geometric attenuation precision. Since the satellite is operational, the geometric relationships with coverage area are different. Then the precision is different from one region to another.

7.4 Satellite Navigation Velocity Measurement Principle

The important mission of RNSS service is to determine the navigational speed for the user. According to the derivative of position to time, solution of navigational speed is theoretically feasible [3]. As for engineering, it is hard to express the instant navigational speed of the user and cannot be used for kinetic characteristic of high-speed carrier. Then, it is suitable to use the change rate of pseudorange, namely the observed value of the derivative of pseudorange to time $\frac{d\rho'}{dt}$. As the change rate of pseudorange is related to Doppler frequency of satellite navigation signal, that is:

$$\Delta f = \frac{1}{c}\frac{d\rho}{dt} \quad (7.25)$$

Therefore, solution of the user's speed can be derived with this relationship expression. In the positioning solution (7.4), when frequency offset exists in satellite and receiver clocks

$$\Delta f = \frac{1}{c}\frac{d\rho}{dt} + \delta f^j - \delta f \quad (7.26)$$

In the formula, δf^j and δf represent the frequency offsets of the satellite clock and receiver clock, respectively. Frequency offset of the satellite clock can be obtained from satellite broadcast ephemeris as a known value. The clock offset of receiver is the value to deduce. However,

$$\vec{\rho} = \vec{r^j} - \vec{r}$$

$$\frac{d\vec{p}}{dt} = \frac{d\vec{r^j}}{dt} - \frac{d\vec{r}}{dt}$$

Written as component form:

$$\frac{d\rho_x}{dt} = \frac{dx^j}{dt} - \frac{dx}{dt}$$

$$\frac{d\rho_y}{dt} = \frac{dy^j}{dt} - \frac{dy}{dt}$$

$$\frac{d\rho_z}{dt} = \frac{dz^j}{dt} - \frac{dz}{dt}$$

$$\frac{d\rho}{dt} = \left[\sum_{i=1}^{3}\left(\frac{dx_i^j}{dt} - \frac{dx_i}{dy}\right)^2\right]^{1/2}$$

Put into formula (7.26)

$$\Delta f = \frac{1}{c}\left[\sum_{i=1}^{3}\left(\frac{dx_i^j}{dt} - \frac{dx_i}{dy}\right)^2\right]^{1/2} + \delta f^j - \delta f \tag{7.27}$$

In Equation (7.27), Δf is the observed quantity, $\frac{dx}{dt}$, $\frac{dy}{dt}$, $\frac{dz}{dt}$, and δf are the parameters to be solved. The rest are known values.

Using 0 as the approximate value in the formula (7.26), the partial derivatives of $\frac{dx}{dt}$, $\frac{dy}{dt}$, $\frac{dz}{dt}$ and δf are solved:

$$\varepsilon_1^j\delta\left(\frac{dx}{dt}\right) + \varepsilon_2^j\delta\left(\frac{dy}{dt}\right) + \varepsilon_3^j\delta\left(\frac{dz}{dt}\right) - \delta\left(\frac{df}{dt}\right) = [F^j]_0 - [\Delta f]_{0b} \tag{7.28}$$

Wherein,

$$\varepsilon_1^j = \frac{\partial\left(\frac{d\rho^j}{dt}\right)}{\partial\left(\frac{dx}{dt}\right)}$$

$$\varepsilon_2^j = \frac{\partial\left(\frac{d\rho^j}{dt}\right)}{\partial\left(\frac{dy}{dt}\right)}$$

$$\varepsilon_3^j = \frac{\partial\left(\frac{d\rho^j}{dt}\right)}{\partial\left(\frac{dz}{dt}\right)}$$

$[]_0$ is the approximate value, observation value of Δf can be obtained by putting 0 in. When j satellites are observed, the least-squares method can be used for solution.

$$X = (A^T A)^{-1} A^T L$$

$$A = \begin{bmatrix} \varepsilon_1^1 & \varepsilon_2^1 & \varepsilon_3^1 & -1 \\ \varepsilon_1^2 & \varepsilon_2^2 & \varepsilon_3^2 & -1 \\ \cdots & \cdots & \cdots & \cdots \\ \varepsilon_1^J & \varepsilon_2^J & \varepsilon_3^J & -1 \end{bmatrix}$$

$$L = \begin{bmatrix} [F^1]_0 - [\Delta f^1]_{0B} \\ [F^2]_0 - [\Delta f^2]_{0B} \\ [F^J]_0 - [\Delta f^J]_{0B} \end{bmatrix}$$

$$x = \left(\frac{dx}{dt} \frac{dy}{dt} \frac{dz}{dt} \delta f \right)^T$$

Apparently, user's speed is obtained according to (7.29) under the condition that the satellite speed is known. Satellite kinematic speed can be calculated from the orbit information of the message data block.

The speed obtained in this way can be estimated with GDOP similarly in principle. This speed calculation is performed when the satellite clock speed change is known. When the satellite clock rate deviates from model variation, this deviation will be mistakenly considered a calculation of the user's total speed. Therefore, solution of the satellite navigational speed still requires a high-accuracy satellite clock.

7.5 Positioning Velocity Measurement Precision

Positioning accuracy is the first thing to be discussed. Based on the positioning principle previously, the main error of satellite navigation positioning comes from three aspects:

1. *Space*: Satellite ephemeris error, satellite clock error, satellite equipment time delay.
2. *User*: Measurement error of user receiver, error of user's calculation.
3. *Signal propagation path*: Troposphere signal propagation time delay, ionosphere signal propagation delay.

Error of space comes from operational control of the system. Only satellite equipment time delay must be ensured by satellite production. Satellite clock error can be weakened further by observation and modeling of operational system. The precision and stability of the satellite clock increases with increasing of the precision of the modeling.

Measuring error of the user receiver is not only related to measuring technique but also related to transmitting power of satellite and the influence of electromagnetic interference on the carrier noise spectral density (c/N_0).

According to the concept of satellite navigation system in this chapter, the discussion of satellite navigation precision will include the discussion of different navigation user, under different navigational stage (condition), and based on the demand of positioning accuracy, then it is economic and feasible to specifically analyze and provide the technological approach of meeting precision.

7.5.1 Positioning Accuracy of Global System

From the demands of the users listed in Tables 7.1 and 7.2, positioning accuracy of a global system is mainly to meet the demand of the user who travels long distance or demand of extending to non-precision approach. The highest precision is at 100 m (95%) magnitude. Slight improvement is made in order to meet the requirement of weapon and space navigation. The global system mentioned here means the global coverage or regional coverage system by only utilizing stable operational non-GEO satellite and no-pseudo satellite technology. The practice shows that under the condition of not considering the rigor integrity, it only relates to selection of the satellite constellation and will be discussed in relevant sections.

7.5.2 Global and Regional Augmentation System Positioning Accuracy

The global and regional augmentation system adds GEO satellites in a non-GEO constellation in the global coverage (or regional application), which both improves the positioning accuracy and meets the requirement of availability, continuity, and integrity. The highest service target is the ß-class I precision approach (CAT-I). The so-called regional satellite navigation system needs to be designed according to the user's demand of regional augmentation, the coverage area precision of which must meet the requirement of the class I precision approach for the lands with airport and islands. The index listed by ICAO should be its standard, namely, horizontal precision: 18.2 m and vertical precision: 7.7–4.4 m. The requirements of continuity risk of $8 \times 10^{-6}/15^{s}$approach and integrity risk of 2×10^{-7}approach must be met accordingly. Warning time is 6 s. Based on the equivalent distance error of the user, UERE = 2.0 m (95%), design of the satellite constellation must allow the following index of the coverage area of the airport in the region to meet the following requirement:

$$HDOP \leq 9$$

$$VDOP \leq 3.85$$

When the GEO satellite navigation message properly arranges the broadcast of short-period and hard real-time messages, such as integrity information, integrity warning information, and so on, multiple GEO satellites can achieve the requirement of reliability and availability.

7.5.3 Global, Regional, and Local Augmentation Positioning Accuracy

This is to meet the CAT-2 user's demand.

This has a short transitional process; it is the augmentation of integrity and precision through ground pseudo-satellite technology based on global and regional augmentation system, the nature of which is utilizing the achievement of regional system to establish landing system for the airport. Generally, it is not appropriate to increase the global or regional burden to meet the requirement of positioning accuracy.

7.6 Distance Difference and Radial Velocity Difference

A distance difference measuring method can be used to get the navigation mission solved, which is based on distance difference measured from user to one or several satellites, essentially, this method is similar to pseudorange measurement; as there is unknown difference existing in pseudorange measurement, it is suitable to adopt this method. The distance difference measuring method is using three differences to four satellites $\Delta R_{ij} = R_i - R_j$, as during the navigation time, unknown difference ΔR is invariable; therefore, pseudorange difference equals to actual distance difference. In order to determine the distance difference, three independent equations are required. Navigation parameter is ΔR_{ij}. Position surface is determined by $\Delta R_{ij} = const$, which is a hyperboloid of revolution with the coordinates of focal point or even supporting points i and j (the centers of antenna phases of the satellite i and the satellite j). The distance between these base points is called baseline of the measuring system. If the distance from each supporting point of the satellites to the user is longer than baseline, it will superpose its own surface – a circular cone on the hyperboloid of revolution near the user point. The apex of the cone superposes the central point of the baseline. Determination of the coordinate precision of the user with this method is equivalent to pseudorange method. The defect of this determination is that it cannot determine the time migration of the user.

Radial velocity difference method can also be used to determine the velocity vector of the user, the nature of which lies in determining three difference values of two radial velocities to the satellite $\Delta \dot{R}_{ij} = \dot{R}_i - \dot{R}_j$. Therefore, the differences can be calculated relative to one satellite or different satellites. Actually, pseudorange radial velocity can be used when calculating the differences.

As for distance difference, the precision of velocity vector components obtained using this method is the same as the precision determined with pseudorange change rate. The advantage of the radial velocity difference is insensitivity to unstable frequency standard and the defect is that the instability of frequency standard cannot be estimated.

7.7 Combined Method

Besides the basic methods of determining user's position listed, there are still a lot of combined methods that adopt the additional measuring devices provided by the user, except for the satellite navigation system. For instance, when the user has height measurement, in those measurements pseudoranges of only three satellites, instead of four satellites, are required. For example, if the elevation of the user on the ocean slightly changes, under the condition that the precision requirement is not that high, the pseudorange of three satellites observed can also be used to solve the user's position.

Other situations using combined methods include combination of simultaneous and proceeding measurements or just the proceeding measurement. For example, replace the simultaneous measurement with velocity difference to determine the coordinates of the user for combination. Replace at least four times the simultaneous measurements of four satellites with two lots of proceeding measurements of two satellites, or four times the proceeding measurement of one satellite, and so on.

7.8 Carrier Phase Difference Method

The user who requires extremely high accuracy can adopt carrier phase difference technology. Such users measure the carrier phase relative to reference station. This carrier distance measurement precision is a percentage of carrier wavelength (typically ≈1 cm). These phase difference values can also be used for dynamic positioning, especially for the approach landing of high grade aircrafts, such as the Category II or Category III precision approaches. However, distance between antenna is usually over 1 wavelength (e.g., GPSL_1 is 19 cm). Therefore, estimated position possesses multi-valuedness as a wavelength integer is unknown. As for practicability, carrier phase difference (such as DGPS) needs to get the complete cycle multi-valuedness ($n\lambda$ or $N_{u\cdot s}\lambda$) solved.

As for carrier phase DGPS, the mobile user receives the carrier phase measurement from the reference station (original or converted), and forms them into a single difference and double difference. Single difference, between user and reference station, eliminates the error source of strong correlation. The differences existing in the ionosphere, troposphere, and satellite clock are completely eliminated as the distance from the user to the reference station is generally approximately 30 km. As for double difference, the satellites that the user receives are used to eliminate the difference between the reference clock and user clock. In order to get "complete cycle multi-valuedness" solved, a further proposal should be made. Therefore, double difference of different times is used for elimination. During two sufficient times, the double difference between satellites provides sufficient information for the solution of three components of position differences $\Delta\bar{r}_u$ between user and reference station. However, double difference observed serial number time of 30 min is generally required. As for survey and mapping user, the later treatment is acceptable.

As for the Categories II/III precision approaches required by aircraft, they not only require eliminating multi-valuedness reliably but also require real-time positioning. There are two ways of real-time determination of carrier wavelength integer. One is using a pseudo satellite, which is a ground transmitter transmitting the satellite navigation signal. It can be operated at a single frequency. For example, working at the GPS L_1 frequency point, this satellite is placed under the approach path. An airplane flies over at a height of 1000–1500 m. Under this configuration, when the airplane is flying over it, the view vector from the airplane to the pseudo-satellite takes a large sweeping angle. Through this big change of angle, complete cycle of multi-valuedness can be solved. This angle change is completed in less than 1 min, which is caused by the aircraft rapidly flying over the pseudo satellite. The success rate of this technology demonstration can reach almost 100% so that centimeter-level three-dimensional positioning accuracy can be provided. This technology is the foundation of the "integrity beacon landing system."

The other one is named the "wide lane," which is a feasible way to simplify multi-satellite search to distinguish the complete cycle multi-valuedness. The so-called "wide lane" uses the beat frequency (347.82 *MHz*) formed by multiplying L_1 by L_2. Its wave length is 86 cm, which is obviously longer than L_1 (19 cm) or L_2 (24 cm). Therefore, code observation can be used to solve the complete cycle numbers of the beat frequency signal. In return, the searching range is significantly reduced in order to get complete cycle numbers solved. Navigation signal of code phase on L_2 is relatively convenient, otherwise, the L_2 carrier needs to be recovered through square technology.

8

Performance Demand and General Design of RNSS

Navigational demand for modern ocean, aviation, aerospace, and land transportation increases each day. In particular, modern war deems satellite navigation as an addition to weapons and there are extremely high expectations of it. More performance requirements and a higher performance index are proposed. These can be divided into three categories:

1. *Essential performance requirement*. This means the essential and basic requirements to finalize en-route navigation, which includes precision, availability, continuity, integrity, and so on.
2. *Value-added performance requirement*. The performance extension of navigation including the realization of a series of value-added requirements such as timing, communication, and user identification by the navigation system.
3. *Integrated performance requirement*. Starting from the entire performance of the system and not only navigation, a series of integrated innovation requirements is proposed, including high-performance index requirements such as system height combination and information integration. Design of a satellite navigation system includes initiation design and engineering design. The so-called initiation design, also known as the top-level design or system architecture, belongs to the project's overall plan. System engineering is the breakdown of the system functional index and solution, and specific indexes of the planning satellite, operational control system, augmentation system, and application system.

8.1 Essential Performance of RNSS

The essential performance of RNSS is to be the performance index proposed by the ICAO for civil aviation navigation: See Table 8.1. Previously, departments in USA and Russia had corresponding standards (see Tables 8.1–8.7). The names and explanations of indexes are different and the scopes specified by the indexes are also different. Up to now, system design of satellite navigation, including Galileo satellite navigation, considered the following indexes as essential.

- Precision
- Availability
- Continuity
- Integrity

GNSS Systems and Engineering: The Chinese Beidou Navigation and Position Location Satellite, First Edition. Shusen Tan.

Table 8.1 Safeguarding requirements for Russian radio aviation navigation.

Mission	Flight Range	Positioning Accuracy (m, rms)	
Airline flight	Over the ocean	5800	30–40
	20 km wide airline	2500	30–40
	10 km wide airline	1250	30–40
	Local airline		
	Category 1	500	30–40
	Category 2	250	30–40
Airport area flight		200	
Top flight (Lifesaving)		1–10	
Non-type preparation landing		50	
Preparation landing			
	Category 1 H = 30 M	4.5~8.5	1.5~2.0
	Category 2 H = 15 M	2.3~2.6	0.7~0.85
	Category 3 H = 2.4 M	2.0	0.2~0.3

1. *Accuracy*
   In the given service area or phase of navigation duty implementation, it is the difference between coordinate parameters of undetermined position of user equipment and actual coordinate parameters; as for aviation user, the reliability is generally 0.95, for example, 2 dRMS (twice standard error.)
2. *Availability*
   Availability is the measurement of working performance probability. Currently, there is no uniform standard for satellite navigation yet. Availability of weapon equipment means the operational degree that can be reached by the system at any random time. Availability of satellite navigation system can be interpreted as the capability of mission performing in user given service area and positioning accuracy at given time. It is defined as the ratio of serviceable time to expected service time:

$$A = \frac{\text{Serviceable time}}{\text{Expected serviceable time}}$$

   As for the users of satellite navigation, they expect 24-hour continuous service. Then, the percentage of rated service time provided in one day can be used to express the availability. Based on this concept, the availability of the system includes:
   a. PDOP value meeting the positioning accuracy requirement in service area.
      When system constellation is selected, inherent availability of the constellation is expressed as satellite deflection CVs, which is the upper limit of system availability.
   b. Reliability of satellite and operational control system and backup strategy.
      Reliability of satellite can be expressed with the probability of normal service provided during service life and is decreased gradually along with the extension of the service time. In order to ensure the availability of the system constellation, a developed satellite transmission backup strategy is formulated to meet the satellite

Table 8.2 Russian index requirements for the descent stage of flight landing.

Level	Channel	Deviation limitation (m)		Height (m)	Availability	Integrity	Problem Solving Probability	Unremittance
		Orientation	Vertical					
1	Internal	±40	±12	60	0.9975	T < 6S	0.95	$1 - 1.4 \times 10^{-6}/15$ s
	External	±121	±37	60	0.9975	$1 \sim 3.3 \times 10^{-7}$	$1 \sim 3.3 \times 10^{-7}$	$1 - 1 \times 10^{-4}/150$ s
2	Internal	±21	±4.6	30	0.9985	T < 2 S	0.95	$1 - 1.4 \times 10^{-6}/15$ s
	External	±64	±14	30	0.9985	$1 \sim 3.3 \times 10^{-8}$	$1 \sim 3.3 \times 10^{-8}$	$1 - 1.4 \times 10^{-5}/165$ s
3	Internal	±15	±1.5	15	0.999	T < 1 S	0.95	$1 - 1.4 \times 10^{-6}/30$ s
	External	±46	±4.6	15	0.999	$1 \sim 1.5 \times 10^{-7}$	$1 \sim 1.5 \times 10^{-9}$	$1 - 1.4 \times 10^{-6}/30$ s

Table 8.3 Safeguarding requirements for Russian radio navigation of marine vessels.

Mission	Positioning Accuracy (RMS, M)	Availability	Integrity[a)]
High seas navigation	1400~3700	0.99	0.99
Offshore navigation	100~60	0.99~0.997[b)]	0.99
Pass gorgc long area, arrival at port	Superior to 20	0.99~0.997	0.99
Port flexibility	8	0.997	0.99
Ground mapping instrument oceanography	0.25~5	0.99	0.9~0.99
Geological prospecting, mining	1~5	0.99	0.9~0.99

a) Allowable report termination time is 1 s to 1 min, subject to the mission of ocean vessels.
b) When value is 0.997, it is relative to large-tonnage vessels.

Table 8.4 Planning requirements for Russian radio navigation.[a)]

Integrity Detection Reference	Airline Flight	Airport Area	Landing Approach	
			Non-Grade	Grade
Availability		0.9996		
Middle value (m)	250~8000	200	50~70	2.0~8.5
Integrity detection probability	0.999		≥ 0.999999	

a) *Reliability*: The probability of reliable position information (conforming and non-conforming) obtained by the user through required precision at the working area and given time as the integrity of auxiliary navigation and value of 0.999 as the agreed navigational integrity.

Table 8.5 Federal Radio Navigation Proposal Requirements of the United States Department of Transportation.

Flight Phase	Ocean	Internal Airline	Airport Area	Non-Grade Landing
Allowable value				
Threshold value, NM	12.6	2.8	1.7	0.3
Allowable delay, S	120	60	30	10
Future value				
Threshold value, NM	5	1	0.5	0.1
Allowable delay, S	30	30	10	6

quantity in the constellation and the constituted PDOP value. This backup strategy is determined subject to planning transmission or transmission as required. The backup strategy of planning transmission is based on the reliability of the satellite. New satellites will be added to the constellation when others have reached the end of their design life, no matter whether they are operating normally or not. However, a backup strategy as required is to supplement satellites to the constellation as

Table 8.6 US Federal Aviation Administration FAA Navigation System Precision Standard.

Flight Phase	Min. Height (m)	Horizontal Precision (m, RMS)	Vertical Precision (M, RMS)
Air route terminal	152	7400	500
Non-precision approach	76.2	3700	100
Category 1 precision approach	30.5	9.1	3.0
Category 2 precision approach	15.2	4.6	1.4
Category 3 precision approach	0	4.1	0.5

Table 8.7 Requirements of the US Radio Technical Commission for Aeronautics RTCASC 159 for the RAIM Integrity Recommendation.

Flight Phase	Door Limit NM	Max. Warning Probability (h^{-1})	Warning Time (s)	Min. Detection Probability of Integrity
Air route	2.0	0.002	30	0.999
Terminal	1.0	0.002	10	0.999
Non-precision Approach	0.3, Horizontal	0.002	10	0.999

actually needed after the performance of satellite is lowered or totally lost. In order to not interrupt the service or lower performance, the numbers of backup satellites in the orbit of constellation is thoroughly schemed, as well as the satellite delivery cycle, ground backup numbers of satellite, transmit cycle, orbit injection opening cycle, and so on. Generally, a backup satellite in orbit is launched in advance on same orbit plane to ensure the availability of constellation. Based on the condition of backup, one RNSS system is better for selecting three orbit planes rather than six, which can save backup satellites and shorten the cycle of constellation control.

Although the availability of ground operational control system can also affect the usability of satellite system, the influence on satellite navigation system is not big as the main backup method and equipment of operational control has sufficient redundancy with the usability of 0.9998 and above.

3. *Continuity*

Continuity is the probability of system service performance of the user in the important navigation phase, which specifically describes the probability that meets precision and integrity without unexpected interruption happening during navigation provided to the user. It can also be expressed with the probability when unexpected interruption is happening. ICAO expresses it with the maximum non-continuity of navigational signal with unit of 1/h, namely the probability of unexpected interruption. See Table 8.1. However, Galileo expresses it with continuity risk and maximum interruption time. See Table 8.2. The definition of continuity risk is same as maximum non-continuity of ICAO. The maximum interruption is expressed with minute. Specific interruption time is still pending and to be confirmed further.

The concept of continuity includes the continuity of positioning and of integrity monitoring. Continuity of positioning is the probability that positioning accuracy is not exceeded. Continuity of integrity detection is that undetected probability is not happening or integrity protection tolerance is not exceeded, which are used in a different navigation phase. For example, the CAT-I precision approach is expressed with continuity of integrity monitoring and determined according to the user's cost by various navigational industries. In order to meet the requirement of higher continuity, ground augmentation technology is used in most cases. There are still some industries that also use the probability that the system cannot meet the requirement of navigation information to describe the continuity, the continuity index of which is even lower.

4. *Integrity*

   Integrity is the probability that the navigation signal fails, it can be specifically described as capability of in-time warning provided by the system to the user when the navigation signal provided by the system deduces positioning accuracy out of tolerance and does not meet the requirement of navigation. It is expressed as integrity risk, warning error threshold, and warning time.

   - Integrity Risk (IR)

     IR = Probability/Time

     This is the probability of a warning caused for whatever reasons during operation. It is naturally required that no warning is expected to happen in a certain phase of navigation no matter what reason it is. For example, during the period of precision approach, any signal warning will cause damage to the navigation user. Even if prompt action is taken, the risk is still big. Therefore, integrity risk is the probability that any error during navigation may cause a positioning error out of the warning threshold error. This index is generally not used for en-route navigation. It is only used in the precision approach phase through thorough augmentation measure and specific facilities.
   - Warning threshold error

     Maximum allowable error of the user position causing a warning in the specific navigation phase. Different navigation phases have difference warning position error values, which are expressed as horizontal and vertical errors.
   - Warning time

     Defined as the time cycle from the time of the failure happening to time of warning information received by user. Different navigation phases of different users have different warning times. Refer to Table 7.1. Currently, only ICAO and a few countries in the world have the specific requirements of these essential performance parameters. CAAC announced the implementation of the specifications and requirements of ICAO.

An essential performance requirement is the basic prerequisite performance requirement of finalizing interregional and international navigation. Designated navigation safeguards are provided by a country or a region for international navigation, which is the periodic outcome of peaceful development of countries. In order to bring its own navigation facilities into an international common navigation safeguarding system, the demands of advanced navigation carrier and personal safety must be met. The US

Government constantly adopts peaceful development and force control to implement global hegemony. The so-called peaceful development method is the adoption of advanced science and technology. For example, satellite navigation technology is widely used in traffic management of national and global aviation and aerospace, planning and management of ground transportation informatization, exploration and excavation, communication power network synchronization, scientific surveying and mapping, and other scientific experiment research. The purpose of controlling economic technology fields of other countries is achieved through civil use service. Therefore, civil use is the extension of military use. The heavier the responsibility of civil use carried for the international society is, the more opportunity there is, the stronger the political and economic control power is. Therefore, in the satellite navigation field, enhance the proportion of civil use and promote the international cooperation of civil use to both propel the peaceful development and improve the power of solution for international conflict. Military and civil uses are two aspects of enhancement of national strength, neither of which should be neglected. Certainly, essential navigation performance is also the necessary navigation performance of military user.

8.1.1 Value-Added Performance of RNSS

Value-added performance of RNSS refers to performances such as timing, positioning, and awareness reporting in addition to positioning service.

The satellite navigation system has become the most advanced time comparison method in the world. Due to the development of the satellite navigation system, interregional time synchronization accuracy has been improved to 20–50 ns. GPS has become the main measure of time synchronization for every country. The China Compass Navigation Experiment System has reached two-way timing precision of 20 ns, which is gradually being used in the synchronization of communication networks. If a nationwide power synchronization is realized, namely, the efficiency of power network is improved by several percentage points, it is a giant contribution made to energy saving and environmental protection. Therefore, to provide a convenient and diverse timing service is one of the methods to improve the value-added performance of satellite navigation.

From the viewpoint of the user's demand for navigation, demand for a user's position report is growing, which is the essential condition of realizing aviation and aerospace traffic control, as well as ground transportation control. Therefore, the positioning report is the valve to increase the value of the satellite navigation system, which makes overall efficiency multiply. The US Government takes position reporting as an important navigational demand in the course of constructing the modernization of GPS. The new concept of NAVCOMM has been proposed. The Chinese satellite navigation system has realized the integration of positioning, timing, and communication. Preliminary awareness reporting of users is realized. Actually, the awareness report of the user contains more extensive implication, including the more extensive position report involving thermal imagery position awareness, microwave remote sensing position awareness, optical imaging position awareness, and so on, which will be realized gradually with the requirement for high-dimensional satellite navigation performance.

8.1.2 High-Dimensional Performance of RNSS

Demand for high-dimensional satellite navigation performance comes from the comprehensive demand of satellite navigation. Besides the essential performances and value-added performances mentioned previously, performances, such as anti-interference, anti-deceptive, security, autonomy, repellency, signal acquisition, and so on, are intensively reflected in the concept of navigation warfare in the modernization process of GPS, with the goals of (1) ensuring the authorized use of GPS signal, (2) preventing the use of unauthorized user, and (3) slightly influencing the civil user outside the military conflict zone.

High-dimensional performance is reflected outstandingly in the performance requirement of military use. This new military demand comes from the US Army's negation of traditional military science and experience of precision engagement. Long-term military occupation of one region constitutes an enormous burden to the US Government. Stronger hostility is generated. It is also contradictory to democratic politics. However, precision engagement replaces the regional control by properly delivering the military strength and materials to form abrupt combat strength. It is proved by the practices of Kosovo, Afghanistan, and Iraq that this strategy and tactics will be developed continuously. Active information control rights can obtain meaning the battlefield situation is consistent with combat dynamic planning to victories with minimum limit strength. Based on the information control right, navigational demanding concept includes: ensure precise position awareness; ensure military strength synchronization in the overall process of combat; ensure satisfactory integration of navigation capability under the C4I structure in the globe and battlefield. According to the demands for military and civilian use before, the GPS of the future will be changed to a positioning system with the following functions:

1. Ensure global continuous positioning service capability for civilian use.
   - Eliminate the influence of military use on global service for civilian use.
   - Service for military use only controls navigation signals at war.
   - Maintain the so-called backward capability of existing civilian and military uses. Realize smooth transition and multi-measures and multi-equipment redundant service.
2. Provide navigation for aviation and space traffic control.
3. Improve the high precision navigation signal for ground vehicle-mounted and vessel positioning.
4. Enhance the battle field awareness capability of armies and allied forces, including capability of identification of friend or foe.
5. Protect from the traditional and untraditional interference signal and electronic deception and maintain the existing electronic countermeasure capability.
6. Accommodate the evolution and adaptation of backward capability.

Satellite navigation system will develop continuously from simple position, velocity measurement to multi-functional demands such as positioning and velocity measurement, timing, position awareness report, tracking and monitoring, accident report, and an interactive display of the battle field situation.

8.2 Mission and Procedure of the General Design

The US satellite navigation science and application industry has divided the general design of GPS modernization into architectural design and engineering design 11. The top-level design of satellite navigation systems is compared to system architectural design because general design is just like building design to accommodate a user. Firstly, it aims to help the user determine the key point of the demand and acceptable performance, composition, scheduling and technical risks, political and legal limitations, as well as the organizational management structure of the project. The nature of architectural design is to build up an interfacing relationship with multiple structures, various performances, and engineering systems. The main functions of space-based navigational system are positioning, velocity measurement, timing, and position information reporting, which are accomplished through the user's equipment, satellites in the space, and operational control on the ground. The design of system is restrained by external conditions such as frequency and bandwidth, administrative policy for civilian use, relationship of international community, threat of interference and disturbance, available budget, planning and scheduling, and so on. All of these must be demonstrated and solved at the stage of architectural design. However, successful architectural design is established by simplifying high dimensional demand and high level solutions. The reduction of dimension is obtained through the balancing and matching between user's demand and design restraint as well as the proper analysis of current technological level and industry tendency. The basic procedure of architectural and engineering designs is shown in Figure 8.1. Basically, six steps are taken to get the

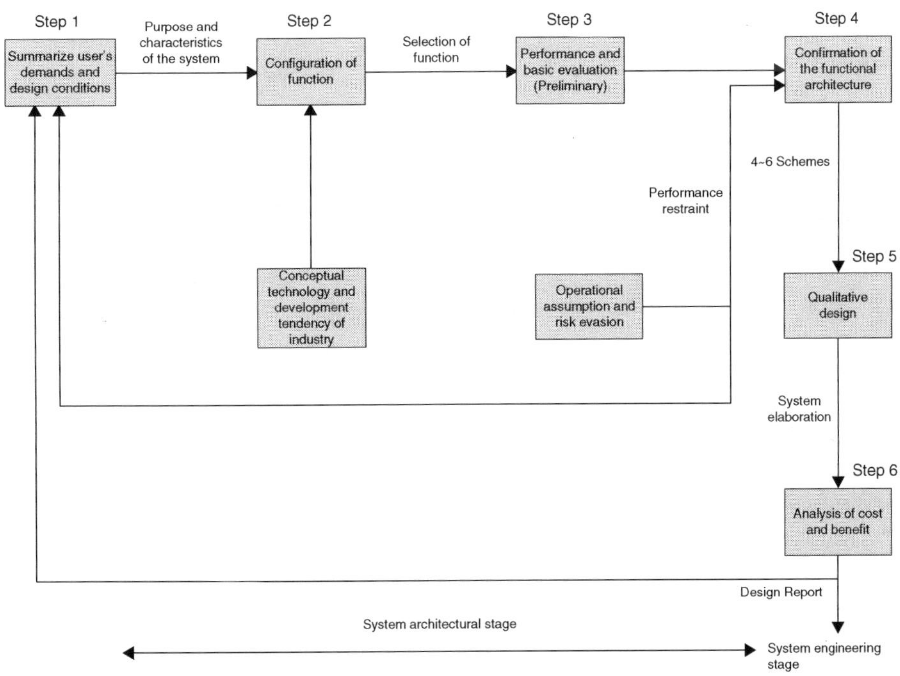

Figure 8.1 Block diagram of system architecture and engineering procedures.

general design accomplished. Steps 1–4 are the system architectural design. The rest are engineering design.

Step 1: summarize user's demands and design conditions

Through comprehensive analysis, purpose and performance of the system are elaborated accurately to highlight the service object, service mode, and service business of the system.

Step 2: configuration of function

Functions of the system under the restraint of user's demands and design conditions are listed. Options of function configuration of system are proposed. During this process, technology and industrial fundamental conditions involved in the system are analyzed in depth. The industrial base of so-called conceptual technology includes high power satellite technology and device level, large diameter figurative satellite antenna design, fabrication, implementation and control technology, and programmable integrated digitalized receiver technology. Equipment and technology of ground difference augmentation system analyze these industrial foundation levels. The development potential, direction, and speed are sufficiently assessed. As it takes around 10 years to set up a satellite navigation system, the first life time is also around 10 years. The current level shall be considered, as well as the development tendency for 20 years to make the system have sufficient vitality and value-added opportunity.

Step 3: performance and basic evaluation

According to different function configurations, a performance and cost evaluation report is produced accordingly. At this stage, a suitable operational assumption is designed. Since the development tendency of technology and industry of the 20 years to come is estimated in Step 2, there is one clear operational assumption for the system engineering establishment. Measurement of risk evasion shall be taken accordingly for financial budget and technological risks, and the operational assumption is taken as the restraint conditions of function design.

Step 4: confirmation of the functional performance of architecture

According to the initial performance and cost evaluation, the functional performance must be confirmed considering design conditions such as operational mechanism, risks, and so on: 4–6 selections are iterated to 2–3 selections.

Step 5: qualitative design

The so-called qualitative design includes system constellation configuration, quantity, cost and lifetime of satellites, basic scheme, main method and equipment configuration of operational control, ground augmentation scheme, and establishment mode that may be produced. The systematic elaboration is made with respect to the system composition, cost input, construction cycle, and application benefit.

Step 6: cost-benefit analysis

Through the analysis of the total cost of construction, including the cost of assurance conditions and the benefit and cost recover that may be generated, the overall scheme report is figured out. If the result is not satisfactory, iterate according to Steps 1 and 2 until the optimal cost-benefit ratio is maintained and the general design of system and the establishment operation mechanism are confirmed.

After the general design is done, the design of engineering overall implementation plan is carried out. Specific solutions for engineering construction are proposed for the development of soft and hard equipment.

8.3 Mission and Procedure of Engineering Design

System engineering design aims to finalize the general implementation scheme including space satellite, ground operational control, and user, in which ground operational control plays the role of kernel and the link stipulates the interfacing relations of the three systems in the functional configuration of each part. Refer to Figure 8.2 for the basic procedure

Step 1: determination of basic composition of system
This consists of space, ground operational control, and the user.

Step 2: confirmation of satellite navigation system and interfacing relationship of system
Indexes are assigned according to general design to perform the general design of operational control system and the satellite navigation system. Operational control systems formulate the scheme of system time synchronization based on the system mission, as well as the schemes of satellite orbit determination, ground augmentation, and satellite-ground TT&C and propose the requirements of corresponding functions and indexes of the satellite. The corresponding schemes of satellite time synchronization, ground augmentation, and general TT&C and interfacing relationships are implemented. According to orbit determination, identify or formulate the schemes of inter-satellites link synchronization and pseudo-range measurement.

Step 3: design of general satellite scheme
The general scheme design mentioned previously is accomplished according to satellite orbit, payload requirement, and satellite-ground measurement and control interfaces. Missions of carrying and transmission are finalized.

Step 4: general design of application system
According to the satellite navigation signal provided by the system, dynamic performance design of application system and designs of various users' application systems and user receivers are finalized.

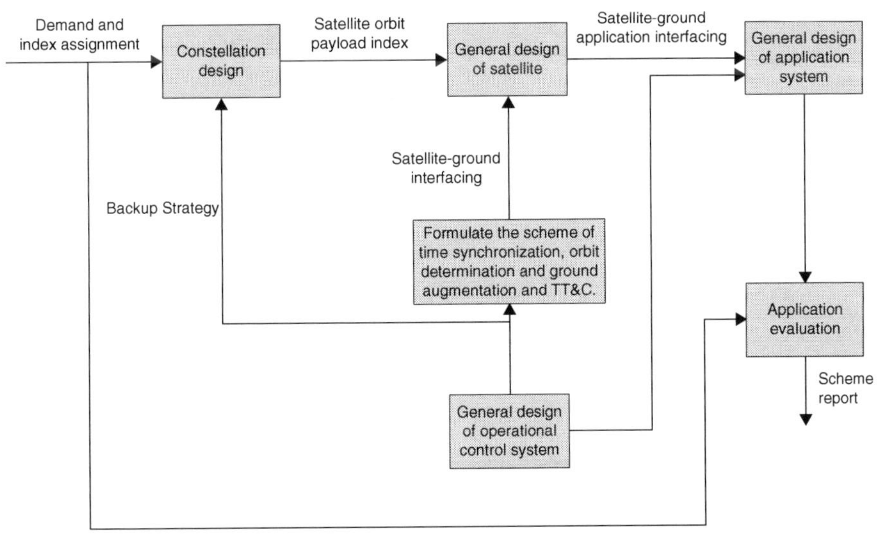

Figure 8.2 Engineering design of the system.

Step 5: application evaluation and final engineering design report of system

According to the results of expectant efficiency and engineering application design, the engineering design evaluation is made to draft the system engineering design report.

9

System Design of Satellite Navigation

9.1 System Design Principle and Content

Satellite navigation originated in the early 1960s when the United States and the Soviet Union built the first-generation global satellite navigation system with a navigation constellation composed of LEOS (low earth orbit satellite) (altitude of about 1000 km), very high frequencies (150 MHz, 400 MHz) as operating frequencies, and Doppler measurements as the basic observations, namely the US Naval Navigation System Transit and the Soviet Union's CICADA. At the same time, the US Naval Research Laboratory (NRL) organized the Timation plan, and Air Force SAMSO organized the 621B plan. In 1973, the US Defense Department merged these two plans; consequently, the basic system and development plan of Global Positioning System (GPS) was established where the satellite at an altitude of about 20,000 km was taken as the navigation constellation, L frequency band (1559~1610 MHz, 1215~1260 MHz) as the navigation signal frequency and pseudorange as the observed quantity. Meanwhile, the Soviet Union designed a similar global navigation system, GLONASS. The system evolved so rapidly and completely and has left a profound lesson and inspiration.

The satellite navigation industry in China started relatively late; out of basic consideration for national conditions, in the 1980s, the building of the first-generation satellite navigation began. The satellite radio determination system was used with Geostationary Orbit (GEO) satellites as the navigation constellation, and S, L-frequency bands as operating frequencies to complete the regional satellite navigation system covering land and sea of our country. On this basis, we completed the enhancement test of GPS satellite navigation signal accuracy and integrity, and comprehensively carried out the test, demonstrating the building of China's satellite navigation system. We began building the second-generation satellite navigation system in 2004, participated in the discussion and coordination of the work of the International Telecommunications Union (ITU) on the frequency of the Radio Navigation Satellite Service (RNSS), and reported the building plan under the network name of Compass and the system name of Compass Navigation System to the International Telecommunication Union. We participated in and organized a series of international consultations to create an international environment for the building of China's satellite navigation system. The building of the system, proceeding from historical experience and development trend of high-dimensional performance of the system, has overcome the impact of overthrowing an old system and creating a new one as well as proposing a series of design thinking and solutions such as the smooth transition of the system, independent system, international compatibility, solving urgent

GNSS Systems and Engineering: The Chinese Beidou Navigation and Position Location Satellite, First Edition. Shusen Tan.

needs, and looking forward to the future. The best integrated solution combining the Radio Determination Satellite Service (RDSS) and Radio Navigation Satellite System (RNSS) was selected to achieve high-dimensional needs of NAVCOMM, and conditions for the interactive display of satellite navigation and user identification, tracking monitoring, and user location were created.

9.1.1 System Design Principle

In order to achieve long-term user needs and interests and to create stable and viable service efficiency, the satellite navigation system is designed in accordance with the following principles.

9.1.1.1 Adhering to the Long-term Duty and Continuity

Proceeding from sustainable development potential, this pays attention to the continuity of the system service mode, avoids chop and change or expedient technical measures, and treats the decision to overthrow an old system and create a new system with caution. There are a number of satellites with a wide range of users in the satellite navigation system that is costly and technologically sophisticated. In particular, the operating frequency involved in the high-dimensional requirements of satellite navigation positioning not only has complex relations with International Satellite Navigation System, but also frequently interlaces with international mobile communication systems, traditional radio astronomy, aeronautical radio navigation, radar, and so on, and also has great influence on domestic traditional electronic countermeasure systems and other radio systems built. With the rapid development of radio information systems today, it's easy to forgo and even destroy old ones, thus, it is difficult to form overall efficiency at the key moment.

9.1.1.2 Adhering to Coordination and Integration

Humankind has entered a civilization period of harmonious development; as mentioned before, satellite navigation and other information systems are closely related, while people's demands are growing. From the perspective of whole benefit, the development of satellite navigation systems coordinates with the development of communication, radar, and other application systems, and has the potential to integrate with communications, surveillance, detection, remote sensing, resources, and other information. Nowadays, space and frequency resources are not the only the important resources and objects competing for interest, but also become the objects of competition in all walks of life. Only with coordination can we live together and develop jointly. Selecting the system in accordance with international rules coordinated among the majority of countries in the world also reflects a need to open up the market and achieve the economic globalization.

9.1.1.3 Laying Emphasis on Economy and Technical Feasibility

The economy is an important factor in determining a system, and this does not only consider the funding for building a basic system, but also takes life-cycle cost as a measurement. The program of taking small satellites groups at medium and low orbit as a navigation satellite constellation, for example, is desirable when seen from the economy of each satellite, but not necessarily reasonable judging by the life-cycle cost. Simulation results show that the number of satellites in a global satellite navigation system of medium and high orbit with a height of 20,000 km is 24, while that of a low orbit global satellite navigation system with a height of 1000 km is four times of the former. Due to the high movement speed relative to the user, the viewing angle decreases with the reducing of the satellite height. The number of observable satellites in the user's field of view is reduced, and the performance of autonomous integrity also decreases; as a result, the difficulty of the user terminal increases, and the use cost and risk increases. This is one of the reasons why navigation satellite orbit has risen from 1000 km of transit to 20,000 km.

Regarding the aspect of technical feasibility, we not only see the current level, but also consider the development trend in at least 20 years in the future. The building of a system takes about 10 years, while the first life cycle is about 10 years. It's necessary for technical feasibility to consider the current restricted reality and development trends, and the influence on system security of navigation countermeasure system in the high-dimensional performance and the technical vitality.

9.1.1.4 Security and Competitiveness

Today, the social needs and competition are very strong; satellite navigation is taken as an important infrastructure from the perspective of both peaceful development and military struggle. The objectives and programs for Europe to develop Galileo placed security and competitiveness in an important position.

9.1.2 System Design Content

With increasing high-dimensional demands of the user, the content included in the system increases. At the current stage, the following should generally be included.

- *Service mode and content*
  At present, any one satellite navigation system adopts the two modes of public and authorized service. With different modes, there are differences in the content of the services. The content of the service is no longer satisfied with solving the positioning problem of where I am, but also needs to solve where I am at a certain time as well as finding out more about where the users are. The position report has become a universal requirement. Logistics management, military strength, and military delivery also understand the implementation effect of the mission. So, from a single positioning need, service contents gradually extend to the timing, communication, identification (monitoring), location reporting, and interactive display all in one.
- *Selection of satellite orbit and constellation*
  In order to reduce the number of system satellites, improve the stability of the orbit, and increase the user's viewing angle, a higher satellite orbit shall be chosen; to enhance received power of the user and improve the anti-jamming performance of navigation signals, a lower satellite orbit shall be chosen; and in the case of constant satellite transmitted power, the strength of navigation signals is inversely proportional to the square of the orbital height. In the case of the same signal power on the ground, the effective radiated power of a 1000 km LEO satellite is 400 times (26 dB) lower than that of a 20,00- km satellite.
- *Navigation signal frequency and code modulation mode*
  To adapt to a high accuracy and reliable service under complex conditions, the multi-band and large bandwidth of navigation signals is a basic condition. In order

to make full use of the frequency spectrum resources and extend the service mode, it has become a common method to modulate navigation signals by adopting a variety of modulation and coding methods within the specified bandwidth. For instance, QPSK+BOC modulation is adopted at the same time, and the service of different accuracies and signal quality will be offered to three kinds of authorized users.

- *System confrontation and competitive strategy*
  With complicated international relations and market competition, an independent navigation system is established to ensure stable operation at any time and under any competitive conditions, and the system needs high anti-spoofing and anti-jamming performance. This is necessary to ensure the global market and maintain the interoperability and compatibility with the influential international popular satellite navigation system in order to be truly invincible. The confrontation and competitiveness of the system is fully embodied in the choice of navigation frequency, the distribution of service modes, and all aspects of signal coding and modulation.

9.2 Service Mode and Content

There are two kinds of divisions and methods for service modes. One is divided by user objects; the other is divided based on the positioning principle and performance. GPS positioning is divided into authorized users and non-authorized users according to the user objects: authorized users use precise positioning service PPS, while non-authorized users use standard location service SPS, and the selective availability (SA) can be carried out according to national security needs. Galileo, subject to user objects, also provides services basically in an authorized and non-authorized way. See Table 9.1 for details of the class of service.

The service class is defined as:

OAS:
: Open access areas/global services used by the public, classified as unimportant class, with no impact on security or commerce. It provides basic information and non-commercial services.

SAS:
: Secure, controlled services. Local/regional/global business, with impact on the safety of life; high risk may be caused due to data errors, classified as important class.

CAS:
: Non-secure, controlled access to local/regional/global services. Classified as an important class, with impact on business performance and commercial operations; it is potentially dangerous to safety, but not directly. Safety procedures can mitigate its hazards. Commercial use may need to ensure availability and may raise liability and related loss issues.

GAS:
: Limited, secure business under the control of government/military departments. This can be divided into the following two types according to the positioning principle and performance:

  - RNSS service mode
    The users can complete positioning and speed measurement by themselves in accordance with the traditional principle of satellite radio navigation services.

Table 9.1 Galileo Business Characteristics [7].

Business	OAS	SAS, Security Applications	CAS, Non-Security Applications		GAS (not mentioned in the business case)
User group	Public entertainment	Air and maritime transport	Ground transport: highways, railways	Urban taxi businesses	Military security, government, emergency
Importance	Not important	Security being the key	Important commercial use	Important commercial use	Important
Market	Very large	Small	Large	Small	Unknown
Coverage area	EU + global	Local + EU + global	EU + global	Local + EU	EU
Potential income	No direct income	Low, recyclable costs	Medium, recyclable costs	Low, recyclable cost	Low, government subsidies
User interface	Simple, low cost, no standard	Suitable for advanced MOPS	Commercial grade	Commercial grade	Dedicated
Safety certificate	None	Yes	None	None	None/Yes
Access	Open	Air transport: No difference Maritime transport: Authorized control	Authorized control (commercial use)	Authorized control (commercial use)	Authorized control (military use)

There is no need to transmit a response signal. The user must observe four or more satellites to continuously obtain three-dimensional position, three-dimensional speed, and the corresponding time information. Positioning needs a number of satellites, which costs much in terms of system overhead.

- RDSS service mode

  The positioning principle is: through the response of the user, the control systems other than the user measure the distance between the satellite and the user, calculate the location, and then give the location results to the user. It has position reporting capability with good confidentiality, which is characterized by a small number of satellites required, and the user simply needs to observe two satellites. The system cost is low, but instantaneous speed measurement cannot be completed.

The integration of these two services into one system is a comprehensive choice, which does not only complete the continuous real-time positioning, speed measurement, but also takes into account the user location report and short message communication, to form abilities such as user identification and monitoring, formation management, scheduling command, and high-risk environmental work monitoring.

9.3 Satellite Orbit and Constellation Selection

RNSS orbit selection has gradually matured. The basic principle of orbit selection follows the general principle of system design, namely achieving stable accuracy, availability, continuity, and integrity. It is also required in order to adequately consider the cost of system and technical feasibility. Design content includes satellite altitude, orbital plane and satellite distribution, constellation composition, and methods of measurement and control.

9.3.1 Orbital Altitude

The height of satellite motion is divided into ground altitude and orbital altitude. The satellite ground altitude is the distance from Earth's surface when the satellite is running in orbit. It varies as the geocentric latitude changes. We can see from Figure 9.1:

The ground altitude of satellite [8]:

$$H_s = R_s - R'_S \tag{9.1}$$

Wherein,

R_s is radius vector of satellite, $R_s = OS^i_n$
R'_S is the geocentric distance of sub-satellite point; and in Figure 9.1, $R'_S = OS_1$

$$R'_S = 6356755.625 + 21277.285\cos^2\varphi(\text{m}) \tag{9.2}$$

φ is geocentric latitude of satellite, approximately equal to geographic latitude.

Orbital altitude of satellite is the average value of the ground latitude of perigee and the ground latitude of apogee. It is shown by the following formula.

$$H_{sm} = \text{a} - \frac{1}{2}(R'_{SP} + R'_{SA}) \tag{9.3}$$

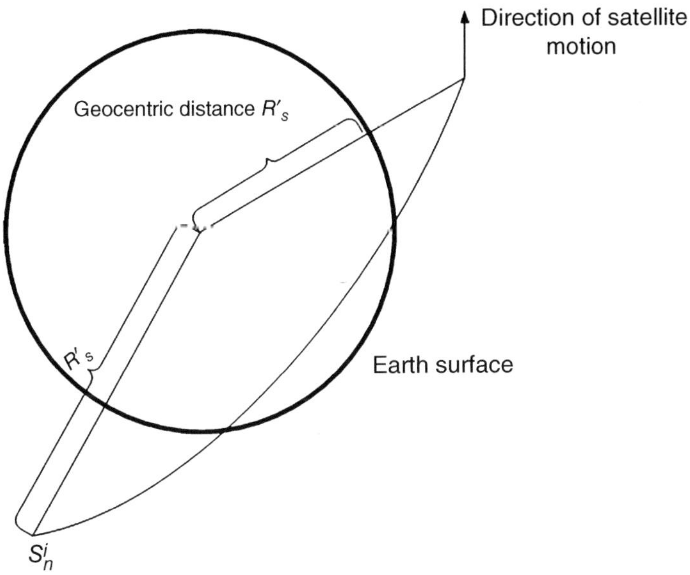

Figure 9.1 Schematic diagram of satellite altitude.

a is the semi-major axis of satellite orbital ellipse; R'_{SP} is the geocentric distance of sub-satellite point at perigee; R'_{SA} is the geocentric distance of sub-satellite point at apogee; H_{sm} is the height of satellite orbit.

The greater the height of satellite orbit is, the larger the angle of visibility α is. When the beam coverage angle β of satellite signals is the Earth coverage beam, we can see from Figure 9.2:

$$A = 180° - \beta\text{-}2\ E_S \tag{9.4}$$

$\beta = 2\arcsin\frac{R_e}{\mathrm{Re}+Hsm}$; R_e is Earth's radius, $R_e = 6371.1$ (km).

Figure 9.2 Angle of visibility of satellite.

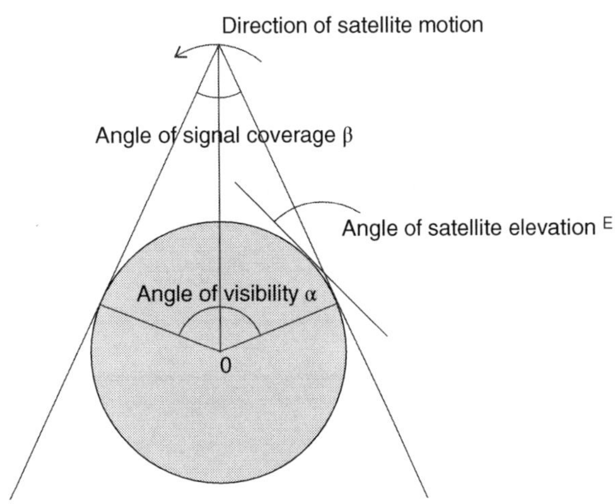

For the GPS satellite with a height of 20,200 km, $H_{sm} = 20{,}200$ km, the angle of visibility of the ground user to satellite is $\alpha = 152.26°$.

When $H_{sm} = 36{,}000$ km, the angle of visibility is $\alpha = 162.7°$.

Therefore, high orbit is beneficial for a user to increase the angle of visibility. For the same number of satellite constellations, the larger α is and the greater the number of observed satellites. This is not only beneficial to improve geometric figure and location accuracy, but also improves the autonomous integrity of the user receiver (RAIM). But when selecting satellite altitude, it is also important to consider the satellite regression period and the track of sub-satellite point as well as the difficulty of effective load; and when the orbit is higher, the power flux density, transmitting power, and satellite total power increase in square of height in order to reach the ground.

9.3.2 Track of Sub-Satellite Point and its Effect on Measurement and Control Plan

At any moment in time t, the crosspoint of the connecting line from satellite S to Earth's center of mass O and Earth's surface, is called the sub-satellite point. The connecting lines of all sub-satellite points are called track of sub-satellite point as shown in Figure 9.3, S1, S2, and S3 are sub-satellite points, and their connecting lines are tracks of sub-satellite points. Ideally, the track of a sub-satellite point is a large circle falling on the Earth's surface. However, because of Earth's rotation and precessional motion of orbital plane, the ground track that the satellite goes through each time is not repeated. When the earth rotates from west to east, on the ground, the ground track of satellite shows the westward movement by circle. This characteristic means that the measurement and control station located on Earth has the chance to see the satellites at any orbital plane, providing convenience for ground measurement and control. The track of sub-satellite point of GPS satellite is shown in Figure 9.4:

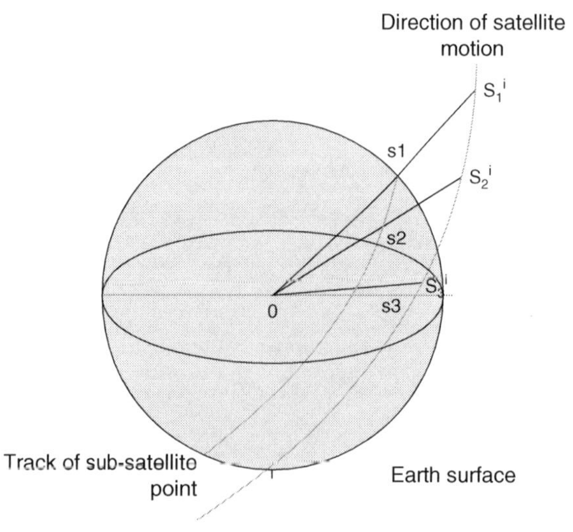

Figure 9.3 Track of sub-satellite point.

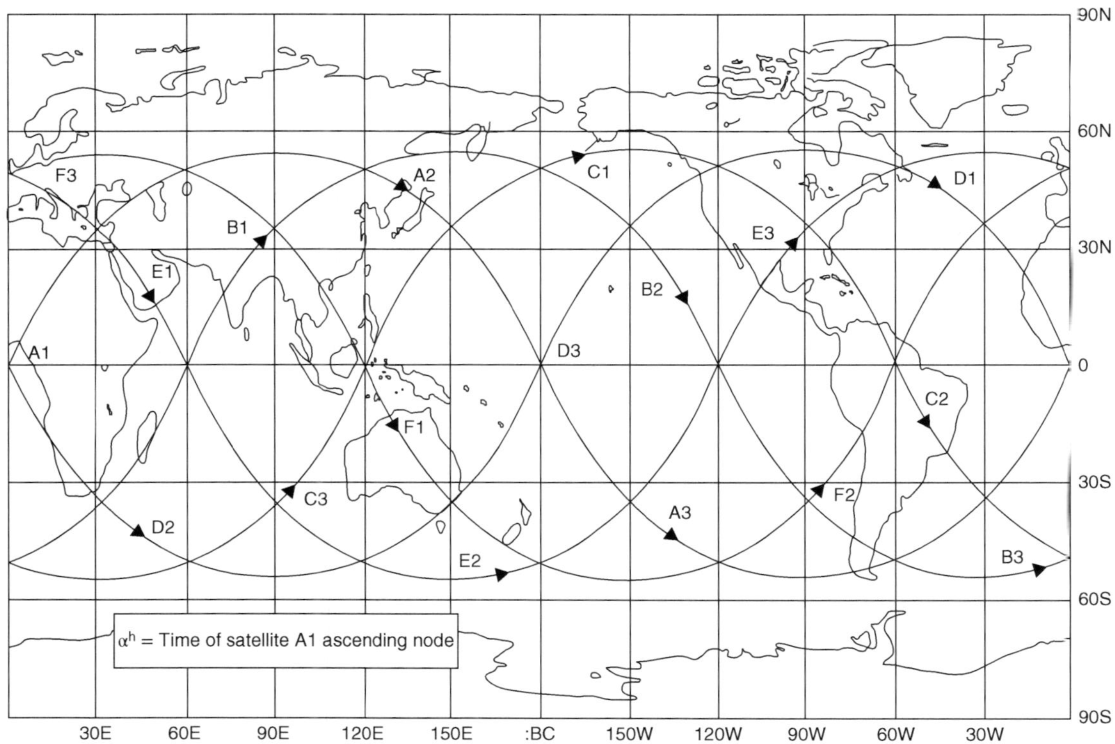

Figure 9.4 Example of a track of GPS sub-satellite point.

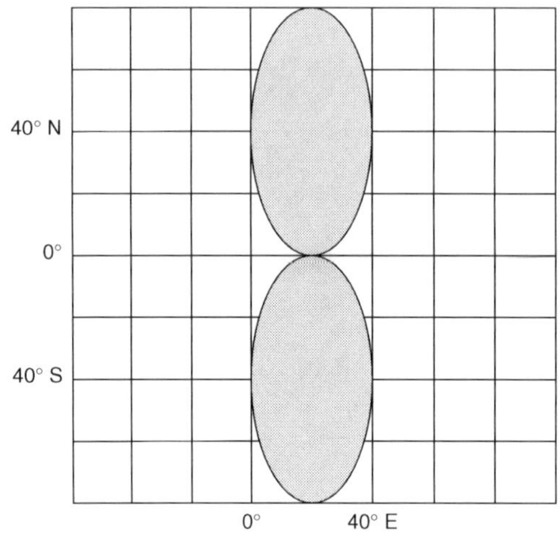

Figure 9.5 Track of the sub-satellite point of IGSO.

In ground track of two satellites, one traverses the equator at the longitude of 0° (the longitude of northbound node is 0°), and the other satellite is at 90° interval in the same orbital plane.

It is different for satellites in Earth's Inclined Geosynchronous Satellite Orbit (IGSO). The track of sub-satellite point is "8" through a certain cross-point of equator as a fixed track, as shown in Figure 9.5.

So, at the ground measurement and control station, only the fixed IGSO satellites can be seen. To observe all globally-covered IGSO satellites, implementing the global station distribution scheme is necessary. Otherwise it is needed to complete satellite measurement and control, satellite time synchronization, and track measurement through inter-satellite links, so IGSO satellites at a height of 36,000 km, like the Geosynchronous Earth Orbit (GEO) satellite, are only used for regional coverage, it is difficult to complete global coverage.

9.3.3 Orbital Plane and the Number of Constellation Satellites

The orbital plane is another parameter to describe satellite orbits, which are described with the right ascension of ascending node Ω and orbit inclination i. The prior Transit navigation system adopts the polar orbit of $i = 90^0$. GPS and other modern satellite navigation systems adopt the inclined orbit of $0 < i < 90°$. In order to achieve the coverage for middle and low latitude regions, i is relatively low, generally $45 \sim 55°$. To achieve the coverage for the middle and high latitude regions, $i = 55 \sim 65°$ is selected.

For the selection of global positioning constellations, the theory and practice show that the Low Earth Orbit (LEO) satellite constellation with an altitude below 2000 km is improper. The analysis indicates that if the number of the Walker constellation satellites with an altitude of 2000 km is four times that of an altitude of 20,000 km, the cost and maintenance expense of the system will soar [12]. Although reducing the total power of satellite is inadvisable, for an MEO Walker constellation with an altitude of 20,000 km, whether the number of satellite is 24, 27, or 30, adopting three trajectory planes has the highest availability, and under the condition of PDOP = 6, availability can reach up to 99.9%.

So, how to design the number in constellations? The answer is still from the demand analysis. In terms of the current reality, in the near future, there will be possibly four global navigation satellite systems in the world, which are, respectively, GPS in the US, GLONASS in Russia, Galileo in Europe, and The China Compass Navigation Satellite System (CCNSS). Global performance can be completely achieved by Receiver Autonomous Integrity Monitoring (RAIM). The so-called receiver autonomous integrity refers to the integrity of receipt signal detected by the receiver for the observed quantity and positioning results from four residual satellites, thereby removing the observed results of the faulted satellite signals and getting the needed positioning accuracy as well as integrity. For this purpose, at least five satellites need to be observed so as to make the integrity judgment. If, at the same time, six satellites are observed, the signals of one faulted satellites can be removed, and if seven satellites are observed, the signals of two faulted satellites can be removed, and so on. If the satellite number of each constellation in the four global systems is not less than 24, the total satellite number that the user can observe equals to 24 (6/system × 4 systems). The requirements of waypoint navigation and airport terminal navigation can be completely met. Even if one or two systems do not provide the service, 12~18 satellites also can be observed, with reasonably high accuracy and integrity.

Thus, the design should focus on regional area service. Starting from the principle of the optimal price performance ratio, it is appropriate to fix the goal of regional area service on the precision approach of Class I (CAT-I). According to the standard of the International Civil Aviation Organization, it can be concluded as:

Accuracy (m): Horizontal: 18.2
(95%)
Vertical: 7.7~4.4
The maximum non-continuity: $8 \times 10^{-6}/15$ s
The maximum non-integrity: $2 \times 10^{-7}/$ approach
Warning time (s): 6
Availability: 0.9975

Because CAT-I service concerns life security, the service based on authorization or commercial system is appropriate. Like the global service, although the user has 24 usable satellites, there is still a national or regional system that gets legal and economic responsibility guarantee to offer support. The International Civil Aviation Organization also has a stricter responsibility guarantee constraint to evaluate whether a nation (region) joins in the system of GNSS. The regional system represents national sovereignty and service ability, and is strictly designed.

The China Compass Navigation System adopts a Geostationary Orbit (GEO) satellite (or IGSO satellite) to enhance navigation performance so as to achieve the capability to meet CAT-I. User integrity is completed with two kinds of redundancy means: Receiver Autonomous Integrity Monitoring (RAIM) and System Integrity Monitoring (SIM). It can also be referred to as Ground Integrity Channel (GIC). So-called ground integrity monitoring aims to establish the Integrity Monitoring Station (IMS) in the service region using a ground receiver with special performance, and collects the navigation signals of all satellites within the sight line. The Central Station (CS) makes the integrity judgment for the satellite signals, and then the signals are broadcast to the users through the GEO satellite. Now the number of satellites needed for two kinds of integrity is to be analyzed.

1. Total number of SIM satellites
   The total number of the constellation satellites that adopt SIM for integrity monitoring mainly meets CAT-I accuracy and availability. The integrity guarantee is mainly for the GEO satellite. The design principle of providing the number of integrity GEO satellites is to make any airport within the region ensure broadcast integrity of two channels, namely at least to ensure every airport in the whole region receives integrity signals from two GEO satellites, and the observation elevation angle of each satellite is larger than 25°. Therefore, China's regional system shall have 4~5 GEO satellites to be used for SIM integrity information broadcast.
   For the considerations of accuracy and availability, it is important to make the continuous 24-h PDOP design for the whole region on the basis of the number of MEO satellite plus the number of GEO satellites. Galileo design analysis shows that, for one global system composed of 30 MEO satellites, the minimum number of satellites observed by a user within the field of view is 8 (superior to the probability of 99.99%) [10], which is consistent with the results of modern GPS and China's analysis. Plus, at least two GEO satellites can be observed. The total number of satellites that a user can observe will be up to 10. If China Compass chooses five GEO satellites plus 27 MEO satellites plus three IGSO satellites, China's region can undertake a CAT-I mission in an entirely independent manner.
2. Number of RAIN satellites
   The number of SIM satellites mentioned previously shows that, because of the addition of the number of GEO satellites, the number of observable satellites judged by RAIM receiver autonomous integrity also increases. When the China Compass Navigation System adopts the scheme of 4~5 GEO satellites plus 30 MEO satellites, the number of satellites observed by a user within the field of view is at least 10 (2 + 8). And when China Compass Navigation System adopts the scheme of 4~5 GEO satellites plus 27 MEO plus three IGSO satellites, the number of satellites observed by a user within the field of view is nearly equal to that of before, and the availability is partly improved.

The redundant backup of these two kinds of means will create necessary and sufficient condition for China Compass Navigation System to enter into world GNSS. When designing the constellation of the Galileo system, the conclusions are as follows [10]:

1. To achieve the performance indexes of middle and high level, at least 24 satellites are needed. The effect of satellite altitude on performance index reduces with the increase of the number of satellites. When the number of global constellation satellites is ≥27, the contribution of satellite altitude to accuracy can be ignored. See Figure 9.6, Figure 9.7, and Figure 9.8.
   Assumptions of UERE are presented in Tables 9.2 and 9.3 as follows:
2. The constellation scheme of 30 MEO satellites is excellent, and the constellation design of Walker 30/3/1 is selected as the best scheme. When the semi-major axis is a ≥ 25,000 km, both vertical and plane accuracy can be superior to 5.5 m (availability is superior to 99.7%). See Figure 9.9.
3. For further improvement of availability, it is important to increase back-up satellites in orbit. It is not necessary to conduct constellation modification.

The comparison between the following two Walker constellations of Galileo and the GPS of current 24 satellites is as shown in Table 9.4. These two constellations are:

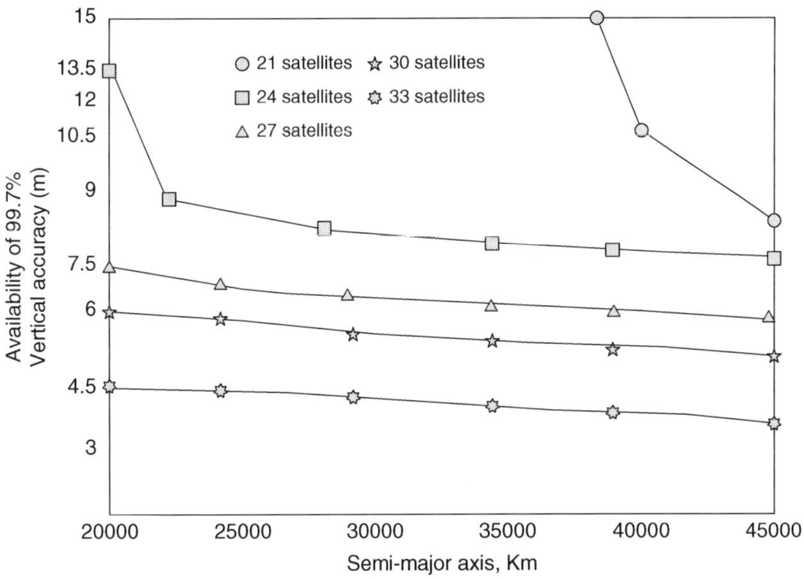

Figure 9.6 Relationship of vertical location accuracy (99.7% availability) and the number and altitude of MEO satellites.

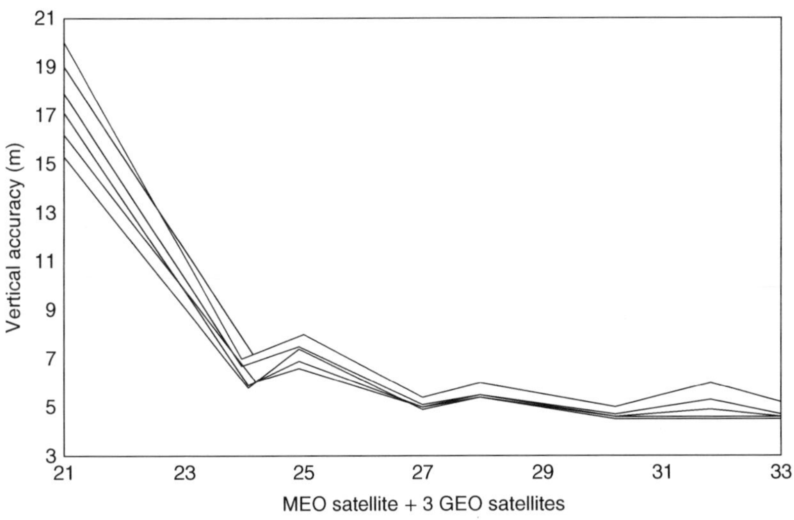

Figure 9.7 Relationship of vertical accuracy and the number and altitude of satellites (with 3 GEO satellites, each curve corresponding to a different altitude: from 20,500 to 30,500 km, with a step length of 2000 km).

- 24/6/1; angle of tilt (58°), and the altitude is same as that of GPS. (Orbital parameters are provided by ESA);
- 24/3/2; angle of tilt (57°), and the altitude is same as that of GPS.Simulation region is (180°E ~ 180°W) × (75°S ~ 75°N)

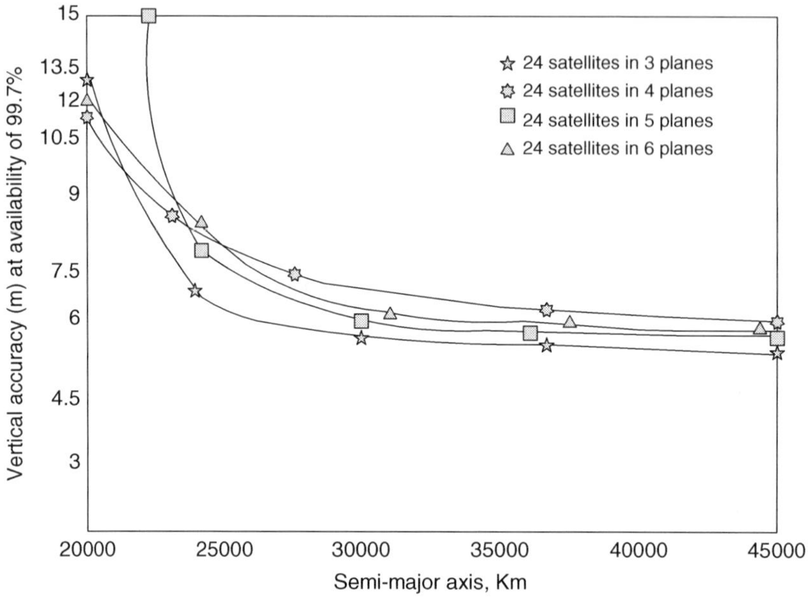

Figure 9.8 Relationship of vertical accuracy and the number and altitude of planes (MEO 24 satellites).

Table 9.2 UERE of the first action.

Angle of elevation (°)	5	10	20	30	45	90
UERE (m)	4.27	2.2	1.24	0.96	0.79	0.69

Table 9.3 UERE of the second action.

Angle of elevation	5	10	15	20	30	40	50	65	90
OAS UERE (m)	3.25	2.05	1.65	1.4	1.3	1.2	1.0	0.9	0.9
CAS1 UERE (m)	2.4	1.45	1.1	0.9	0.8	0.75	0.7	0.7	0.7
CAS2 UERE (m)	2.3	1.3	1.05	0.9	0.8	0.75	0.7	0.7	0.7

The UERE is calculated according to the first action.

The comparative result shows that the current GPS index differs a lot with CAT-I requirement. The constellation design is not the best, so it is necessary to optimize 24-h worst vertical accuracy.

9.3.4 Selection of Types of Satellite Orbits

This chapter has so far comprehensively discussed the impact of orbital altitude on the number of satellites, constellation availability, accuracy, and here the selection for the types of satellite orbits is specifically proposed, which is considered based on the

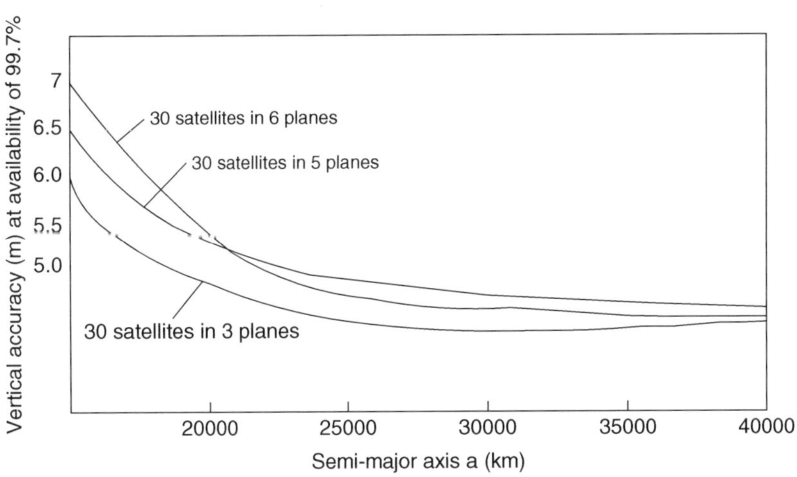

Figure 9.9 Relationship of vertical accuracy and the number of planes and the satellite altitude (30 satellites).

Table 9.4 Walker/GPS performance comparison.

The worst vertical accuracy within 24 h (m)	GPS	24/6/1	24/3/2
	Maximum: 19	Maximum: 11	Maximum: 6.6
	Average: 6.23	Average: 5.54	Average: 4.95
Average vertical accuracy within 24 h (m)	Average 3.63	Average 3.66	Average 3.66
Availability (4 m perpendicular)	Minimum 24%	Minimum 43%	Minimum 44%
	Average 72.9%	Average 69.6%	Average 73.8%
	Maximum 91%	Maximum 96.8%	Maximum 95.8%
Availability (6 m perpendicular)	Minimum 90.6%	Minimum 80%	Minimum 94.9%
	Average 98.1%	Average 96.9%	Average 98.6%
	Maximum 99.5%	Maximum 99.5%	Maximum 99.3%

integrity of satellite navigation. The statistical analysis for the number of MEO satellites shows that if the number of MEO satellites is enough (see Figures 9.6, 9.7, and 9.8), it can meet not only the CAT-I accuracy requirement, but also have enough usable satellites to meet the integrity requirement. The user can select healthy satellites from 6~12 observable satellites, but the workload of Receiver Autonomous Integrity Monitoring (RAIM) of users certainly will be increased. The user receiver performs an alarm report within 6 s without any help, but the estimates for the integrity (UE estimate) of satellite signal and the overall error UERE of satellite signal during the propagation process cannot have adequate means. It is often required that the ground control system provides the integrity prompts. So, the only help of the so-called Ground Integrity Channel (GIC) can ensure three kinds of integrity indexes (accuracy error, integrity probability and warning time) of the user. The use of the GEO satellite does not only strengthen accuracy, but also enhances the 99.7% availability of pure MEO constellation to the 99.75% availability needed by the CAT-I I precision approach, particularly contributing to integrity warning time and probability. There is no GEO satellite for GPS and GLONASS, so the completion of CAT-I navigation requirement is not adequate. Even if the number of satellites increases to 30, the integrity of CAT-I still needs to be strengthened. Galileo designed the GEO satellite several times in the design process, but the scheme of GEO satellite participation was still abandoned in 2005, so the completion of CAT-I navigation mission similarly needs a GEO satellite to strengthen integrity. CCNS, on one hand, takes integration of navigation and communication into account and on the other, also considers meeting integrity warning time and warning probabilities at different levels. The addition of a GEO satellite is beneficial in improving the navigation integrity in China. The purpose of full utilization of resources is achieved.

This is the reason why the best regional navigation mission constellation is the global constellation and GEO satellite improvement.

9.4 Signal Frequency and Modulation Coding Mode

9.4.1 Selection Principle of Navigation Signal Frequency

Navigation signal frequency has gone through selection and coordination from VHF, UHF, to L frequency (extended C frequency included) from the earliest Transit Satellite Navigation System to today's mature Global Navigation Satellite System. The common principles followed are as follows:

1. A high-frequency available for selection so that there is larger frequency bandwidth for higher ranging accuracy.
2. Relatively low atmospheric propagation loss to keep the Effective Isotropically Radiated Power (EIRP) of the satellite low as possible so as to reduce total satellite power consumption and weight.
3. Strong vegetation penetrating ability.
4. Minimum frequency interference between systems.
5. Mature microwave components available for selection.

Considering the maturity of components at that time, two-frequency signals of 150 MHz and 400 MHz have been chosen to conduct ionosphere propagation delay verification since the Transit. Up until now, the current ITU continues to maintain the effectiveness of these two traditional satellite navigation frequencies. In order to ensure that more systems can have reliable navigation frequency, the World Administrative Radio Conference in 2000 (WRC-2000) conducted a series of efforts for a new navigation spectrum.

9.4.2 Navigation Frequency Recommended by the ITU

The concept of radiolocation can be traced back to 1912 when the International Telecommunication Union (later known as International Union of Radio Telegraphy) at that time defined radiolocation as the business to determine the position, velocity and other characteristics of an object with propagation characteristics of radio wave as a means. The later-developed radar positioning, Doppler positioning, and pseudorange positioning were all included in this category. This definition was a useful reference for other radio services, but was not scheduled for clear radio frequencies at that time.

Due to the rapid development of satellite technology in the 1960s, navigation technology rose such as the Transit and the цикада Doppler. Satellite communication has become commercial reality because of the efforts of AT&T and the new-built COMSAT Company. Based on this development, an Administrative Radio Conference was held by International Telecommunication Union in 1963 to determine which frequency should give way to satellites. As the aerospace industry was still very young then, it failed to form a satisfactory concept. Until 1971, the World Administrative Radio Conference (WARC-ST71), space communications, drew up the general concept and management system for satellite communications. And several important concepts were formed: First, member states agree that every country has the right to "equally occupy" Geostationary Satellite Orbit (GSO) location and satellite frequency band; second, the satellite communication system launched subsequently must avoid causing interference with systems launched earlier; and third, try to accommodate all satellite business that comes to mind by a series of definitions. The concept of using satellite resources to determine the position, velocity, and attribute characteristics of an object was determined and established as the Radio Determination Satellite Service (RDSS). The frequency meeting (WARC-79) held in 1979 determined the frequency for Radio Navigation Satellite System (RNSS) of Aeronautical Mobile-Satellite Service (AMSS). In the countries, AMSS accepts its frequency arrangement in the legal form of Civil Aviation Authority, yet the RNSS accepts its frequency arrangement in the legal form of United States Federal Government, which was used for the planned Global Position System (Navstar GPS). As no country has RDSS service requirements, no frequency was assigned to the RDSS until 1985 when the United States Federal Communications Commission formally recognized the RDSS technical standards based on spread spectrum modulation. A frequency of 2483.5~2500 MHz is for Satellite Radio Determination (air-to-ground) and a frequency of 1610~1626.5 MHz is for Aeronautical Radio Navigation and is also assigned to ground-to-air service of satellite radio determination. During the Compass Navigation test conducted by China in the 1990s, China tried hard to build RDSS frequency rules in the World's Third District, due to the impact of MSS Service, for 2483.5~2500 MHz and 1610~1626.5 MHz frequency bands, RDSS can only acts as the secondary service, and MSS acts as its main service.

WRC-2000 proposed that Satellite Navigation Frequency must be added. China supported the following suggestions:

1. Agree to add 1164~1215 MHz and 1260~1300 MHz as air-to-ground frequency of satellite radio navigation.
2. Agree that 1300~1350 MHz is the upload injection frequency of Satellite Radio Navigation.
3. Agree to add 5010~5030 MHz as air-to-ground frequency of radio navigation.
4. Conduct negotiations on power density with frequency declaration network operators by working groups.
5. 5. Agree that the shared research on 1610~1626.5/2483.5~2500 MHz used for IMT-2000 mobile satellite service by ITU-R group has not completed yet.

Currently, the built-up GPS system and GLONASS systems have basically taken up two frequency bands of 1559~1610 MHz and 1215~1260 MHz, and parts of newly

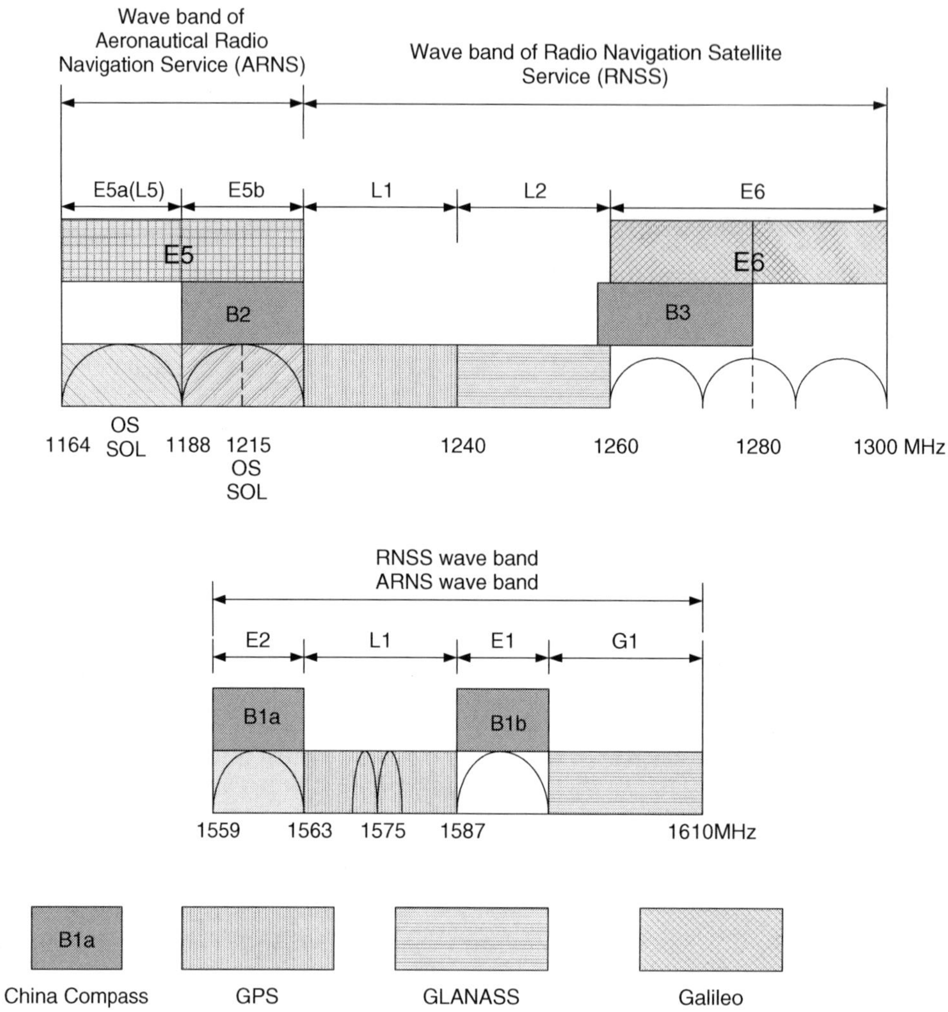

Figure 9.10 Occupation condition of navigation frequency.

added frequency bands have been declared after the WRC-2000 Conference. Galileo and China Compass have declared 1559∼1610 MHz, 1215∼1260 MHz, and have added new navigation frequency bands. The actual basic spectral allocation of systems is shown in Figure 9.10.

Working frequency of the Compass navigation system at a second-step development strategy is:

B_{1A} : 1561.098 ± 2.046 MHz
B_{1B} : 1589.742 ± 2.046 MHz
B_2 : 1207 ± 12 MHz
B_3 : 1268 ± 12 MHz

Working frequency of the Galileo system is:

Open Service (OS): Providing location and time information for all the global users for free in the E5a, E5b, and L1 signals.
Commercial Service (CS): Providing additional commercial encrypted data in the E6 signal based on OS standard service.
Safe-of-Life Service (SOLS): Attaching SISA information based on standard service in the E5a, E5b, and L1 signals.
Public Utility Service (PRS): Adopting encrypted code and encrypted data in the E6 and L1 signals.

The signals are planned as follows in Tables 9.5 to 9.7.

It can be seen from the frequency design of China Compass and Galileo that they have quite good compatibility. The design conditions of Compass/Galileo's compatibility are quite sufficient, but attention shall be paid to power distribution and modulation code coordination.

Table 9.5 E5 signal.

Service	OS-SOL			
	Data channel	Pilot channel	Data channel	Pilot channel
Carrier frequency	1176.45 MHz		1207.14 MHz	
Data/symbol rate	25 bps/50 bps	N/A	125 bps/250 bps	N/A
UMRP (10°)	–155 dBW		–155 dBW	
Modulation mode	Alternating BOC modulation			

Table 9.6 E6 signal.

Service	PRS	CS	
		Data channel	Pilot channel
Modulation mode	BOCcos (10, 5)	BPSK (5)	BPSK (5)
Power ratio	50%	25%	25%
Code rate	5.115 Mbps	5.115 Mbps	5.115 Mbps
Data/symbol rate	To be determined	500/1000 sps	N/A
Supported service	PRS	CS	CS
UMRP (10°)	–155 dBW	–158 dBW	–158 dBW

Table 9.7 E2–L1-E1 signal.

Service	PRS	OS-SOL	
		Data channel	Pilot channel
Modulation mode	BOCcos (15, 2.5)	BOC (1, 1)	BOC (1, 1)
Code rate	2.5 × 1.023 MCPS	1.023 MCPS	1.023 MCPS
Data/symbol rate	To be determined	125/250 SPS	N/A

9.4.3 Signal Frequency and Bandwidth Selection

As the air-to-ground frequency resources available for satellite navigation are short, the subsequent system must not cause interference toward GPS and GLONASS established earlier, and has to share frequency with Galileo at the same time, so the frequency selection for China Compass Navigation System (CCNS) is never easy. Design conditions and working content for signal frequency and bandwidth selection are as follows:

1. Never cause frequency band overlap with GPS and GLONASS. Pay attention to the band-to-band interaction between CCNS and GPS, GLONASS. Conduct related interference analysis calculation and coordination.
2. Try to reduce overlap in the frequency arrangement with Galileo. Conduct interference calculation toward co-frequency multiple access users within the user field. Coordinate the total number of satellites within the constellation and the power spectral density of the corresponding satellite toward the ground.
3. Based on available bandwidth, design CCNS spreading code rate (chip), and estimate the anti-jamming capability of CNSS navigation signal and achievable ranging accuracy and information rate.

The following is the calculation of mutual interference taking CNSS $B_1 = 1589.742 \pm 2.046$ MHz as an example:

GPS C/A code and P code, CCNS B_{1B} all adopt QPSK modulation, and their power spectrum is distributed in accordance with $\left(\frac{\sin x}{x}\right)^2$, $X = 2T_b\text{F}$, and T_b is the width of chip, its density of main lobe is within ± 1 frequency doubling, accounting for 92% of total power. The first minor lobe is within $\pm 1 \sim \pm 2$ frequency doubling, 12 dB lower than the main lobe (accounting for 6% of total power). It can be seen that:

1. The power caused when GPS C/A code falls into CCNS B_{1B} is negligible.
2. The power caused when GPS P code falls into CCNS B_{1B} is less than 6% of total power of signal (at least 12 dB lower than total power).
   Suppose the power of P code signal of each satellite on the CCNS receiving end is −163 dBW, then the maximum interference caused by 12 GPS satellites can be calculated as:

$$\begin{aligned} P_{ma} &= 10 \times \log_{10}(\text{N}) + P_s - 12 \text{ dB} \\ &= 10 \log_{10}(12) + (-163 \text{ dBW}) - 12 \text{ dB} \\ &= 10.8 - (175 \text{ dBW}) \\ &= -164.2 \text{ dBW} \end{aligned}$$

While the noise power on the CCNS receiving end is

$$P_o = \text{B } n_0 = \text{BKT}$$

B is CCNS receiver bandwidth 4.092 MHz = 66 dBHz
K is Boltzmann's constant −228.61 dBW/Hz
T is receiver noise temperature 300 K = 24.77 dBk

$$p_0 = -137.8 \text{ dBW}$$

P_{ma}'s joining increases the total noise power by 0.03 dB, which is negligible.

3. The power caused when CCNS B_{1B} signal falls into GPS P code bandwidth is less than 1% of B_{1B} signal power, at least 17 dB lower than the main lobe. It will not cause harmful interference to GPS.

Let's analyze the interference of GPS M code toward CCNS B_{1B}. If GPS adopts BOC modulation at 1575.42 MHz, two separate main lobes of BOC modulation are right between the main lobe and the first side lobe of P code (see the subsequent analysis). This causes more serious interference toward CCNS B_{1B} than P code does, and must be analyzed in accordance with the actual design of GPS M code.

The interference between GLONASS and CCNS B_{1B} signals can be analyzed with the same method. The difference is that GLONASS adopts frequency division modulation satellite identification method. In the frequency band of $L_1 = 1602 \sim 1165.5$ MHz, the working frequency of multiple satellites has the following relationship:

$$f_n = \mathrm{f} + \Delta f_1$$
$$\mathrm{F} = 1602\,\mathrm{MHz}$$
$$\Delta f_1 = 0.5625\,\mathrm{MHz}$$

Working frequency of the first satellite is 1602 + 0.5625 MHz=1602.5625 MHz

Spread-spectrum code rate of P code is 5.115 Mbps, and that of C/A code is 0.5115 Mbps, so $f_{1p} = 1602.5625 \pm 5.115$ MHz

Its main lobe difference with CCNS B_{1B} is 5.6595 MHz

The subsequent satellite frequency gradually keeps away from CCNS B_{1B} signal as shown in Figure 9.11. As the interference gets weaker, the interference calculation of multiple users must be conducted one satellite to another. Analysis shows that: The interference among B_{1A}, B_{1B}, B_2, B_3 signal frequencies selected by CCNS and each signal of current GPS, GLONASS and the future Galileo system must not cause significant damage. However, if the implementation power of certain frequency signal increases under conditions of conflict, it must be further analyzed in detail in accordance with the size of enhancement. So confrontation in war navigation is long and complex. The basic point of confrontation is that its own navigation signal has an independent frequency width and large bandwidth. The selection of multiple navigation frequencies is also one of the most effective methods.

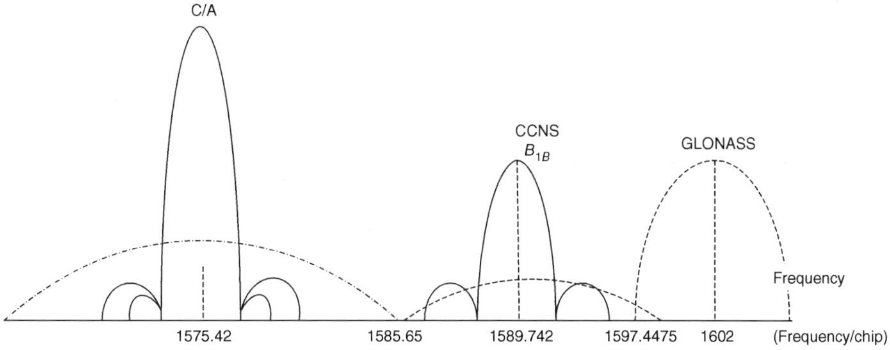

Figure 9.11 Mutual spectral interference between CCNS and GPS.

9.4.4 Satellite Multiple Access Identification and Ranging Code Design

The current two global satellite navigation systems adopt two different satellite multiple access identification modes separately. GPS adopts CDMA (Code Division Multiple Access) identification, while GLONASS adopts FDMA (Frequency Division Multiple Access) identification. The Russian system underlines its superiority of strong anti-interference capacity, for narrow-band interference, it is impossible to affect all observable satellites within the view field. Currently, it is quite easy to produce broadband man-made malicious interference, so its superiority is not quite outstanding. On the contrary, the complex frequency system of the receiver adds its complexity. And as the high frequency of GLONASS L_1 exceeds 1610 MHz, it causes harmful interference toward radio astronomy. Consultative Committee of International Radio (CCIR) organization suggests that Russian GLONASS L_1 frequency band shall be compressed by half, and can only be designed with the two satellites opposite to the Earth adopting the same frequency. In this way, there will be a series of problems to be solved. Galileo and CCNS both adopt Code Division Multiple Access Mode (CDMAM). This navigation coded signal designed in accordance with CDMA can complete the satellite identification and is also the basic of precise pseudorange measurements; for general pseudo-random ranging code (PR), the missions and features of its signal coding design are:

1. Ensuring good satellite identification, strong autocorrelation of PR code and weak cross-correlation of each PR code within code family.
2. Ensuring the pseudorange measurement accuracy and bit error rate of information transmission.
   The chip rate of PR code determining the accuracy of the ranging measurement has been determined by the bandwidth discussed previously. The bit error rate of information transmission is determined by information rate and information coding scheme.
3. With the ability of anti-interference and anti-multipath.
   This is related to the repetition period (code length) of PR code, as the repetition period gets longer, the spread frequency and gain gets higher, the anti-interference capacity will get stronger.
4. With very short capture time.
   To capture signals rapidly, its code period shall be short, which is contradictory to point (3). There are two ways to solve this problem; one is to adopt QPSK modulation, make I branch the short code, and Q branch the long code, and guide the capture of Q branch with I branch. The other reliable way is to research the direct capture scheme of Q branch long code.

In accordance with these requirements, QPSK modulation is adopted generally in the design of multiple satellite signals, to guide the capture of Q branch long code with the rapid capture of an I branch with a short period and to complete the acquisition of a satellite to obtain the exact time-*priori* information of the local code, making the time of local code moment and receiver code moment approach rapidly and in synchronization to 1 ms to several ms, so as to complete the direct capture of other satellite Q-branch long codes.

I branch is generally called the coarse or civilian code, and is often called C code. Q branch is arranged with higher code rate and longer code period; it is often called precise

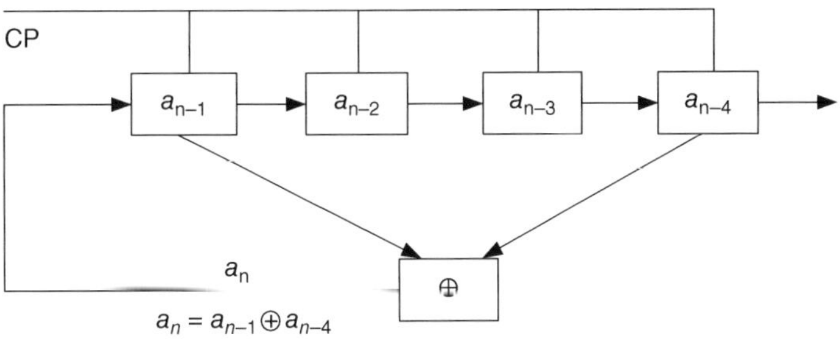

Figure 9.12 The m-sequence register with linear feedback.

ranging code, or P code. C code is generally the pseudo-random code of m sequence, or the pseudo-random code of Gold sequence.

The m-sequence code has good pseudo-random characteristics and is easy to generate. It is a very desirable pseudo-random code, and is short for the longest linear feedback shift register sequence. It is generated by a shift register with linear feedback. The diagram of the m-sequence generator is shown in Figure 9.12.

Characteristic polynomial of the m-sequence is

$$f(x) = c_0 + c_1 x + c_2 x^2 + \cdots + c_n x^n \tag{9.5}$$

The necessary and sufficient condition of the sequence generated by feedback shift register is that its characteristic polynomial f(x) is a primitive polynomial, the period length of m sequence generated by an n-level linear feedback shift register is

$$P = 2^n - 1$$

The character of the m-sequence can be summarized as follows:

1. In an m-sequence period , the number of "1" is only one more than "0."
2. The number of runs with the length of K accounts for 2^k of the total number of runs, and runs mean the number of consecutive "1" and consecutive "0" in the sequence.
3. Add an m sequence m_p and an m sequence m_r module 2 produced by the delay of any number of times of m_p together, and the result is also the sequence m_s produced by certain delay of m_p,
   That is $m_p + m_r = m_s$
4. The autocorrelation function of the m sequence is a double valued function:

$$R(i) = \begin{cases} 1 & i = 1 \\ -1/p & i = 1,2,\cdots p-1 \end{cases} \tag{9.6}$$

   It has good autocorrelation characteristics.
5. The autocorrelation function of m sequence can be seen as an impulse function. As the autocorrelation function and the power spectral density function form a pair of Fourier transforms, so the power spectral density of m sequence tends to white noise with the increase of period length P and code rate.

Although the m-sequence has good pseudo-random characteristics and related characteristics, the number of sequence is quite small, so the *Gold sequence* is generated.

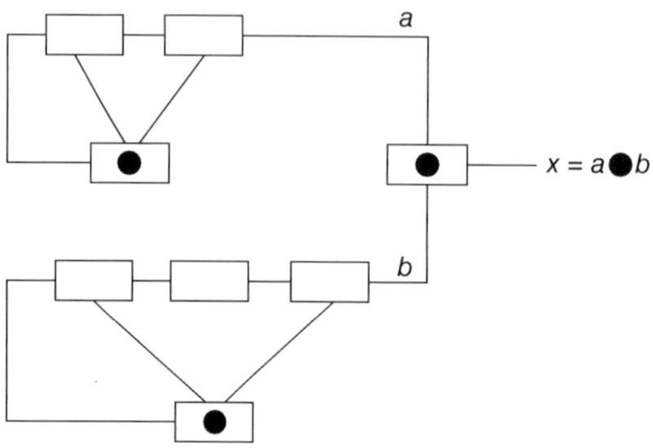

a: 011011011011011011011   0110110

b: 111010011101001110100   1110100

x: 100001000110010101111   1000010...

Figure 9.13 Gold code sequence.

It inherits the advantages of the m-sequence, and the sequence number is much larger than that of the m-sequence. As it is the composite code of the m-sequence, it is generated by bitwise module 2 with the addition of the preferred pair of m sequence with the same length, the same rate but different sorting. A pair of m-sequence with n-grade can generate 2^n+1 Gold code, this code generator is with simple structure, and easy to achieve, so it is widely used in engineering. GPS C/A code is an excellent Gold code, as shown in Figure 9.13.

GPS C/A code is the short code designed for coarse capture; its length is 1023 bit, code rate is 1.03 Mbps, and its last time is 1 ms; the C/A code of each satellite in GPS is the Gold code generated by module 2 and composition of two 1023-bit pseudo-noise code $G_1(t)$ and $G_2(t)$. So, the period of module 2 overweight is also 1023-bit, and can be expressed as:

$$XG(t) = G_1(t)\ G_2[t + N_i(10\ T)] \tag{9.7}$$

In this formula, N_i represents the number of code element of the phase offset between G_1 and G_2. Attention shall be paid to that the code element width of C/A code is 10 T second. There are 1023 different offset values N_i, so there are 1023 different codes of this kind. G_1 and G_2 codes are all generated by linear shift register with the longest level of 10. In order to synchronize with the X_1 epoch, G_1 and G_2 shift registers are all set to "1." The tap position is determined by polynomials of two generators, that is

$$\begin{aligned} &G_1 : G_1(X) = 1 + X^3 + X^{10} \\ &G_2 : G_2(X) = 1 + X^2 + X^3 + X^6 + X^8 + X^9 + X^{10} \end{aligned} \tag{9.8}$$

The period of Gold code is 1 ms, so each data bit has 20 epochs of C/A code. 50 bps data clock is the epoch synchronization of C/A code and X1.

Figure 9.13 shows that the GPS C/A generator is composed of two 10-level feedback shift registers. Shift register is driven by 1.023 Mbps clock. The feedback taps of G_1 are at level 3 and level 10, and the feedback taps of G_2 are at level 3, 6, 8, and 10. Various delay offsets are generated by the G_2 register taps at certain point, and the two sequences module 2 are added to obtain the G_2 sequence delay form (cyclic summation characteristic). G code should be the 20 Hz frequency divided by 1000 bps epochs to obtain a 50 bps data clock. And all clocks must be synchronized by X_1.

The design of satellite navigation ranging code must focus on the realization of precise ranging. The design requirements can be summarized as follows.

1. It will be an ultra-long period code, or at least longer than the required time delay of ranging. The best shall be a non-cycle code to prevent fraud or theft. It is a short cycle code only when it is acted as the guidance code.
2. The code rate (chip rate) shall be high enough to satisfy the ranging accuracy, but it is limited by the navigation frequency and frequency bandwidth.
3. The autocorrelation function is of two electrical levels, when the sequence is compared with itself, it at maximum value, and when it is compared with its shifted sequence, the secondary peak is uniform and significantly smaller.
4. The cross-correlation function values of different address codes shall be as small as possible so as to meet the processing requirements of multi-satellite signals in the receiver.
5. The direct capture time of ranging code shall be as short as possible.
6. The number of ranging codes in the sequence family shall be quite a few so as to satisfy the address coding of multiple satellites in constellations.
7. The structure of code is simple and easy to achieve and change.
8. It has a large linear complexity, so it is not easy to misappropriate, or be deceived.

According to these conditions and taking comprehensive consideration, the common sequences are m sequence, Gold sequence, Bent sequence, compound code, and so on.

GPS precise ranging code (p code) is a long-period code with the period of a week. Its variation block diagram is shown in Figure 9.14.

The codes of each satellite do not overlap in a period. The GPS p code of satellite *I* is obtained by multiplying two PN codes $X_1(t)$ and $X_2(t + n_i t)$, in which, the period of X is 1.5 s or 15,345,000 code points, and the period of $X_2(t)$ is 15,345,037 or 37 standard code points longer. The two code sequences are reset at the same time (initial phase) to start a week. The code rate of X_1 and X_2 is 10.23 Mbps. And the clock is in the same phase. So, the *P* code can be expressed in the form of the product code here:

$$XP_i(t) = X_1(t)X_2(t + n_i T) \qquad 0 \le n_i \le 36$$

The $X_1(t)$ and $X_2(t)$ in the formula is binary codes with the value of ± 1, and XP_i (t) is reset at the beginning of the week. The delay between X_1 (t) and X2 (t) of each satellite is n_i clock cycles, and each clock cycle is $1/10.23 \times 10^6$ (seconds). See Figure 9.14. Both the X_1 and X_2 codes are generated by multiplying the two different 12-level linear feedback shift register pairs. The two register pairs can be marked as X_1A & X_1B and X_2A & X_2B

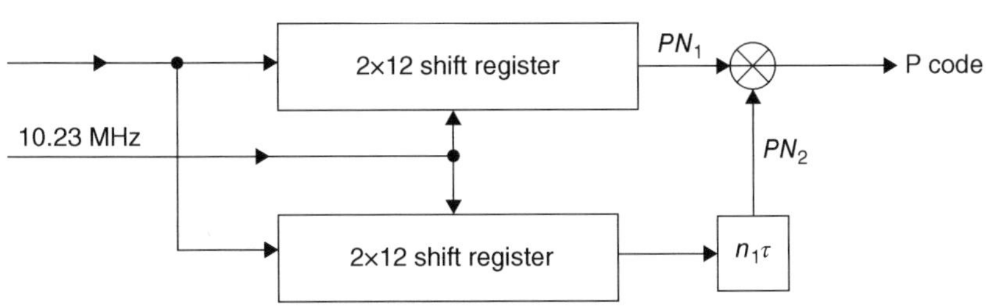

Figure 9.14 Variation block diagram of a GPS p-code generator.

separately, and its polynomial is defined in GPS-ICD-200 as:

$$X_1A: \ 1 + X^6 + X^8 + X^{11} + X^{12}$$
$$X_1B: \ 1 + X + X^2 + X^5 + X^8 + X^9 + X^{10} + X^{11} + X^{12}$$
$$X_2A: \ 1 + X + X^3 + X^4 + X^5 + X^7 + X^8 + X^9 + X^{10} + X^{11} + X^{12}$$
$$X_2B: \ 1 + X + X^3 + X^4 + X^8 + X^9 + X^{12}$$

These polynomials determine the feedback tap positions of four 12-level shift registers, X_1A, X_1B, X_2A, and X_2B, in Figure 9.14. Period of the code generated by the 12-level maximum length shift register is $2^{12-}1 = 4095$. If two code generators are truncated to cycle and cause a relative original period of ≤4095, the period of product codes is the result of multiplying two periods, which is approximately 1.6×10^7, the two product codes of the GPS have been truncated, making the relative original periods of X_1 and X_2 be 15,345,000 and 15,345,037, respectively. Equally, multiplying X_1 code and X_2 code can obtain a new code. The reason why the cycle change of X_2 code is 37 more than that of X_1 code is that the value of n_i is from 0 to 36, so there are 37 pseudo-random P codes. In order to further enhance the complexity, Y code is generated by the module 2 sum operation with stream cipher W. W code further improves the accuracy of the anti-spoofing code. M code will be used in the subsequent BLOCK II F satellite to further improve the performance of military receivers. In order to ensure the continuous use of the original C/A code, P (y) code, M code shall share frequency with them through Binary Office Carrier (BOC) modulation. The initial long code of the Galileo system is 2,097,151 bits ($2^{21-}1$), which may be truncated to accommodate the coded/carrier/multiplexed data set.

In order to further improve the performance of navigation warfare, the US Military prevents the enemy within the war area from using the GPS C/A code service (SPS), it shall cast interference and cause the lead capture of C/A code to P(y) code invalid. The direct capture of P(y) code must be realized. In 1998, the Chairman of the Joint Chiefs of Staff issued a SAASM order requiring that the GPS PPS (i.e., P (y) code or M code precise service) receivers shall contain SAASM after October 1, 2002. This SAASM is a dedicated module with hardware anti-tampering performance, and it is consisted of multiple ICs. It shall complete a series of data signal processing such as the generation, capture, security, and so on of the military code, respectively. The Key Data Processer (KDP) in the SAASM stores and processes all the keys to complete the security algorithm. KDP is provided not by the manufacturer, but by the government, so its hardware design is not open to the manufacturer of the receiver. Therefore, the US government built a set of KDP loading and installation facilities (KLIF), which is the only place to load confidential software. The generation and management process of SAASM is as follows: After the PPS receiver manufacturer (SAASM as the core) obtains KDP from the relevant governmental department, the manufacturer integrates it with other ICs and designs and manufactures the SAASM module of the company. Then this module is sent to KLIF and to load the secret file of the black key software to SAASM. Black key system is very confidential, it can be distributed by non-confidential means, and then be retrieved by the manufacturer to load application software and finish the whole generation process of SAAS M. The SAASM-based PPS receiver shall be defined as non-confidential device after passing the GPS JPO review and approval. When the SAASM based GPS receiver is connected to Host application equipment (HAE), such as INS compound guidance equipment, KLIF needs to re-enter its image so as to register the ultimate purpose of the SAASM receiver, while HAE needs to be licensed by GPS JPO. The key to the process is to send the module to KLIF, what software is loaded into SAASM? Is there effective control? What is the longest period of effective control? The short effective control period is helpful to strengthen confidentiality. But if the period is too short, it may cause inconvenience for use of the user receiver. So, the key technology for black key is to handle the security intensity without adding the trouble of injecting aging parameters. Therefore, it is appropriate to design a *super long code.*

Basic scheme for a super long code: The so-called super long code is a compound code with super long code period. There are two basic generation methods, one is the compound method, and the other is the parameter control method.

The generation of super long compound code is as shown in Figure 9.15. It is generated by the module 2 with the addition of two long codes P_1 and P_2. P_1 and P_2 have the same code rate, but P_1 is controlled by aging parameters.

The generation scheme of parameter control method is as shown in Figure 9.16. P_1 code is the control parameter of generation and control of P_2 code, and P_2 code will generate a precise super long code. There is no essential difference between these two schemes and both can complete the generation of super long precise ranging code in order to reduce the spending of software and hardware generated, it is necessary to design the control parameters carefully.

We have discussed the generation of satellite's super long precise ranging code earlier. While there are dozens of satellites for one constellation, how is the long code allocated at different working frequencies of each satellite? What is the basis for this allocation? There are two schemes for the allocation of super long precise ranging codes at different

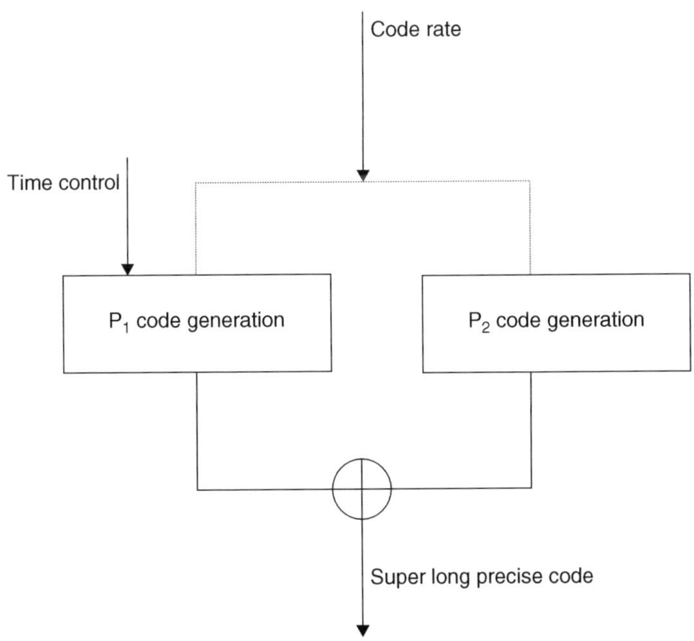

Figure 9.15 Block diagram of super long code generated using the compound method.

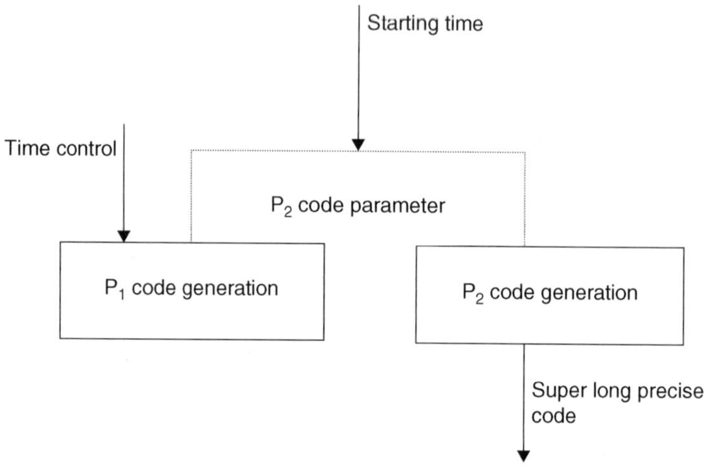

Figure 9.16 Block diagram of super long code generated using the parameter control method.

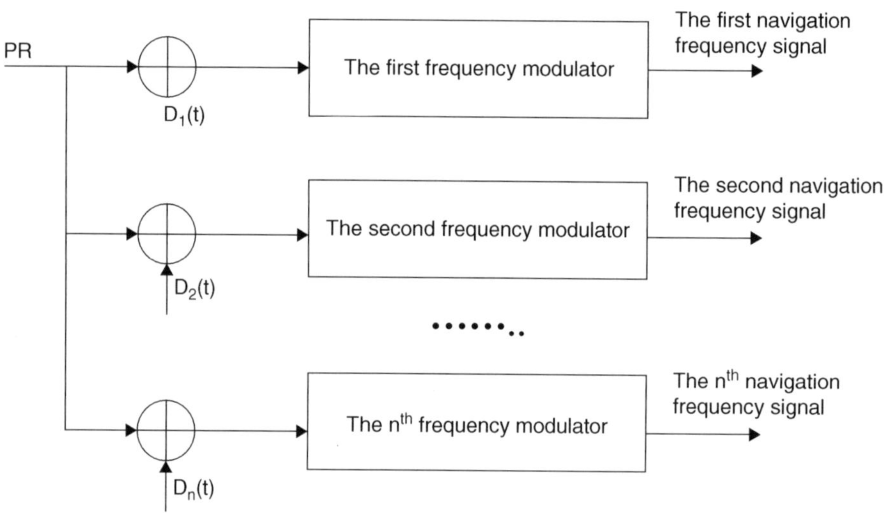

Figure 9.17 Allocation scheme for the same PR code.

working frequencies of each satellite. The first is the allocation scheme for codes of the same length, as shown in Figure 9.17. In order to identify different satellites, the precise ranging codes PR are different, but the PR codes of different frequencies of one satellite are the same. The other allocation scheme is for codes of different lengths, as shown in Figure 9.18; each satellite has different PR codes for working frequencies. The first scheme causes less spending toward satellite load and will help to reduce it, because the generation of super long code costs a lot. The scale of a super long code is within 300~400 thousands of units. It also reduces the spending of the user receiver. But once the user acquires the PR trace of one frequency, it can finish the trace of other frequencies. Currently, the GPS satellite adopts the generation scheme of the same P code. The different codes tracing of L_2 frequency have been realized with its characteristics.

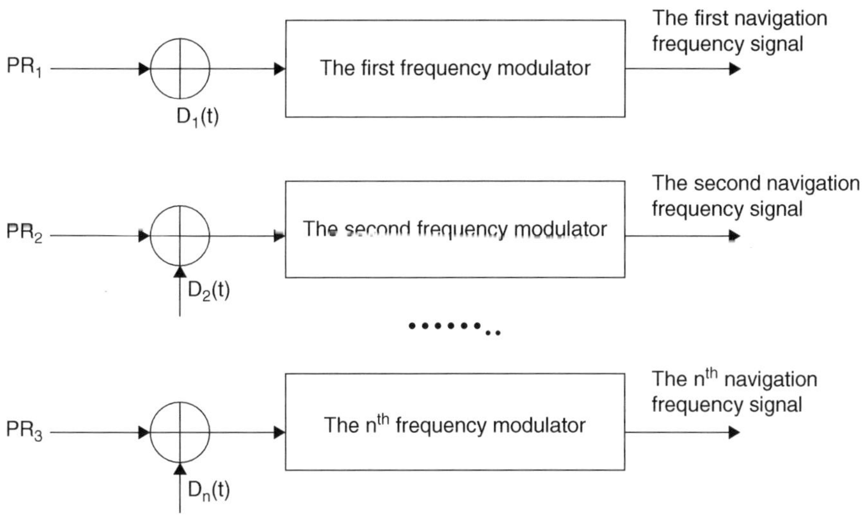

Figure 9.18 Allocation scheme for different PR codes.

For the second allocation scheme, various navigation signals of each satellite are totally independent, and the navigation messages D(t) are totally different, which is helpful for performance allocation and management of various different authorized users.

9.4.5 Navigation Signal Modulation Methods

There is not much difference between navigation signal modulation and communication signal modulation in principle [13, 14]. As the navigation frequency resource is quite rare, all major countries have established their satellite navigation systems. What is more, because of various commercial and military reasons, the authorized use of distribution toward navigation signals is more elaborate and complex so there are several modulation schemes within the same basic bandwidth to provide different service content and accuracy for different authorized users in accordance with phase and spectrum relationship. QPSK+BOC modulation has become the universal choice of navigation signals, while GPS adds the M code of BOC modulation for separation of the M code and C/A code at frequency, so it can be used for interference of the C/A code at war time without influencing M code navigation signals.

Binary phase shift keying (BPSK) modulation: Suppose the spreading code is C(t), and the carrier frequency is ω_0, then the phase-modulated wave can be expressed as

$$S(t) = A\cos[\omega_0 t + \varphi c(t)] \tag{9.9}$$

Wherein, φ is the phase modulation index, if specified in the spreading code sequence,

When $C(t) = 0, \quad \varphi c(t) = \pi \times 0 = 0$

When $C(t) = 1, \quad \varphi c(t) = \pi$

Then this kind of modulation is called binary phase shift keying. And this binary phase shift keying signal can be expressed as

$$S(t) = \begin{cases} A\cos w_0 t & \text{When } c(t) = 0 \\ -A\cos w_0 t & \text{When } c(t) = 1 \end{cases}$$

As in the spread spectrum sequence, (0, 1) and (1, −1) have the same structure. So BPSK signal can be expressed with balance modulation signal as:

$$S(t) = Ac(t)\cos \omega_0 t \tag{9.10}$$

Wherein, c(t) is the spread spectrum code sequence expressed with "+1" and "−1". Quadrature Phase Shift Keying (QPSK) and Offset Quadrature Phase Shift Keying (OQPSK) modulation: the QPSK spread spectrum signal can be expressed with the following formula:

$$\begin{aligned} S(t) &= \sqrt{p}c_1(t)\cos[\omega_0 t + \theta_d(t)] \\ &+ \sqrt{p}c_2(t)\sin[\omega_0 t + \theta_d(t)] \\ &= a(t) + b(t) \end{aligned} \tag{9.11}$$

Wherein, $\theta_d(t)$ represents data phase modulation, and $c_1(t)$ and $c_2(t)$ are independent orthogonal spreading codes and suppose they can only be ±1. $c_1(t)$ and $c_2(t)$ codes are identical in width and synchronous in time. The block diagram of normal QPSK modulator is as shown in Figure 9.19 that also shows the phase relationship between QPSK modulation signals. As $c_1(t)$ and $c_2(t)$ are binary sequences of numbers of ±1, in the QPSK modulation, the signal change of $c_1(t)$ and $c_2(t)$ happens at the same time, so there are the following four situations: (1,1), (1, −1), (−1,1), and (−1, −1).

Therefore, QPSK signal phase can be changed into 0°, ±90°, or 180°.

In the QPSK spread spectrum modulation, the data of in-phase channel and the orthogonal channel can be different and this kind of modulator is called a dual channel modulator. As shown in Figure 9.19, the code rates of $c_1(t)$ and $c_2(t)$ can also be different, powers of in-phase channel and orthogonal channel are also different, and this is the modulation of satellite navigation signals. The navigation signal of QPSK modulation

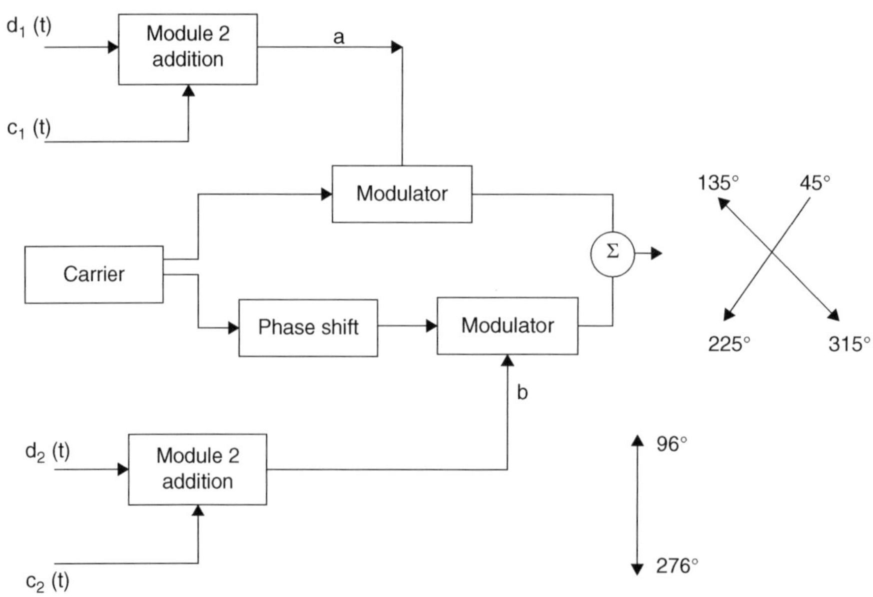

Figure 9.19 Block diagram of a normal QPSK modulator.

can be expressed as

$$S(t) = \sqrt{2p_1}d_1(t)c_1(t)\cos[\omega_0 t + \varphi 1] + \sqrt{2p_0}d_1(t)c_2(t)\sin[\omega_0 t + \varphi 2] \quad (9.12)$$

p_I is the power from I branch to the signal.
$C_1(t)$ is the coarse ranging code, in the GPS, it is called C code or C/A code with its code rate as 1.023 Mbps.
$d_1(t)$ is the data of modulation at C code, and the data rate can be 50 bps or higher.
$C_2(t)$ is precise ranging code and is called P code in GPS, it can also be expressed as P(t). Its code rate is 10.23 Mbps.
$d_2(t)$ is the data of modulation at P code.

From this discussion, we can see that there are only two phases (O° and 180°) in BPSK modulation. Every code element in binary sequence can correspond to two possible phases. There are four possible phases for quadrature phase shift keying signal and it can be seen as the sum of two BPSK signals whose carriers are orthogonal to each other. If the width of binary code elements is T_b, this binary sequence can be divided into odd number sequence and even number sequence through data extractor, and the element width of each branch T_C shall be extended as $2T_b$. The odd number sequence shall be sent into Q branch channel through delay T_C and conduct BPSK modulation toward carrier wave $\sin^{\omega_0}t$; the even number sequence shall be sent into I Branch channel to conduct BPSK modulation toward $\cos^{\omega_0}t$, then the QPSK signal is obtained by adding these two binary phase shift signals together. Naturally, it is conceivable that the OQPSK signal can be obtained by delaying the odd number code sequences by $T_b = T_c/2$ and conducting this operation.

9.4.5.1 Binary Offset Carrier (BOC) Modulation

Binary offset carrier modulation (i.e., BOC modulation technology) should be adopted in order to fully utilize the energy distribution in the limited bandwidth to achieve the purpose of transmitting a plurality of navigation signals. GPS adopts QPSK to arrange C/A code of civilian navigation signal and arrange precise ranging code signal (i.e., P code navigation signal) within the same $\pm 12MH_Z$ bandwidth. In order to further improve the performance of the military code, a new M-code navigation signal is planned to arrange in this bandwidth. This M-code navigation code shall adopt BOC modulation and their corresponding power spectra can be described in Figure 9.20.

BOC modulation is a subcarrier modulation type that uses bipolar non-return-to-zero square wave subcarrier modulation pseudorandom sequence. The period of pseudorandom sequence is quite long, the width of code is T_P, and rate of spreading code is F_P, with the value of ±1, subcarrier is a square wave with the period of T_S and frequency of f_s(Hz), and its value is also ±1. During modulation, the edges of the random sequence correspond to the edges of the square wave. The envelope of modulation signal for a general offset carrier is

$$S(t) = e^{-i\theta} \sum_k a_k v_{nT}(t - knT - t_0)C_T(t - t_0) \quad (9.13)$$

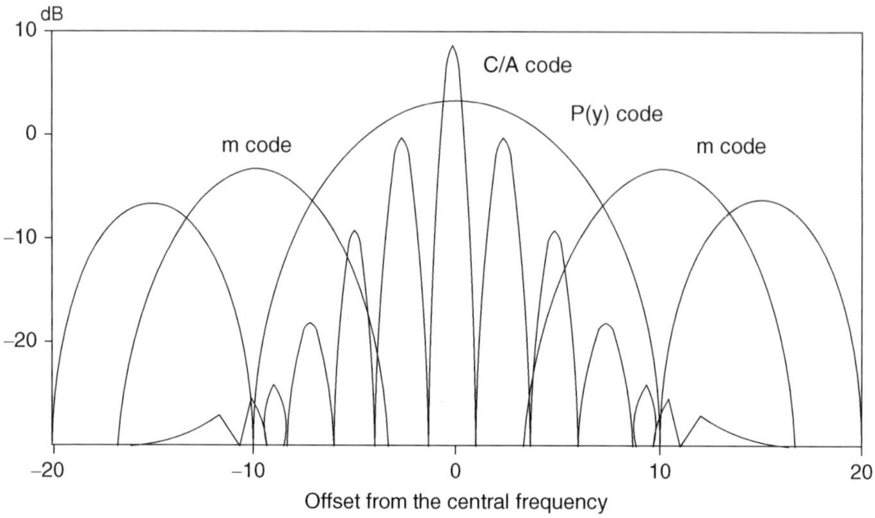

Figure 9.20 Power spectrum relationship between m code and C/A code, P code.

In this formula, $\{a_k\}$ is the unit amplitude, navigation message of the random phase; $C_T(t)$ is the subcarrier with the period of 2T; $V_{nT}(t)$ is the spread spectrum signals of square wave or other special waveform with the width of code element identical to nT; n is the number of subcarriers in half period of code element from $V_{nT}(t)$; θ and t_0 represent the offset of subcarrier phase and time separately.

Offset carrier modulation can also adopt the sinusoidal wave as the subcarrier. The use of sinusoidal wave subcarrier and low pass filter can provide a good spectrum capacity. The block diagram of signal generation is shown in Figure 9.21.

The generation of navigation data, spread spectrum signals and subcarrier adopts the same clock to ensure that the signals are synchronous. Envelope will change when spread spectrum signals pass through the low pass filter. Such modulation is called linear subcarrier modulation LOC and its amplitudes are continuous. It can be expressed as LOC (f_S, f_p, B) where f_S represents the frequency of subcarrier $C_T(t)$; $f_S = \frac{1}{2T}$; f_P represents the rate of precise ranging code; $f_p = \frac{1}{nT} = \frac{2}{n} f_S$; B represents the bandwidth of spread spectrum signals through low pass filters.

The spread spectrum signal is low pass signal, and the subcarrier is sinusoidal signal, so the baseband current is linear.

When the negative carrier $C_T(t)$ in the formula (9.13) is a square wave, and $V_{nT}(t)$ is the square wave signal with the width of code element as nT and not passing the filter, the envelope value of offset carrier signal is constant. At this time, the signal modulation should be BOC modulation. The binary offset carrier signal can be expressed as BOC(f_S, f_P), where f_S represents the frequency of negative carrier, f_P represents the spreading code rate and the time relationship is determined by (9.14) and (9.15). The block diagram of BOC signal generation is as shown in Figure 9.22. Different from the generation of LOC signal in Figure 9.21, all baseband signals in Figure 9.22 are binary numbers and can be realized through binary logic circuit. As a modification of

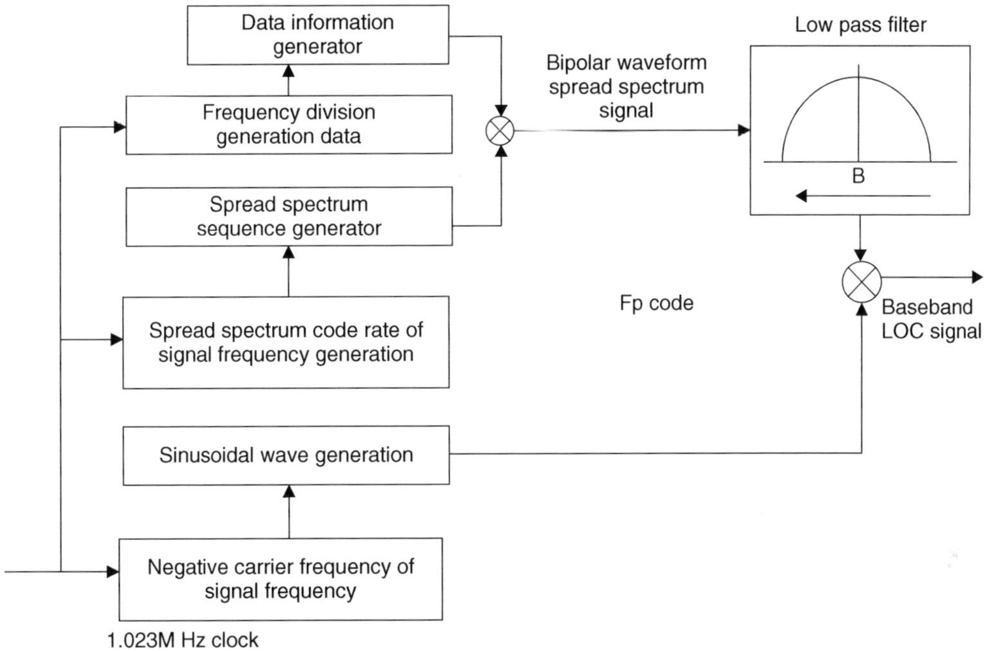

Figure 9.21 Block diagram of a linear subcarrier signal generator.

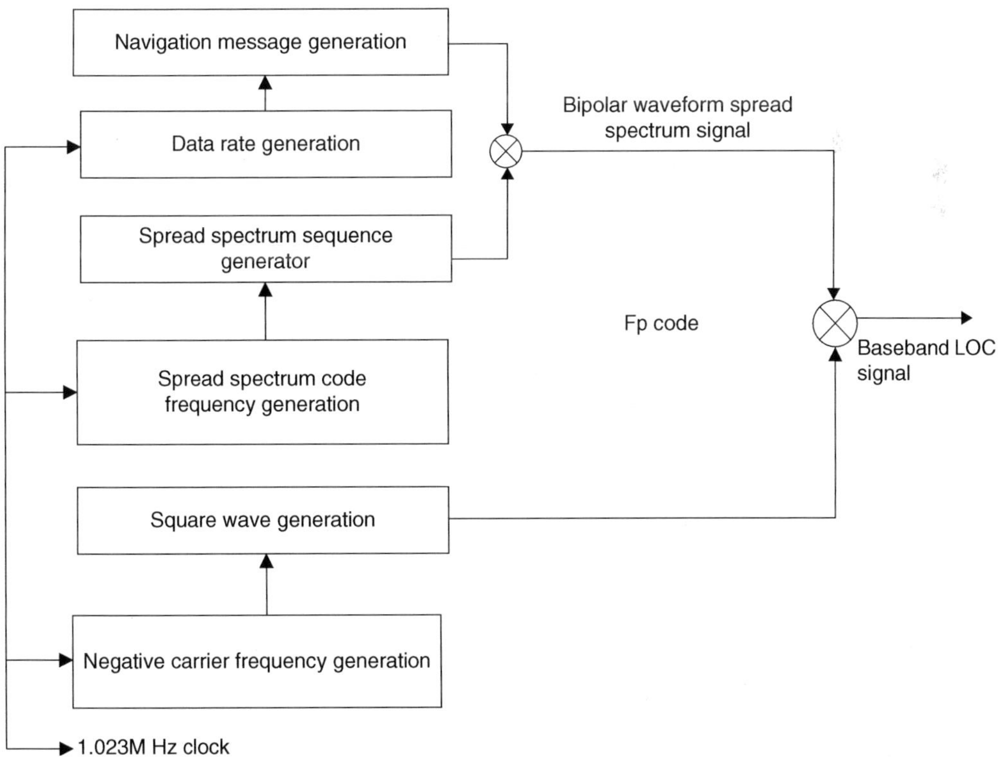

Figure 9.22 Block diagram of BOC signal generation.

formula (9.13), a BOC modulation complex envelope can be expressed as

$$S_{BOC}(f_S, f_P) = e^{-i\theta} \sum_k a_k q_{nT}(t - knT - t_0) \ n \text{ is an even number} \tag{9.14a}$$

$$S_{BOC}(f_S, f_P) = e^{-i\theta} \sum_k (-1)^k a_k q_{nT}(t - knT - t_0) \ n \text{ is an odd number} \tag{9.14b}$$

Wherein,

$$q_{nT}(t) = \sum_{m=0}^{n-1} (-1)^m V_T(t - mT) \tag{9.15}$$

$q_{nT}(t)$ comprises several half-period square waves, that is to say, there are n times of alternation between +1 and −1. When n is even number, $q_{nT}(t)$ is the balanced signal (mean value is zero). BOC modulation can be seen as the extension of Manchester modulation, there are more than one time of alternation between +1 and −1 within a spread code; when n is 2, the BOC code is Manchester code. By taking the BOC (10.5) signal as an example, it is the spread spectrum sequence that the signal adopts the spread-spectrum code rate of 5.115 Mbps to modulate a 10.23 MHz square wave subcarrier, BOC(10, 5) is short for (10.23, 5.115). Normally the BOC (r, m) represents the square wave subcarrier of the r × 1.023 Mbps spread spectrum modulated by the spreading code rate of m × 1.023 Mbps. The waveform is shown in Figure 9.23.

The spread spectrum sequence is +1, −1, +1, +1, −1.

It can be seen from Figure 9.23 that, the essential feature of BOC signal is that the amplitude is constant, which is convenient for the implementation of digital circuits. Each chip of the BOC (10, 5) signal is multiplied by two full cycles of the square wave subcarrier, and this is equivalent to a direct sequence spread spectrum modulation with a BOC modulation signal.

Normally, the designed BOC signal energy is mainly concentrated on the edge of the 24 MHz GPS signal bandwidth, while the signal energy at the center of the frequency

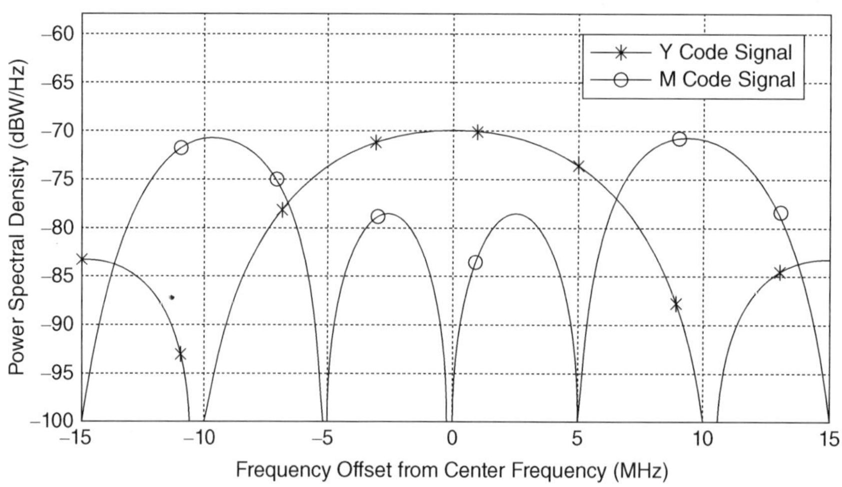

Figure 9.23 BOC (10, 5) baseband signal.

Table 9.8 Examples of main parameters for BOC.

Modulation	Subcarrier Frequency f_S (MHz)	Spreading Code Frequency f_P (MHz)	Width of Spread Spectrum Code Element (ns)	Twice of the Ratio of Subcarrier Frequency to Spread Spectrum Code Rate, n
BOC (8, 4)	8.184	4.092	244.4	4
BOC (9, 3)	9.207	3.069	325.8	6
BOC (10, 2)	10.23	2.046	488.8	10
BOC (10, 1)	10.23	1.023	977.5	20
BOC (5, 5)	5.115	5.115	195.5	2

band is small. The main parameters of the BOC signal at 24 MHz bandwidth are listed in Table 9.8.

9.4.6 Selected Error Correction Coding of Navigation Message

The navigation message is the basic data providing positioning and navigation to users and the users can calculate the clock error, ephemeris, almanac of the satellites, ionosphere correction parameters, differential correction parameters and system integrity information, and other broadcast information by receiving this message. Such data is loaded to the satellite by the operational control system and then broadcast to the users by the satellite. There is some other information, such as satellite integrity information, which is generated directly by the satellite and broadcast to the users. These data have various repeat periods for the users to meet the needs of different data and different response periods. For integrity information, it is generally updated once per second, for ephemeris and almanacs the new period can be 30 s, or even 12.5 min. As message has its time limit and different update period, the error rates of different messages have different importance for users. So, the error detection and error correction coding of the navigation message can be different. Contradiction with high spread spectrum gain and strong error correction ability is required to meet the needs of rich navigation message, high information speed, and strong anti-interference ability. In the following, GPS navigation message is taken as an example to introduce the basic content of the navigation message and methods of error correction coding. Table 9.9 lists the requirements and contents of GPS navigation data.

This GPS navigation message does not contain wide area differential correction signal and ground integrity monitoring information. It does not meet the required navigation performance conditions, so Federal Aviation Administration of the United States (FAA) has put forward the Wide Area Augmentation System (WAAS) to provide differential corrections and short cycle (1 s) integrity information to the users.

The GPS message transmits the navigation message in a data string of 50 bps, and the spreading code rate of the P (y) code is 10.23 Mbps. The spread gain is 10.23 Mbps/50 bps = 53 (dB). It shows when the interference signal is 40 dB (10,000 times) higher than the signal level, after the relevant despreading, there is 13 dB for the signal to noise ratio output by the receiver, meeting requirements of the demodulation of navigation messages. This means that at the 400-W transmitter interference of 200 km range, GPS P (y) code

Table 9.9 Requirements and content of GPS navigation data.

No.	Requirements	Content
1	Exact position when the satellite transmits signals	Use the satellite ephemeris expressed with corrected Kepler's model in the geocentric inertial (ECI) coordinate system and convert it into the earth center earth fixed (ECEF) coordinate system.
2	Exact time when the satellite transmits signals	Correction of satellite clock error model and relativity.
3	Select the general satellite location required by the satellite based on GDOP value	Satellite almanac describes the general position, time, and healthy status of the whole GPS constellation.
4	Satellite's Estimated Ranging Deviation (ERD) provided to the authorized user, which belongs to long-period integrity information	Provided in the form of Navigation Message Correction Table (NMCT) in the navigation message.
5	Time transmitting information	Conversion data of GPS time to Universal Time Coordinated (UTC).
6	Conversion time to guide to capture P code with C/A code	Send the number of sub-frame to keep the 1.5-s period (X_1 sequence) tracing toward P(y) code within a week, and assist the capture of P (Y) code
7	Ionosphere correction data	Approximate model expressed with the relation between ionosphere and user position and time
8	Satellite signal data quality	User Ranging Accuracy (URA).

can normally demodulate the navigation message and ensure the normal navigation and positioning of the GPS receiver. The interference power is calculated as follows:

$$P_J = \left(\frac{\Delta f}{\Delta F} - m\right)\frac{P_S G_S}{G_J}\left[\frac{R_J}{R_S}\right]^2$$

Δf represents spread spectrum code rate, 10.23 Mbps
ΔF represents navigation message code rate, 50 bps
m is demodulation threshold (allowance included), with the value of 13 dB.
G_J represents interference transmitter gain, with the value of 3.16 dB.
G_S represents satellite antenna transmit gain, with the value of 16 dB.
P_S represents signal power of satellite intended to receive, with the value of 30 W.
R_J represents the distance from interference source, 200 km.
R_S represents the distance from the satellite, 20,000 km.

When M is 13 dB, the message error rate is rather low.

To analyze the coding requirements for navigation messages, by taking GPS as an example, the definition and arrangement of navigation data are listed and the relevant materials can be referred to.

In the navigation message coding schemes, the coding scheme that has strong error detection ability and is easy to achieve must be selected. Cyclic Redundancy Check (CRC) plus forward error correction scheme is a good choice, as it has advantages of strong error correction ability, small redundancy rate and being easy to achieve. If a 24-bit check code is selected; that is, CRC24 check plus forward error correction coding scheme, its error detection capability is as follows:

1. CRC24 can detect all two-bit errors (when the error interval is smaller than $N = 2^{23-}1 = 8,388{,}607$).
2. CRC24 can detect all odd bit errors.
3. CRC24 can detect all burst errors with the length less than 24.
4. When the length of burst error is b = 25, the undetectable probability of CRC24 is $P_H = 2^{-23} \approx 1.19209 \times 10^{-7}$, and when the length of burst error is b > 25, the undetectable probability of CRC24 is $P_H = 2^{-24} \approx 5.96046 \times 10^{-8}$.

A block diagram of the CRC24 coding scheme is shown in Figure 9.24.

Channel coding is composed of 24-bit CRC and half code rate of forward error correction convolutional code. This coding can increase the spread spectrum gain by 10 dB.

9.4.7 Advocates of the Compass Operators Toward Satellite Navigation Frequency Compatibility and Compass Signal Structure

Under the background of shortage in satellite navigation bands, the WRC-2000 specified 1164 MHz~1215MHz as the shared frequency bands for radio navigation and satellite radio navigation, which can help bring more satellite radio navigation businesses, but it still can't satisfy the surging needs. Thus, with frequency compatibility placed on agenda, compatibility and interoperation have been hot topics among global satellite navigation systems. The opinions of Chinese Compass operators are as follows: The so-called compatibility refers to that when two or more satellite navigation systems work at the same time, their disturbance to each other will not exacerbate the performance; the partial overlapping of navigation signal frequency of systems is inevitable but feasible and beneficial. The definition is helpful to equal and fair allocation of more satellite navigation businesses. The so-called interoperation is that if users accept the shared services of different systems, the improvement in navigation accuracy, continuity, serviceability, and integrity is superior to the cost spent in access to many systems. The design of Compass navigation signals is compatible with GPS and Galileo for higher navigation performance. Its major parameters are listed in Table 9.10. The signal structures in the table form the foundation for China to participate in discussions of ITU-related international telecommunication rules and bilateral international coordination, and one of the input drafts submitted to the ITU-R WP4C conference by Compass operators in September 2008. During

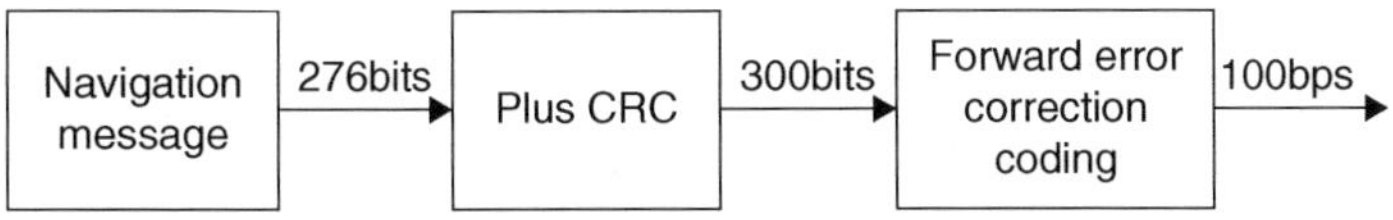

Figure 9.24 Block diagram of CRC coding.

Table 9.10 Signal structures of Compass global system.

Navigation Signal	Carrier Frequency/ MHz	Code Speed/ (Mb/s)	Modulation Method	Navigation Data Speed/(b/s)/ Symbol Rate (sps)	Service Type
$B1_D$(data channel)	1575.42	2.046	BOC(14.2)	50/100	Authorized
$B1_P$(guidance channel)				No	
B1-C_D(data channel)	1575.42	1.023	MBOC(6,1,1/11)	50/100	Open
B1-C_P(guidance channel)				No	
$B2_D$(data channel)	1191.795	10.23	AltBOC(15,10)	50/100	Open
$B2_P$(guidance channel)				No	
B3	1268.52	10.23	QPSK(10)	500bps	Authorized
B3-A_D(data channel)	1268.52	2.5575	BOC(15,2.5)	50/100	Authorized
B3-A_P(guidance channel)				No	

Note: B1-CD: Adopt BOC(1,1) modulation method;
B1-CP: Adopt MBOC modulation, power spectrum density is

$$MBOC(f) = \frac{29}{33}BOC_{1,1}(f) + \frac{4}{33}BOC_{6,1}(f)$$

B1-C power spectrum density is

$$s(f) = \frac{1}{4}BOC_{1,1}(f) + \frac{3}{4}MBOC(f) = \frac{10}{11}BOC_{1,1}(f) + \frac{1}{11}BOC_{6,1}(f)$$

the WPC4 conference, PDNR was upgraded to DNR and was passed at the SG4 conference. The ideas of member countries were collected through letters after the conference.

Signal design features of Compass can be summarized as follows:

1. The authorized service signals take into account the shortage of current navigation frequency resources, GPS, GLONASS, Galileo, and frequency spectrum distribution so that the disturbance to each other will not deteriorate their performance.
2. The frequency spectrums of Compass' open navigation service signals and authorized service signals are separated to avoid harmful disturbance.
3. The open service signals of Compass have the same center frequency, spreading code rate, modulation methods, and message composition formats as those of GPS and Galileo, and highlight preeminent interoperation.
4. Compass global signal structure contains the navigation signals of Compass' current working satellites, featuring inclusivity and continuity.

9.5 Time Standard and Timing Pattern of Satellite Navigation

Satellite navigation system presents the motion trail of satellites and users by time, and the satellite navigation can solve the users' location, speed, and time. The satellite navigation system is the space and time benchmark describing the cosmic events. Satellite navigation time system is certainly an important element of satellite navigation system.

9.5.1 Satellite Navigation Time System (SATNAVT)

In each satellite navigation system, the time synchronization between satellites, satellite and ground control system, satellite, and users shall be maintained according to the time of this system. GPS satellite motion is described according to GPST time system, the time system of each system is guaranteed by specific time frequency to keep the time scale of radio transmitting signal in each satellite precise and synchronized. Their time synchronization accuracy directly influences the users' position accuracy. A satellite clock time synchronization error of 1 ns causes a pseudorange error of 0.3 m. The pseudorange error multiplied by PDOP factor (PDOP≥1) is the resulting position error. SATNAVT of satellite navigation system takes the atomic time as metric to maintain the continuity of navigation. The time synchronization of Universal Time Coordinated (UTC) is converted by a set of parameters. The leap seconds only occur after the conversion of UTC.

SATNAVT maintains the stability of its time system by atomic clock groups with high accuracy and high stability. Atomic clock groups of ground control system provide the main running clock (MAST) with the time and frequency trimming parameters to enable the MAST to keep extremely high and long-term stability. The ground clock groups are composed of hydrogen and cesium clocks. The MAST conducts the time comparison with each ground station and atomic clocks of satellites to ensure that all satellite clocks are synchronized with the SATNAVT.

SATNAVT of satellite navigation keeps strict time synchronization relations with Universal Time (UT), Universal Time Coordinated (UTC), and so on.

9.5.2 Universal Time (UT) (Greenwich Mean Solar Time)

This time system includes year, month, day, hour, minute, and second. This time measurement system was adopted by the third session of International Astronomical Union in 1928.

Universal Time is measured by "mean sun" relative to the Greenwich Meridian's hour angle plus 12 h. Due to the change of meridian location caused by wobbling of the Earth's poles, the universal time can be divided to different kinds according to the magnitude of the perturbation factors considered:

UTO: universal time obtained relative to the uncorrected measurement results of inaccurate Greenwich meridian.

UT1: universal time corresponding to average Greenwich meridian with earth polar wobble taken into account, which is a benchmark of time measurement in daily life.

UT2: universal time after season correction for UT1.

UT1R: tidal correction for UT2 again.

UT is equal to the local mean time minus the geographic longitude of observer L, S_G (Greenwich sidereal time) is equal to local sidereal time minus L, east longitude is positive and the west longitude is negative.

9.5.3 Universal Time Coordinated (UTC)

This is usually used in daily life, measured in atomic time and periodically (0.5~2.5 years) corrected by 1 s to ensure the difference of UT-UTC not more than 0.9 s. The UTC signal is transmitted by the radio network. (TA1-UTC) was 25 s on January 1, 1990. The advantages of UTC are that it is the time scale with relatively high uniformity and it is atomic time in nature; it has a proportional relationship with the natural process of solar time (rising and falling).

9.5.4 Julian Period

Sometimes, it is not convenient to use the time system analyzed before to fully express a specified time. It needs four numbers at least: year, month, day, and hour with decimals, so it actually adopts a time measurement system called the Julian Period. It is convenient due to the use of the number of days to calculate the current time (with a small number of mean solar days). All days in the Period are numbered in sequence, and are irrelevant to the year, month, day, and so on in the adopted calendar system.

The Julian Day (JD) is the continuous count of days since the beginning of the Julian Period, calculated from 12:00 on January 1st, 4713 BC (the beginning of the Julian period) to the specified time. The total length of the Julian Period is 7980 years. And one Julian Century contains 36,525 mean solar days.

9.5.5 Timing Method of Satellite Navigation Time (SATNAVT)

We take GPS as the representative to explain the GPST timing method. In the GPS signal format, there are week count W_N and Z count, which are the basic timing periods. In GPS time, the start time is specified at the junction of Saturday midnight and Sunday morning. GPS is based on the UTC time maintained by the US Naval Observatory. The zero point for GPST is specified at the junction of the midnight of January 5, 1980 and the morning of January 6, 1980, after that, the Z count flips for 1024 weeks every 19 years. The UTC time scale takes the time of the Greenwich meridian as a reference point. The difference between GPS and UTC is that GPS has no leap seconds, and UTC time sometimes inserts leap seconds. However, the control system of GPS must keep the difference between GPS and UTC within 1 μs (module 1 s). With the passage of time, the difference between GPS time and UTC time will be a number of total seconds.

Table 9.11 shows the correction relationship between GPST and UTC.

The information includes the parameters that associate the GPST with UTC, and notifies the users of the δt time in the recent past or the near future caused by the leap seconds $\triangle t_{is}$, and the week number WN_{ISF} with the leap seconds into effect. These relationships are established in the polar parameter time.

However, when the user is working at the time that the leap second is about to occur, the second adjustment must be carried out.

The algorithm using the navigation data of sub-frame 4 to determine the relationship between UTC and GPS time is that:

$$t_{UTC} = (t_E\text{-}\Delta t_{UTC}) \quad [\text{module } 86{,}499 \text{ s}]$$

The unit of t_{UTC} is second.

$$\Delta t_{UTC} = \Delta t_{TS} + A_0 + A_1(t_E\text{-}t_d + 604{,}800(WN\text{-}WN_t)$$

In the formula, t_E = GPST estimated by the user's station, on the basis of the factors given in the discussion of the sub-frame clock correction and the effect of the ionosphere and SA to correct $t_{sl.}$

Table 9.11 GPST-UTC clock correction parameters (taken from p. 18 of GPS telex message sub-frame 4).

Parameter	Digits	Scale Factor LSB	Effective Range	Unit
A_0	32[b)]	2^{-30}	---	S
A_1	24[b)]	2^{-50}	---	s/s
$\triangle t_{IS}$	8[b)]	1	---	S
t_d	8	2^{12}	602,112	S
WN_t[a)]	8	1	---	Week number
WN_{ISF}[a)]	8	1	---	Week number
DN[a)]	8[c)]	1	7	Number of day
$\triangle t_{ISF}$	8[b)]	1	---	S

a) Notes:
The symbols DN and WN represent the number of day and the week number, respectively.
b) The parameter with n shall be 2's complement, MSB is the sign bit (+ or -).
c) It is judged as correct.

$\Delta t_{IS} = \delta$ time due to leap seconds.
A_0 and A_1 = constant and first order term of the polynomial.
T_d = reference time of UTC data.
WN = current week number (given by sub-frame 1)
WN_t = UTC base week number.
The number of second for a day is 86,400 s.

The estimated GPS time is the number of second relative to the end/start of the week. The reference time t_d of the UTC data is based on the start time of the week, and the week number of this week is given in the eighth word of p. 18 of sub-frame 4, which represents 8LSB of the week number. The user must pay attention to the truncation characteristics of the week number. Refer to ICD-GPS-200.

With respect to current time of the user, if the effective time of leap second occurs in the past, the relational expression provided here comes into existence, except that $\triangle t_{ISF}$ replaces $\triangle t_{IS}$. The algorithm has an exception, when it occurs at current time during a time frame from DN+3/4 to DN+5/4 (DN refers to number of day), it can be obtained in this time interval by using the following UTC formula: $t_{UTC} = W[\text{module} - (86{,}400 + \triangle t_{ISF} - \triangle t_{IS})]$.

Wherein the $W = (t_{E} - \triangle t_{UTC} - 43{,}200)\ (\text{module} - 86{,}400) + 3200$

It can use the possible week numbers to convert the correct and acceptable leap seconds. The definition of $\triangle t_{UTC}$ given in the last paragraph is applied throughout the transition period. It should be noted that when plus the leap seconds, it will meet the time value in the form of 23: 59: 60. Some user receivers may be designed to approximate the UTC by reducing the time count in running within several seconds after the leap second, thereby quickly reverting to the correct time indication value. Note that the user must conduct a consistent carry or borrow for the count of year/week/day, whenever the leap second is met.

BDT seconds count of China Compass Navigation Time counts from zero at 00:00 a.m. every Sunday. The corresponding second time means the time corresponding to the first pulse rising time of this sub-frame synchronization head. The whole week count WN is 13 bit, its value range is 0~8192, and is the week cumulative amount of BDT. BDT time counts from zero at 00:00 a.m. of January 1, 2006; BDT and UTC time are consistent at 0:00 a.m. on January 1, 2006. BDT repeats after 143 years.

9.6 Navigation Satellite Trajectories and Ephemeris Expressions

The motion of an artificial earth satellite is carried out according to the laws of celestial mechanics on the action of inertia and the gravity [15]. To describe its motion, the geocentric inertial coordinate used is $0X_0Y_0Z_0$ (see Figure 9.25). The origin of the coordinate is located in the center of mass of the Earth. The axis OX_0 lies in the equatorial plane and points to the vernal equinox or Aries r (r = astronomical symbol of Aries); the OZ_0 axis points to the North Pole along the earth's rotation axis. The OY_0 axis forms a right-handed rectangular coordinate system together with the first two axes. Another coordinate system used in satellite radio navigation is the earth-fixed coordinate system 0XYZ, such as PZ-90 in GLONASS and WGS-84 in GPS. The center of the coordinate system is also located in the center of mass of the earth. The 0Z axis coincides with the $0Z_0$ axis of the $0X_0Y_0Z_0$ inertial coordinate system and points to the North Pole P_N along the earth's rotation axis. The 0X axis lies in the intersection line of the earth's equatorial plane and the Greenwich meridian plane (G), and the 0XZ plane determines the position of the zero point where the longitude is normally calculated. The 0Y axis forms a

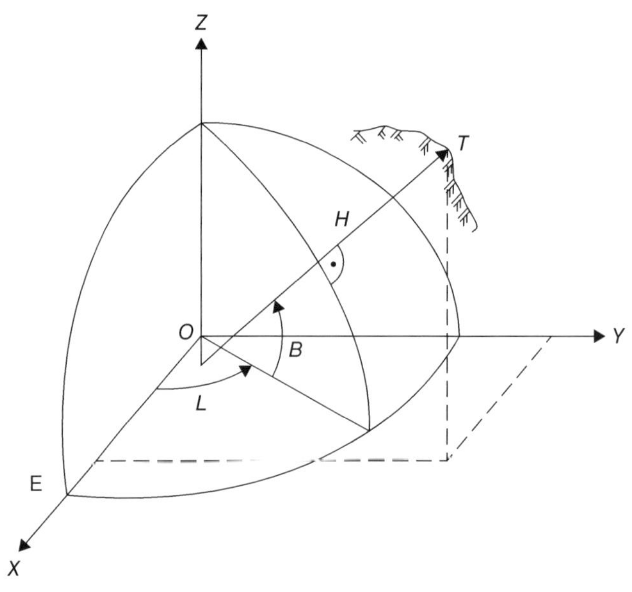

Figure 9.25 Geocentric coordinate system.

right-handed coordinate system together with the first two axes. The angle ϕ_r of 0X and $0X_0$ axis corresponds to Greenwich sidereal time.

The information about satellite motion published in the user navigation message is formed in the fixed-earth coordinate system, and the user equipment calculates the user's own coordinate in this coordinate system at the secondary information processing stage. However, the user is largely concerned with coordinate systems such as altitude, latitude, and longitude associated with the geodetic coordinate system.

The unperturbed orbit equation of the navigation satellite in the inertial coordinate system follows Newton's second law. The motion of the satellite's center of mass in the inertial coordinate system $0X_0Y_0Z_0$ is described by the following formula:

$$mg = F \tag{9.16}$$

The so-called no-perturbance motion means the motion under the Earth's gravity center only, wherein m is the mass of satellite, g is the centripetal acceleration vector, and F is the gravity vector. In accordance with the law of gravitation, for the gravity,

$$F = KMm/r^2 = \mu m/r^2 \tag{9.17}$$

In the formula, $K = 6.672 \times 10^{-11}$ m^3/kg^2: the gravitational constant; m is the mass of Earth, $m = 5.974242 \times 10^{24}$ kg; r is the distance from the center of Earth to the satellite; μ is the geocentric gravitational constant of the Earth; $\mu = KM = 3.9860044 \times 10^{14}$ m^3/s^2;Considering the following relation:

$$g = d^2r/dt^2$$

In this formula, the time derivative is the total derivative in the inertial coordinate system. Therefore, the motion equation (9.16) has the following form:

$$m\frac{d^2r}{dt^2} = F$$

The projection of space trajectory of the satellite no-perturbance motion on each axis of inertial coordinate system $0X_0Y_0Z_0$ is described by the following equations:

$$\begin{aligned} \frac{d^2x_0}{dt^2} &= -\mu\frac{x_0}{r^3} \\ \frac{d^2y_0}{dt^2} &= -\mu\frac{y_0}{r^3} \\ \frac{d^2z_0}{dt^2} &= -\mu\frac{z_0}{r^3} \end{aligned} \tag{9.18}$$

Wherein x_0, y_0, z_0 are the current coordinates of the satellite (the projection of the radial vector r on each axis).

$$r = \sqrt{{x_0}^2 + {y_0}^2 + {z_0}^2}$$

It is often referred to as trajectory of the satellite motion. In practice, however, the navigation satellites have obvious perturbations. Main sources of the perturbation include gravity, second-order zonal harmonic term, lunar gravity, solar gravity fourth-order zonal harmonic term, solar radiation pressure, gravity anomaly, and so on. These forces vary with the orbital position of the satellite and can be seen as constants in a short time. However, when the spacecraft's position approaches the Moon, the lunisolar gravitation plays a decisive role, while all other forces can be considered as constants in a short time, or ignored. Any residual error shall be expressed.

In summary, the navigation satellite ephemeris is the elliptical orbit of the two-body motion plus periodic perturbation and long-term perturbation. The period of the main periodic perturbation is 5.98305 h. Other perturbations are small that can be expressed as the simple function of time (constants or linear functions) in a short time.

Ephemeris expressions of satellite navigation system are gradually being matured after a lot of comparison, the starting point is that it should be beneficial to adapt to the fast, high accuracy positioning, including the following consideration factors:

1. *The user's first positioning time*
   Obtaining the satellite location parameters with the shortest time.
2. *The user's calculation time*
   In order to calculate the satellite position in the earth-fixed coordinate system, it needs to perform many sine, cosine, square root, multiplication, division, and so on. So, the complexity of calculation shall be considered.
3. *User's storage requirements*
   In use, the user requires to store the positions of the satellites. Whether the storage capacity is large can influence the computing power.
4. *Update rate*
   How long a set of data to be used before conversion will affect the storage requirements of the satellite, and involves how often users collects new parameters.
5. *Repetition interval of the application time*
   The repetition interval directly affects the update rate.
6. *Accuracy*
   The change rate of ephemeris and ephemeris expressions shall meet the needs of user positioning accuracy.
7. *Orbital deviation*
   The ephemeris expression of orbital deviation and the range of parameter variation shall be properly given, and the orbital deviation shall be reduced as much as possible. Before the failure of the satellite, the allowable orbital deviation shall be considered. For orbit with large deviation of the orbit, more information groups are needed for expressing.
8. *Performance reduction*
   After exceeding the application period, whether the ephemeris expressions can be applied shall be considered or if the performance reduction can suddenly appear. In the absence of new parameters, it is better to use the original parameters and carry out navigation according to the performance reduction mode.
9. *Time that the user receives the calendar*
   In order to express the calendar with the common algorithm, how much data is required? To keep the calendar data at a minimum amount, the calendar expressions shall express multiple orbits, but in order to do this, it shall also add a common expression with a precise ephemeris to define the ephemeris in question.

10. *Clarity of expressions*
    It considers whether this ** expression is easy to modulate and decode, and whether this algorithm can be applied to the future design, whether other computers can be used for calculation and other factors.
11. *Compatibility with other systems*
    At present, whether the existing satellite navigation systems in the world uses a similar expression so as to improve compatibility of the system. In view of these criteria, there are several major ephemeris expressions:
    a. *Time polynomial*
       This way has a simple calculation method for user, the shortest processing time and the lowest storage requirements; however, polynomial expressions cannot independently express multiple orbits and all orbits. Therefore, for the calendar calculation, it requires a single calculation method. Therefore, except the shortest processing time, the polynomial expression has no obvious advantages.
    b. *Harmonic expansion*
       Although this way also considers the expansion of the harmonic function around the circular orbit, this expression has no vectorial advantages compared with the Kepler expression. Its main disadvantage is that there is no clear physical meaning, making it difficult to determine the coefficient range of the word length for the data. This coefficient is sensitive to the position on the elliptical orbit and the unknown orbital deviation.
    c. *Kepler parameter expression*
       This method is almost beneficial to all of the aforementioned criteria. One of the advantages is that there is a clear physical meaning, which makes it easy to determine the length of the data even to determine the size of the standard orbital deviation. However, it shall add some parameters to the perturbation of the Keplerian orbit that can be expressed clearly. The selected expression for the perturbation is the polynomial and harmonic coefficient of time. Therefore, when the combination of the two long-term drift terms is considered, Kepler expansion plus perturbation will be a better expression. If the system goes wrong, this expression can use the original parameters to provide a better way to deal with performance reduction. Table 9.12 shows the comparison of the various expressions and the aforementioned criteria.

Parameters of Kepler ephemeris expression and the definition of expression are shown in Tables 9.13 and 9.14.

The parameters in the previous expressions are the result of curve fitting and appear only as Kepler parameters. They can only be used for description of the ephemeris over a period of time, and not suitable for the entire trajectory. However, these parameters describe the real Keplerian trajectory within the precise range of a few kilometers.

If the reference time selected is the perigee time, it usually has only six Kepler parameters. However, if we use this parameter, M_0, t_{oe}, to replace the time of perigee, it needs seven parameters (m_0, t_{oe}, e, $\sqrt{A}$, Ω_0, i_0, ω). The effect of this replacement does not require adding words. The t_{oe} and the estimated time are often only a few minutes difference, which makes the decreased sensitivity of calculation to the time derivative, but also points out the stability of curve fitting for the near-circular orbit.

Table 9.12 Comparison of ephemeris expressions and selection criteria.

	Polynomial	Harmonic Expansion[a)]	Kepler Plus Polynomial	Kepler Plus Harmonic Terms
Sub-frame number	3+	2+	2+	2−
Computation time for users	Short, easy	Sine, cosine	Sine, cosine	Sine, cosine
Storage requirements for users	3+ sub-frame Small-scale computation needs calendar	2+ sub-frame Medium-scale computation is the same as calendar	2+ sub-frame Large-scale computation is the same as calendar	2 sub-frame Large-scale computation is the same as calendar
Update rate	Once/h	Once/h	Once/h	Once/h
Update overlapping	1/2 h	1/2 hour	1/2 h	1/2 h or longer
Accuracy	Error of less than 1 foot	Probably accurate.	Error of less than 1 foot	Error of less than 1 foot
Orbital tolerance influence	Not clear	Not clear	Clear for all orbits	Clear for all orbits
Performance reduction	Sudden	Not known	Critical	Exist after reduction
Clock relativity compensation	Incompatible	Incompatible	Compatible	Compatible
Calendar sub-frame	Incompatibility	1+	1−	1−
Clarity	Not clear	Not clear	The orbit is clear and the perturbation is not clear.	Clear

a) *Estimated features

Table 9.13 Parameters of ephemeris expression [15].

M_0	Mean anomaly calculated according to the reference time
$\triangle n$	Difference of the average motion velocity obtained from the calculated values
E	Orbital eccentricity
$\sqrt{A}$	Square root of the semi-major axis
Ω_0	Ascending node's longitude calculated according to the reference time
i_0	Inclination at the reference time
ω	Angle of perigee
$\dot{\Omega}$	Change rate of right ascension at the ascending node
C_{uc}	Amplitude of cosine harmonic correction term for the argument of ascending node
C_{us}	Amplitude of sine harmonic correction term for the argument of ascending node
C_{rc}	Amplitude of cosine harmonic correction term for the orbit radius
C_{rs}	Amplitude of sine harmonic correction term for the orbit radius
C_{ic}	Amplitude of sine harmonic correction term for the orbital inclination
C_{is}	Amplitude of cosine harmonic correction term for the orbital inclination
t_{oe}	Ephemeris reference time
AODE	Ephemeris data age

In GLONASS, the broadcast ephemeris is directly expressed by the positional parameters and the velocity parameters X, Y, Z, $\dot{X}$, $\dot{Y}$, $\dot{Z}$ in the Earth-fixed coordinates for the satellite. These data are updated 15 min before pre-positioning; the update rate is generally time/0.5 h, and is twice as fast as GPS data update rate. Therefore, in GLONASS published ephemeris values, it contains only the time difference with other systems a and time difference drift rate a_1, without change rate a_2 of drift rate of time difference over time.

The scale factors for ephemeris parameters are shown in Table 9.15.

Table 9.14 Definition of ephemeris expressions.

$\mu = 3.986005 \times 10^{14}\, {}^{m^3}/_{s^2}$	Gravitational constant of the Earth (WGS-84)	
$\dot{\Omega}_e = 7.2921151467 \times 10^{-5}\, {}^{rad}/_{s}$	Rotational velocity of the Earth (WGS-84)	
$A = (\sqrt{A})^2$	Semi-major axis	
$n_0 = \sqrt{\frac{\mu}{A^3}}$	Calculation values of the mean motion	
$t_k = t - t_{oe}$	Time calculated from the beginning of ephemeris	
$n = n_0 + \Delta n$	Mean motion after correction	
$m_k = m_0 + nt_k$	Mean anomaly	
$m_k = E_k - e \sin E_k$	Kepler equation	
$\cos V_k = (\cos E_k - e)/(1 - e \cos E_k)$	True anomaly V_k	
$\sin V_k = \sqrt{1-e^2} \sin E_k / (1 - e \cos E_k)$		
$\phi_k = V_k + \omega$	Argument of ascending node	
$\delta_{U_k} = C_{us} \sin 2\phi_k + C_{uc} \cos 2\phi_k$	Correction for the argument of ascending node	Only consider the second- order harmonic perturbation
$\delta_{r_k} = C_{rc} \cos 2\phi_k + C_{rs} \sin 2\phi_k$	Radius correction	
$\delta_{i_k} = C_{ic} \cos 2\phi_k + C_{is} \sin 2\phi_k$	Inclination correction	
$U_k = \varphi_k + \delta U_k$	Argument of ascending node after correction	
$r_k = A(1 - e \cos E_k) + \delta r_k$	Radius after correction	
$i_k = i_0 + \delta i_k$	Inclination after correction	
$x'_k = r_k \cos U_k$	Satellite position in orbital plane	
$y'_k = r_k \sin U_k$		
$\Omega_k = \Omega_0 + (\dot{\Omega} - \dot{\Omega}_c)t_k - \dot{\Omega}_c t_{oe}$	Longitude of ascending node after correction	
$x_k = x'_k \cos \Omega_k - y'_k \cos i_k \sin \Omega_k$	Satellite position in geocentric earth-fixed coordinate system	
$y_k = x'_k \sin \Omega_k - y'_k \cos i_k \cos \Omega_k$		
$Z_k = y'_k \sin i_k$		

a) *t refers to the system time at the signal transmission time, that is system time after correcting the transmission time (distance/velocity of light).

Table 9.15 Scale factors for ephemeris parameters.

Parameter	Digit a	Scale Factor LSB	Effective Range	Unit
IODE	8	–	–	–
C_{rs}	16	2^{-5}	–	m
Δn	16	2^{-43}	–	Semi-circle
M_0	32	2^{-31}	–	Semi-circle
C_{uc}	16	2^{-29}	–	Radian
E	32	2^{-33}	0.03	–
C_{us}	16	2^{-29}	–	Radian
$(A)^{1/2}$	32	2^{-19}	–	
t_{oe}	16	2^{4}	60 478	s
C_{ic}	16	2^{-29}	–	Radian
Ω_0	32	2^{-31}	–	Semi-circle
C_{is}	16	2^{-29}	–	Radian
i_0	32	2^{-31}	–	Semi-circle
C_{rc}	16	2^{-5}	–	m
ω	32	2^{-31}	–	Semi-circle
$\dot{\Omega}$	24	2^{-43}	–	Semi-circle/s
$\dot{Z}$	14	2^{-43}	–	Semi-circle/s

10

Design of the Satellite Navigation Operation Control System

10.1 Mission and Composition of the Satellite Navigation Operation Control System

The satellite navigation system consists of three parts: the navigation satellite in the space segment, the operation control system in the ground segment, and the user segment. The navigation satellite is the coordinate reference of the user positioning. The operation control system in the ground segment completes the networking and maintenance of satellite coordinates, including satellite injection control, satellite precise orbit determination and constellation maintenance, satellite time synchronization, correction of navigation signal propagation delay, and detection and prediction of navigation signal integrity. The operation control system is generally composed of the master control station, the time synchronization station, the injection station, and the navigation signal monitoring station. The master control station, the center of the operation control system, completes the processing of observation data of satellite navigation signals in constellation, including the calculation of satellite time synchronization, satellite precise orbit and satellite navigation signal integrity, and correction and prediction of signal propagation delay. A high-performance atomic clock system and a system time benchmark are provided to maintain system-wide time synchronization. With high accuracy satellite-ground measurement data, the system coordinate reference is established and maintained. The time synchronization station measures the time synchronization between the satellite and the ground station, thus completing the measurement and prediction of synchronization parameters among satellites. The injection station completes the injection of satellite ephemeris, ephemeris, clock error, control parameters, ionosphere correction parameters, and navigation signal integrity parameters, as well as the user-receiver-related aging parameters. The navigation signal monitoring receives and processes satellite system navigation signals and satellite system remote signals via a multi-band monitoring receiver and the satellite telemetry and telecontrol system on the ground, to obtain the precise orbit determination data and integrity determination data, which are sent to the master control station for comprehensive processing to obtain different injection parameters.

A laser station is set up at the master control station and the time synchronization station in order to further verify and improve the time synchronization accuracy of the system and the satellite orbit determination accuracy; and the satellite-ground time synchronization measurement and the two-way distance measurement are performed by laser. In addition to the laser station for satellite-ground two-way distance measurement,

GNSS Systems and Engineering: The Chinese Beidou Navigation and Position Location Satellite, First Edition. Shusen Tan.

the satellite shall be equipped with laser receiving and ranging equipment, the so-called satellite-ground time synchronization measurement completes on-board laser pseudorange measurement through received laser signals and the satellite clock time scale, thus obtaining satellite-ground time synchronization parameters, and checks the accuracy of wireless two-way time synchronization, with time synchronization accuracy better than 100 ps. The China Compass system with this ability means that the measurement of satellite-ground system time synchronization accuracy can be greatly improved from the current 10 ns.

Based on the means and equipment, GPS, GLONASS are slightly different in principle in terms of the system composition mentioned before. GLONASS calls the basic system of time and frequency the phase monitoring system, and the navigation signal monitoring equipment the monitoring equipment, and is involved in the satellite-ground time synchronization measurement and orbit measurement by using response radar, thus calculating time synchronization and satellite orbit measurement separately. It has more advantages than GPS unified calculations in terms of observation data acquisition, error separation, improvement of calculation accuracy, and reducing the complexity of the calculation system. The China Compass treats the time synchronization and satellite orbit determination, respectively, by using more advanced laser and wireless satellite-ground two-way time comparison data, which not only improves accuracy but also reduces processing difficulty.

10.2 Satellite Time Synchronization and Timing

The purpose of satellite time synchronization is to obtain an accurate satellite clock error parameter, which is the difference between the time scale of each satellite navigation signal parameter and the navigation system reference time. The error is provided to the user by the satellite broadcast parameters, thereby reducing the positioning error due to the pseudorange positioning principle. Timing technology is one that the satellite navigation system provides time synchronization services to the user. The satellite time synchronization of the operation control system is solved by means of the satellite-earth time synchronization. Due to different satellite data and orbit height, all the satellite time synchronization of a navigation constellation cannot be completed by the ground synchronization station, therefore, time synchronization between stations is introduced. We'll discuss each of them separately next.

10.2.1 Method of Satellite-Ground Time Synchronization

10.2.1.1 Method of Satellite-Ground Pseudorange Time

The satellite transmits navigation signals marked by time t_s under the control of high accuracy atomic clock, the ground station receives and restores satellite navigation signals, and then compares them with the local clock on the ground to complete the pseudorange measurements, and the pseudorange is:

$$\rho_g = r_0 + (t_s - t_g)c \tag{10.1}$$

The ground stations transmit uplink ranging signals marked by time t_g under the control of the ground atomic clock, the satellite receives and restores uplink ranging signals, and then compares them with the satellite clock to complete the pseudorange measurements, and the pseudorange is:

$$\rho_s = r_0 - (t_s - t_g)c \tag{10.2}$$

$$\text{Since } t_s - t_g = \Delta t_{sg} \tag{10.3}$$

Which is very small, the relative position between the satellite and the ground and signal transmission path is basically the same. Therefore, there is a possibility that the signal transmission path is unchanged. Then (10.1) and (10.2) are subtracted to obtain the following relationship:

$$\Delta t_{sg} = t_s - t_g \tag{10.4}$$

$$\Delta t_{sg} = \frac{1}{2c}(\rho_g - \rho_s) \tag{10.5}$$

Equation (10.5) is the relationship where the satellite-ground time difference is found through the satellite-ground two-way pseudorange measurement.

The satellite-ground two-way pseudorange time synchronization error includes:

1. Pseudorange measurement error. This is related to the spread-spectrum code rate of signals, with rate, error capacity. When the spread-spectrum code rate is 5 Mcps, the pseudorange measurement error can be up to 0.5 ns.
2. Ionosphere error. This is related to selected downlink navigation frequency f_N, and the uplink transmitted signal frequency f_u when applied to the two-way satellite-ground time synchronization. When the frequency difference of both signals within the L band is less than 70 MHz, the path in the satellite-ground uplink and downlink is basically the same. The ionosphere delay error is less than 0.5 ns.
3. Multipath error. Both the receiving antennas on the satellite and ground can adopt large aperture antennas with strong directivity, and the multi-path can be 0.3 ns in general.
4. Troposphere error. When the uplink and downlink signals travel through the same troposphere, the troposphere delay difference can be negligible. Thus, the satellite-ground two-way pseudorange synchronization error is:

$$\begin{aligned} m_{st} &= \frac{1}{2}\sqrt{m_{rs}^2 + m_{gr}^2} + m_{ion} + m_w \\ &= \frac{1}{2}\sqrt{0.5^2 + 0.5^2} + 0.5 + 0.3 \\ &= 1.05\ ns \end{aligned} \tag{10.6}$$

m_{rs} : Satellite receiver ranging error, 0.5 ns.
m_{gr} : Ground receiver ranging error, 0.5 ns.
m_{ion} : Ionosphere transmission error, 0.5 ns.
m_w : Multipath error, 0.3 ns.

10.2.2 Method of Inter-Station Time Synchronization

The means available for inter-station time synchronization includes: Two-way Satellite Time Transfer (TWSTT), two-way satellite common view, and Geostationary Orbit (GEO) satellite common view.

10.2.2.1 Method of Two-Way Satellite Time Transfer

This method transmits the remote station's ranging signals via the same GEO satellite transponder based on the ground station. The clock error among stations is obtained after two sides measure the pseudorange. The process is as follows: The ground stations A, B send ranging signals to each other at their own clockface time T_A and T_B, respectively, these two signals are received by stations A and B, respectively, at their clockface time T_B' and T_A' after being transmitted by the satellite, thus measuring two pseudorange delays with reference of the local clock. If $T_A \approx T_B$, two pseudorange delays are directly subtracted, which is the clock error between two stations. The key technique is to obtain the pairing of pseudorange measurements, that is, to approximate the time difference between T_A and T_B, and in other words, to obtain the inter-station clock error by repeated iterations.

The error sources include: pseudorange measurement error, satellite transponder and other equipment delay error, ionosphere delay error, and so on. The measurement error can be controlled at 1 ns, the error of forwarding delay and equipment time delay can be ignored, and the ionosphere propagation error is 2.65 ns by using c-band. If the ku band is used, the transponder can be better than 1 ns. Therefore, the accuracy of two-way satellite time transfer can be better than 1 ns. The test results in Beijing and Urumqi on September 1 and September 12, 2004 were 1.32 and 1.81 ns, respectively.

10.2.2.2 Method of Two-Way Satellite Common View

This method is a by-product when the two-way satellite-ground radio method performs satellite-ground time synchronization. As shown by the two-way satellite-ground radio method, when two time-synchronization stations, A and B, on the ground, perform satellite-ground time synchronization measurements on one satellite at the same time, two clock errors between the satellite and station A and between the satellite and station B are obtained, and subtracting the two satellite clock errors directly is the clock difference between the two stations. Of course, the inter-station and inter-satellite phase synchronization and pseudorange matching application will be considered in actual operation. According to the similar error analysis previously, time synchronization precision of 2.0 ns is also obtained.

10.2.2.3 Method of Two-Way GEO Satellite Common View

This method completes time synchronization between stations based on the method of Radio Determination Satellite Service (RDSS) two-way time alignment. The principle of RDSS two-way time alignment is that the central control system sends inquiry signals to the time station A (the user) required to be synchronized via the GEO satellite transponder, the time station A receives the inquiry signals and transmits the response signals, and then the central control system receives these signals and performs the time difference measurement to station A.

Finally, the time difference between the central station and the user station A is calculated, and the clock differences of the different user stations are derived from the clock errors of each station and the central station. The precision of two-way time alignment is better than 20 ns when the measurement error of China Compass-1 is 12.5 ns. When the measurement error is further improved, it is estimated that the GEO satellite two-way common view method will achieve a synchronization accuracy of 10 ns.

10.2.3 User Timing Service

The satellite navigation system provides timing services to the user in two ways: two-way and one-way.

The so-called two-way timing servicc means that a synchronization accuracy of 10 ns can be obtained through the satellite two-way time alignment method in the RDSS function. The so-called one-way timing service method is that the user receives signals of a navigation satellite to calculate the time difference between the local clock and the satellite clock by measuring pseudorange and on the premises of obtaining satellite clock error, orbit parameters, ionosphere propagation delay parameters, and observation station position, and to obtain an accuracy of timing services of better than 50 ns. The one-way timing service method is the basic method for the satellite navigation service to provide time synchronization services, and it's difficult to use in the time synchronization of motion carriers from the principle, while the two-way timing service can meet the needs of timing services for motion carriers.

10.2.3.1 Prediction Model for the Satellite Clock Error

This prediction model is established based on the following two reasons: Firstly, the calculation of integrated atomic time is non-continuous and non-real-time, which is conducted once only when the atomic clock runs for a fixed time interval (1 day or 1 month); secondly, the comparison between the satellite clock error is not continuous, real-time, because the ground time synchronization station is unable to trace the satellite in time to complete the time synchronization measurement. Then its synchronization error must be maintained by the atomic clock during the period other than measurement periods. The error of the satellite clock varies with the clock speed and clock drift. Therefore, the satellite clocks must be modeled with satellite clock speed and clock drift as reference in order to enable users to get accurate satellite clock errors. From these reasons, the satellite clock model must be established based on the performance of satellite atomic clock, synchronization comparison method, and parameter values. The basic methods to calculate the clock error include the polynomial fitting method, Kalman filtering method, AR model method of time series analysis, and so on. If the second-order polynomial is used for least squares prediction, the time prediction model can be generally expressed as

$$t(t) = a_0(t_0) + a_1(t_0)(t - t_0) + a_2(t_o)(t - t_0) + \Delta \tag{10.7}$$

Where a_0 is the clock error at time t_0; a_1 is the clock speed at time t_0. a_2 is the clock drift at time t_0, and $t-t_0$ is the clock error caused by the degree of uncertainty.

Then the prediction error of satellite clock error can be expressed as

$$M_t^2 = m^2{}_0 + m^2{}_1(t - t_0)^2 + m^2{}_2(t - t_0)^4 + \sigma_y^2(\tau)(t - t_0)^2 \tag{10.8}$$

Where m_0 is the error of a_0, m_1 is the error of a_1, m_2 is the error of a_2, and σ_y (z) is the frequency stability.

When the satellite clock indicators are determined, the predication accuracy of clock errors within different predication time intervals can be calculated according to formulas (10.7) and (10.8). For example:

The indicator of radio frequency port of satellite atomic clock is selected as:

Accuracy 5×10^{-10}

Drift rate	1×10^{-13}/day
Stability	1×10^{-11}/day
	3×10^{-12}/10 s
	1×10^{-12}/100 ns
	3×10^{-13}/10,000 s

It can be calculated that the error caused by stability in 10 h is 10.8 ns, and the error caused by drift rate is 1.5 ns. The total error is better than 12 ns.

10.3 Correction of Navigation Signal Spatial Propagation Delay

This includes the two parts of troposphere and ionosphere propagation delay. The correction of troposphere propagation delay has been discussed in the satellite positioning project. We will mainly discuss the correction of ionosphere propagation delay here.

When the satellite radio signals pass through the ionosphere, due to different electron content in the propagation path, the different refractive index produces a different degree of dispersion, the signal propagation path varies in degree, resulting in changes in the propagation path delay, affecting the pseudorange measurement accuracy, which will cause errors in the satellite orbit determination, time synchronization, user positioning, and speed measurement. Therefore, the accurate correction of ionosphere propagation delay is one of the key technologies in satellite radio navigation.

Ionosphere propagation refractive index is:

$$\mathrm{NP} = \left(1 - \frac{N_e e_t^2}{4\pi^2 f^2 \varepsilon_0 m_e}\right)^{1/2} \tag{10.9}$$

Where: f is the operating frequency of radio signals; N_e is the electron density (number of electron/m^3); e_t is the amount of electron charge, $e_t = 1.6021 \times 10^{-19}$; ε_0 is vacuum dielectric constant, $\varepsilon_0 = 8.85 \times 10^{-12}$; m_e is electron mass (kg), $m_e = 9.11 \times 10^{-31}$.

Expand (10.9), and omit quadratic small terms, then

$$\mathrm{Np} = 1 - 40.28\frac{N_e}{f^2} \tag{10.10}$$

The additional distance of the propagation path due to the refractive index, when the radio signal passes through the ionosphere

$$\begin{aligned} \sigma_\rho &= \int^s (\mathrm{Np} - 1)ds \\ &= -\,40.28\,\frac{1}{f^2}\int^s \mathrm{Neds} \end{aligned}$$

Set N_Σ as the total number of electrons on the propagation path of the radio waves, and

$$\mathrm{N}_\Sigma = \int^s \mathrm{Neds}$$

Then

$$\sigma_\rho = -\,40.28\,\frac{N_\varepsilon}{f^2} \tag{10.11}$$

Define σ_ρ as

$$\sigma_\rho = \rho - \rho_0 \tag{10.12}$$

ρ is measured pseudorange
ρ_0 is real space distance

According to formula (10.12), if dual band f_1, f_2 are used for pseudorange measurement, then

$$\begin{aligned} \rho_{f1} &= \rho_0 + \delta\ \rho_{f_1} \\ \rho_{f2} &= \rho_0 + \delta\ \rho_{f2} \\ \delta\rho_{f_1} &= -40.28\frac{N_\Sigma}{f_1^2} \\ \delta\rho_{f2} &= -40.28\frac{N_\Sigma}{f_2^2} \end{aligned} \tag{10.13}$$

From these two formulas, there is

$$\delta\rho_{fz} = \delta\rho_{f1}\frac{f_1^2}{f_2^2}$$

For the observation, there is

$$\rho_{f1} = \rho_0 + \delta\rho_{f_1}$$
$$\sigma_{f_2} = \rho_0 + \delta\rho_{f_2}$$

These two formulas are subtracted, so

$$\begin{aligned} \sigma_\rho &= \rho_{f_1} - \rho_{f_2} = \delta\rho_{f_1} - \delta\rho_{f_2} \\ &= \delta\rho_{f_1}\left(1 - \frac{f_1^2}{f_2^2}\right) \\ &= \delta\rho_{f_1}\left(\frac{f_2^2 - f_1^2}{f^2{}_2}\right) \end{aligned}$$

Namely, $\delta\rho_{f_1} = \sigma_\rho \frac{f^2{}_2}{f_2^2 - f_1^2}$

Substitute the amount of correction into the observation equation, the true value of the observation can be obtained

$$\rho_0 = \rho_{f_1} - \sigma\rho\frac{f^2{}_2}{f_2^2 - f_1^2} \tag{10.14}$$

Formula (10.14) is the basic model of the dual band ionosphere correction. The ranging accuracy, namely, the minimum interval between two operating frequencies can be designed according to the ionosphere correction residual. When f_1 = 1.56 GHz, f_2 = 1.21 GHz, the pseudorange measurement accuracy is 1 ns, the dual band ionosphere correction residual is 2.49 ns. When the frequencies of 1.27 GHz and 1.21 GHz are used, the dual band ionosphere correction residual is 10.6 ns. The greater the frequency difference interval is, the higher the ionosphere correction accuracy is. Therefore, the frequency selection of GPS, GLONASS, Galileo, and China's satellite navigation positioning compete for the two operating frequencies between 1559~1610 MHz and 1215~1260 MHz. While GPS takes advantage of a strong domestic frequency coordination mechanism to seize 1176.45 MHz as L_5 navigation frequency, and passes through the 1176.45 MHz international coordination by using the demands of China and EU on the new operating frequency. Cooperated with the 1575.42 MHz operating frequency, the ionosphere correction accuracy of 2.26 ns was obtained by the pseudorange measurement accuracy of 1 ns.

The high accuracy navigation receivers tend to use a dual band working mechanism that allows ionosphere correction accuracy to achieve the desired results. In order to reduce the cost of users, the single band work is often needed, and then how to improve the ionosphere correction accuracy? There are two methods to use. One is to establish a single band ionosphere correction model; and the other is to broadcast ionosphere correction parameters based on regional grid for the user. In wide area differential systems, it is exactly the ionosphere grid form that is used for real-time broadcasting of ionosphere correction parameters.

10.3.1 Ionosphere Model Correction for Single Band Users

GPS provides the ionosphere correction model for single band users. This method can correct 75% of the ionosphere path delay. According to formula (10.13), when the total electron content N_Σ is $1 \times 10^{18}/m^2$, the additional ionosphere delay by using frequency $f_1 = 1560 \times 10^6$ Hz is:

$$\delta\rho_{f_1} = -40.28\frac{1 \times 10^{18}}{(1560 \times 10^6)^2}(m) = 16.55(\mathrm{m})$$

75% is corrected, implying that the ionosphere delay residual corrected by the model is:

$$16.55 \times (1 - 75\%) = 4.13(\mathrm{m}).$$

GPS describes the ionosphere delay parameter corrections in different observation directions (azimuth and elevation) by using eight broadcast parameters: $\alpha_0, \alpha_1, \alpha_2, \alpha_3$ and $\beta_0, \beta_1, \beta_2, \beta_3$. They are the cubic coefficients of the vertical time delay amplitude value and the cubic coefficients of the model period respectively, as defined in Table 10.1.

The algorithm needs to calculate the elevation angle E and azimuth A of the GPS satellite, the observation time t according to the user's approximate geographic latitude φ_u and longitude λ_u, and combined with α_n, β_n parameters, to calculate the ionosphere delay correction parameters. The unit for the angle measurement is semicircle, the time is second, and the steps are as follows:

1. Calculate the geocentric angle ψ

$$\psi = 0.0137/(\mathrm{E} + 0.11) - 0.022 \text{ (Semicircle)}$$

2. Calculate sub-ionosphere latitude ϕ_I

$$\varphi_I = \varphi_u + \psi \cos A \text{ (Semicircle)}$$

If $\varphi_I > +0.416, \quad \varphi_I = +0.416$; and if $\quad \varphi_I < -0.416, \quad \varphi_I = -0.416$

Table 10.1 GPS ionosphere parameters [14].

Parameters	Bits	Scale Factors	Effective Range	Unit
α_0	8*	2^{-30}		Seconds
α_1	8*	2^{-27}		Sec/semicircle
α_2	8*	2^{-24}		Sec/(semicircle)^2
α_3	8*	2^{-24}		Sec/(semicircle)^3
β_0	8*	2^{11}		Seconds
β_1	8*	2^{14}		Sec/semicircle
β_2	8*	2^{16}		Sec/(semicircle)^2
β_3	8*	2^{16}		Sec/(semicircle)^3

Note: Parameters marked with * match up with Figure 10.2, and the most significant bit (MSB) is the sign bit (+ or –).

3. Calculate sub-ionosphere longitudeλ_I

$$\lambda_I = \lambda_u = \psi \sin A / \cos \varphi_I \text{ (Semicircle)}$$

4. Determine the geomagnetic latitude of the sub-ionosphere position when directing at each observation hour of GPS satellite

$$\phi_m = \phi_I + 0.064 \cos(\lambda_1 - 1.617) \text{ (Semicircle)}$$

5. Determine the local time on the sub-ionosphere

$$\mathrm{t} = 4.32 \times 10^4 \lambda^1 + GPST\,(\mathrm{s})$$

If t < 0, t = t + 86 400; and if t > 86 400, t = t – 86 400,

6. Convert to the oblique time delay, and calculate the tilt factor F

$$\mathrm{F} = 1.0 + 16.0 \times (0.53 - \mathrm{E})^3$$

7. Calculate x, and calculate the ionosphere time delay

$$X = \frac{2\pi(t - 50\,400)}{\sum_{n=0}^{3} \beta_n \phi_m{}^n}$$

If $|x| > 1.57$,

$$\mathrm{T_{ion}} = \mathrm{F} \times [5 \times 10^{-9}]$$

Or else

$$\mathrm{T_{ion}} = \mathrm{F} \times \left[5 \times 10^{-9} + \sum_{n=0}^{3} \chi_n \varphi_m{}^n \times \left(1 - \frac{x^2}{2} + \frac{x^4}{24}\right)\right]$$

10.3.1.1 Ionosphere Correction in Regional Grid

The so-called grid is an ionosphere correction map 350 km above the ground. Each electrical IGP on the grid corresponds to the vertical ionosphere delay at that point; the user receiver obtains the ionosphere vertical delay of the IPP of the ionosphere puncture site by calculating these delays, thus obtaining the pseudorange delay correction t_{ion} in the propagation direction. The size of the grid cell is determined by the degree of ionosphere correlation, the number of reference stations to obtain grid parameters, the spacing, the size of the coverage area, and the message capacity assignable by the GEO satellite ionosphere broadcasts. It is usually 5 × 5 ° in the middle and low latitudes, 10 × 10° in the middle and high latitudes (55~75°), and 15 × 15° in the high latitudes (75° or more), which can meet the above requirements. The relationship between ionosphere correlation properties and actual distance is as shown in Table 10.2.

In the standard correlation, the east-west correlation distance is about twice the north-south correlation distance. Its correlation with the latitude is weak, the north-south interval is 1800 km, and east-west distance is 3000 km. The appropriate resolution with a spacing of 400~500 km is selected in the grid point design. In order to facilitate the broadcast in the satellite message, the grid point broadcasting adopts the unified number of north and south, from west to east, as shown in Table 10.3.

Table 10.2 Basic relationship between ionosphere correlation properties and distance.

Correlation distance	100~400 km	400~1500 km	1500~3000 km	3000~5000 km	5000~10,000 km
Correlation coefficient	0.95	0.87	0.70	0.50	0.30
Correlation	Very strong	Strong	Standard correlation	Weak	Loose

Table 10.3 Grid point 1GP number demonstration.

B/L	70	75	80	85	90	95	100	105	110	115	120	125	130	135	140	145
55	10	20	30	40	50	60	70	80	90	100	110	120	130	140	150	160
50	9	19	29	39	49	59	69	79	89	99	109	119	129	139	149	159
45	8	18	28	38	48	58	68	78	88	98	108	118	128	138	148	158
40	7	17	27	37	47	57	67	77	87	97	107	117	127	137	147	157
35	6	16	26	36	46	56	66	76	86	96	106	116	126	136	146	156
30	5	15	25	35	45	55	65	75	85	95	105	115	125	135	145	155
25	4	14	24	34	44	54	64	74	84	94	104	114	124	134	144	154
20	3	13	23	33	43	53	63	73	83	93	103	113	123	133	143	153
15	2	12	22	32	42	52	62	72	82	92	102	112	122	132	142	152
10	1	11	21	31	41	51	61	71	81	91	101	111	121	131	141	151

Geomagnetic latitude $B^{m}{}_{IGP}$ and local time t_{IGP}

$$B_{IGP} = 5 + CNUM_{IGP} - int[(NUM_{IGP} - 1)/10 \times 10] \times 5$$

$$L_{IGP} = 70 + int[(NUM_{IPG} - 1)/10 \times 5]$$

$$B^{m}{}_{IGP} = B_{IGP} + 11.6^{0} \cos(L_{IGP} - 291^{0})$$

$$t_{IGP} = 3600 \times (L_{IGP}/15 + UTC)$$

Latitude B_{IGP} and Longitude L_{IGP} on any grid point are calculated by grid point number NUMIGP.

The basic steps to collect grid ionosphere delay data are as follows:

1. Calculate the carrier phase smoothing pseudorange in the monitoring station.
2. Calculate the ionosphere time delay in multiple monitoring stations and the master control station.
3. Calculate the projection function.
4. Calculate the ionosphere zenith delay at pierce points.
5. Calculate the coordinates at pierce points.
6. Calculate the local time at pierce points.
7. Calculate the ionosphere delay at pierce points.
8. Calculate the ionosphere delay at grid points.
9. Fit the ionosphere zenith delay at grid points.
10. Calculate the zenith delay error: GIVE at grid points.
11. User zenith delay UIVE.
12. User ranging path error URVE.

10.4 Determination of Precise Orbit and Satellite Clock Error

Modern satellite navigation systems adopt the medium earth orbit (MEO) satellite, geostationary orbit (GEO) satellite, and inclined geo-synchronization orbit (IGSO) satellite to constitute navigation satellite constellations. The degrees of difficulty of the satellite's precise orbit determination are different, so there are different considerations for orbit determination parameters, schemes of orbit determination station, orbit determination models, and other aspects. For a global positioning system composed of MEO satellites, generally, precise orbit determination is conducted by adopting the scheme of global station distribution, pseudo-range, and carrier phase as measurement parameters, and through the long-periodic observation fitting, such as GPS and Galileo. GLONASS does not belong to the global distribution station, but the scope of its observation station also crosses the range of 180° longitude. China's satellite navigation system mainly depends on domestic stations for orbit determination, so there are great difficulties. To realize the precise orbit determination for MEO satellites, development of satellite-to-satellite links is also required to achieve satellite-to-satellite precise pseudorange measurement and time synchronization.

Orbit determination mainly adopts a kinetic method, the solution of the satellite motion equation adopts a numerical method, and state estimation adopts a batch processing method and sequential method. The satisfied orbit determination accuracy can be obtained through periodic fitting.

For the orbital maneuver happening during the process of satellite motion, such as the maneuver of longitude direction conducted by a GEO satellite about every 2 weeks, the maneuver of latitude direction conducted each month, because mechanomotive force is not easily accurately measured and modeled, the motion track within a certain time quantum after the maneuver has a larger deviation, the transition process will be 24~72 h, and thus, in this time, quantum the effect of satellite orbit on positioning accuracy only can be weakened with the method of orbital modulation. The influencing

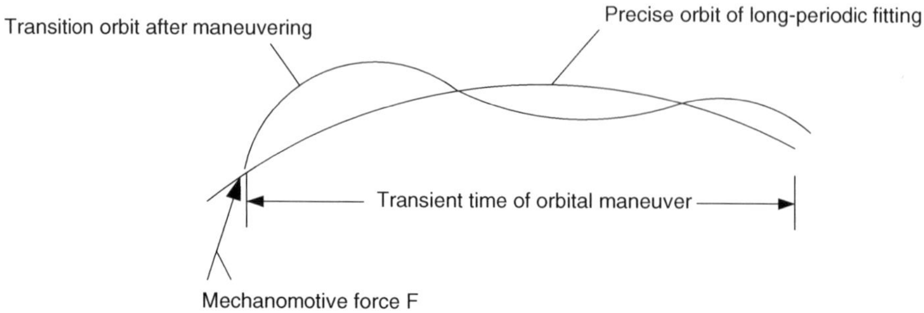

Figure 10.1 Transient process of orbital maneuver.

process of orbital maneuver on precise orbit determination is shown in Figure 10.1, and tends to be stable under new orbital parameters after maneuvering adjustment.

10.4.1 Correction of Satellite Orbit

Precise orbit determination can overcome the effect of satellite maneuver through the satellite orbit correction, which is the basic means adopted by wide area differential positioning, it has more obvious advantages for the system that adopts the satellite-ground two-way radio time synchronization scheme, and can separate the satellite clock error from the satellite orbit clock error. Through broadcasting satellite positions and satellite velocity correction parameters (x, y, z and $\dot{x}, \dot{y}, \dot{z}$) to users, the purpose of providing precise satellite positions can be achieved.

When the coordinate of observation station is known, pseudorange observation data (ρ) is the function of the coordinates of the observed satellites (x, y, z), the satellite-ground clock error (Δt_s) and time t.

$$\rho = \rho = (x, y, z, \Delta t_s, t) \tag{10.15}$$

Δt_s is the satellite-ground clock error, and the time synchronization station together with satellites uses the satellite-ground two-way radio method to complete the time difference of satellites and ground stations.

Pseudorange measurement error caused by synchronous satellite position error is

$$\Delta\rho = \frac{\partial\rho}{\partial x}\Delta x + \frac{\partial\rho}{\partial y}\Delta y + \frac{\partial\rho}{\partial z}\Delta z \tag{10.16}$$

$$\Delta\rho = e_x\Delta x + e_y\Delta y + e_z\Delta z \tag{10.17}$$

In the formula, e_x, e_y, and e_z, respectively, are cosines of three coordinate axes at the satellite observation direction. It is observed that, if the positions of receivers are different, the pseudo-range errors caused by satellite positions will be different, and change along with the satellite observation direction. So, it is impossible to use the pseudo-range correction, and the adoption of the satellite position correction value (δx_i) and the velocity correction (δx_j) is appropriate, among which $j = 1, 2, 3$ corresponding to three coordinate axes.

The fundamental principle and process of calculating δx_i and δx_j are as follows:

1. Obtain the satellite pseudorange (and carrier phase) observation data ($\rho^{(i)}$) during the satellite control by orbit measurement observation station, i is the number of orbit measurement observation station, and $i = 3$.
2. The time synchronization station uses the satellite-ground two-way pseudo-range (or laser) method to obtain the satellite-ground clock error (Δt_s). Because of the orbit measurement station and the time synchronization station, the clock errors (Δt_s) are of the same digit parameter.
3. Adopt the inverted positioning method to calculate the satellite position $x_s(t)$ at moment t

$$\rho^{(i)}(t) = \left[\sum_{j=1}^{3} (x_j^i + x_{sj}(t))^2\right]^{1/2} + \Delta t_s \tag{10.18}$$

4. Work out the satellite position $X_{spj}(t)$ predicted at moment t according to the orbit prediction established by satellite maneuver.
5. Calculate the correction value of satellite position at moment t

$$\delta x_i = x_{s_i}(t) - x_{sp_i}(t) \tag{10.19}$$

6. With the similar method, work out the velocity correction value

$$\delta \dot{x}_i = \dot{x}_{s_i}(t) - \dot{x}_{sp_i}(t) \tag{10.20}$$

10.4.2 Accuracy Estimation

With the examples from setting up stations in China's Kashgar, Urumchi, Chengdu, Sanya, Beijing, Xiamen, and Harbin, orbit corrections for 80°E, 110.5°E, and 160°E are made to conduct accuracy estimation.

Assume that the pseudorange observation error $m_\rho = 0.3\ m$, the satellite-ground time synchronization error $m_{st} = 0.6$ m, the dual-band ionosphere correction residual error $m_{ion} = 0.75$ m. PDOPs of 80°E, 110.5°E, and 160°E satellites are, respectively, 22, 21, and 57.4.

Equivalent distance error $m_R = (m^2{}_\rho + m^2{}_{st} + m^2{}_{ion})^{1/2} = 1.25$ (m)

The position errors are, respectively:

80°E satellite $\delta x = 1.25 \times 22 = 27.5$ m
110.5°E satellite $\delta x = 1.25 \times 21 = 26.25\ m$
160°E satellite $\delta x = 1.25 \times 57.4 = 71.75\ m$

Compared with no orbit correction, the accuracy is obviously improved. Compared with the long-periodic accuracy prediction of 10 h, the accuracy worsens 2.5-fold. The accuracy usage can be reduced to improve availability of the GEO satellite. Certainly, in order to reduce the effect of orbit accuracy deterioration on the system, $\delta x, \delta \dot{x}$ update rates need to be considered. The higher the update rate is, the higher the accuracy that a user can obtain. The design of update rate is not only limited by the system working capacity, such as sampling of observation data, operation of master control station, information update rate, and so forth, but also restricted by the format of satellite broadcast signals. The replacement cycle of $\delta x, \delta \dot{x}$ will be less than 1 s as far as possible.

10.5 Integrity Monitoring and Prediction

Integrity of the satellite navigation system consists of three parts, which are satellite autonomous integrity monitoring, receiver autonomous integrity monitoring (RAIM), and ground integrity channel (GIC). The ground integrity channel is the integrity monitoring embodied in ground operation control system or application system. The focus of this discussion is the missions and implementation of the ground integrity monitoring. And RAIM will be discussed in the user receiver part.

The concept of ground integrity channel (GIC) originates in the scheme that US Federal Aviation Administration provided to the US domestic aviation for non-precision operation in the 1980s. The scheme at that time was to provide integrity information broadcasting to aviation users through four monitoring stations and the geostationary orbit of two segments with L-band communication transponders established in Contiguous US, for the following satellite integrity broadcasting, it has the same working frequency and navigation signal frequency, and it arranges the integrity information uniformly into the navigation messages, such as the cases of WAAS and EGNOS.

This integrity function is included into the operation control system as part of regional mission segment. The implementation of this part depends on the observation of the pseudorange error by the IMS network and is completely independent from the orbit determination process. A real pseudorange error can be observed by an integrity monitoring station marked accurately by position, so it can be compared with the accuracy of orbit prediction; if the observed error does not match the accuracy requirements of the forecast, there may be something wrong in the satellite navigation signals. While error in the space navigation signal consists of satellite clock, orbit deviation, propagation error, and satellite payload distortion, it is hard to separate which causes this error. Providing an effective separation method is not only beneficial to the improvement of positioning accuracy, but also beneficial to the determination of satellite signal integrity.

10.5.1 Satellite-Ground Two-Way Pseudorange Time Synchronization Separated Satellite Integrity

The two-way pseudorange time synchronization is the only method that can separate the clock error and orbit error. According to the principle of two-way pseudorange determination of satellite clock and ground clock, when the accuracy of satellite-ground coarse synchronization meets the satellite-ground relative motion, the satellite-ground clock error can be derived from the formula (10.6):

$$\Delta t_{sg} = \rho_s - \rho_g$$

As being separated from satellite orbit parameters, predicted clock error $\Delta t(t)$ shall be calculated according to formula (10.7) and the model to determine the integrity of the satellite clock according to difference of $\Delta t_{sg} - \Delta t(t)$.

$$\Delta T_{Itg} = \Delta t_{sg} - \Delta t(t) \tag{10.21}$$

It can alarm the satellite clock error when ΔT_{Itg} is beyond the satellite clock prediction accuracy.

10.5.2 Integrity of DLL Related Monitoring Satellite Payload

Non-coherent delay locked loop DLL is used in the satellite navigation system to complete the recovery of satellite signals so as to conduct pseudorange measurement (see Chapter 12 for more details). When the bandwidth of signals is infinite, its normalized pseudorange measurement variance with chip width is

$$\delta^2{}_{DLL} = \frac{B_L d}{C/N_0}\left[1 + \frac{1}{(C/N_0)T(1-d)}\right] \tag{10.22}$$

δ_{DLL} multiplies the width of chip is the standard deviation of pseudorange measurement expressed with time (s), and the result multiplies the speed of light C is the standard value of pseudorange measurements expressed in distance.

B_L is the loop bandwidth, in HZ.
d is the relevant interval after normalization.
T is the accumulated time of correlation processing.
c/N_0 is the carrier noise ratio.

The navigation signal cannot be infinite bandwidth in practical engineering. The crowded navigation frequency can only choose the bandwidth twice of the code rate. While J. W. Betz made some modifications, when the band-pass filter modeling for the receiving channel is a rectangular filter, the approximation formula for variance with only thermal noise is

$$\delta^2{}_{nDLL} = \begin{cases} \dfrac{B_L(1-0.5B_L T)}{2(c/N_0)} \times 2d\left[1 + \dfrac{1}{T(c/N_0)(1-d)}\right] & bd \geq \dfrac{\pi}{2} \\ \dfrac{B_L(1-0.5B_L T)}{2(c/N_0)}\left[\dfrac{1}{b} + \dfrac{b}{\pi - 1}\left(2d - \dfrac{1}{b}\right)^2\right]\left[1 + \dfrac{1}{T(c/N_0)(1-d)}\right] & \dfrac{1}{2} < bd \leq \dfrac{\pi}{2} \\ \dfrac{B_L(1-0.5B_L T)}{2(c/N_0)}(b)\left[1 + \dfrac{1}{T(c/N_0)}\right] & 0 < bd \leq \dfrac{1}{2} \end{cases} \tag{10.23}$$

b is the signal bandwidth with chip width after normalization. Refer to Chapter 12 for more details.

It can be seen that, with different b and d, there are different pseudorange measurement variances. so integrity monitoring can be conducted towards satellite signal distortion to determine the integrity of effective payload.

10.5.2.1 How to Conduct the Distortion Determination of Satellite Signals

Two kinds of correlators are set in the monitoring receiver, one is a wideband correlator where d = 1/2, and the other is a narrow correlator, taking d ≦ 1/16 as an example. For normal receiving signals without distortion, although the relevant characteristics are not standard triangles, the left-right benchmarking is better. The maximum point of correlation peak appears when the receiving code aligns with the local instant code. See Figure 10.2. After linear approximation processing toward relevant results of several narrow correlators, the acquired local instant code coincides with the result of wideband correlation. For abnormal navigation signals with distortion, the results after wide and narrow processing will not coincide and the deviation ΔTc cannot be produced, so whether or not distortion occurs within the navigation signals can be determined.

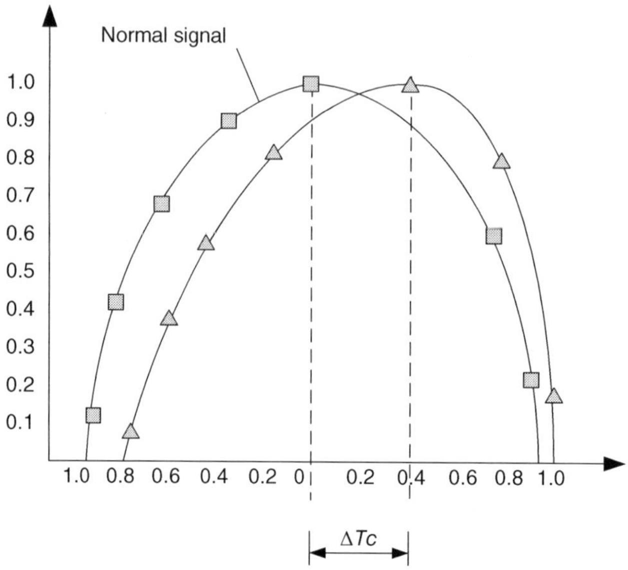

Figure 10.2 Schematic diagram of correlated characteristics.

The F(d) in Figure 10.2 is a normalized correlated value, and d is the normalized correlated interval. It can be seen from Figure 10.2 that this δ_{DLL} does not represent the variance produced by thermal noise.

The distortion of received signals can be caused by the following reasons:

1. The navigation signal launched by the satellite suffers distortion and causes migration of correlative peak, or multiple correlative peaks occur.
2. It is affected by coherent spread spectrum pseudo-satellite signals and causes the correlative curve to be misaligned, for example, affected by multipath interference, it shows deterioration in the leading edge of code waveform.
3. It is affected by RF filter of the receiver and causes distortion in signals. An output filter fault can occur in satellite system and this can be classed as a (1) reason.
4. Affected by strong noise, it makes the correlative peak flat.

Because of the cost, general user receivers can only use the wide correlator, accuracy can be guaranteed when there is no distortion in the signal. When there is distortion within the signal, it cannot be detected. And that is why the monitoring receiver still needs to adopt multiple narrow correlators. The integrity monitoring of satellite signals will be completed by special integrity monitoring stations and central processing systems.

It can be seen from this analysis that the monitoring receiver with wide-narrow correlation processing can complete the integrity determination of signals received. Monitoring stations that have three monitoring receivers working simultaneously can complete the integrity determination of navigation signals under this environmental condition. Excluding the fault reasons of receiver and the multipath effect, the joint determination of multiple monitoring stations can finally determine the integrity of navigation signals, which requires well-set monitoring station network and integrity processing center, and the integrity processing center shall determine the integrity of navigation signals through the observed range-measuring error. Certainly, integrity processing center must conduct determination towards the ionosphere integrity, and this determination can monitor the ΔT_c of three monitoring receivers at the monitoring station and separate the effect of the ionized layer.

10.6 Integration of Operation Control System

The integration of operation control system has two meanings: one is the integration within the system and the other is the integration of multiple navigation systems. Integration within the system means a series of integration of functions and indexes of the same system such as precise orbit determination and orbit adjustment in orbit control period, time synchronization between satellite-ground to satellite-to-satellite and between stations, integrity monitoring, integrity calculation, and so on. It constitutes the whole RNSS service by the published navigation message and signal format.

Integration of multiple navigation systems, or integration beyond the system, is the integration to complete navigation with timing, communication, and the integration of the home satellite navigation system with a navigation system abroad. The second integration will be discussed next.

10.6.1 Combination of RNSS and RDSS to Realize Integration of Three Functions of Navigation, Communication, and Identification

A successful satellite navigation system must realize the high integration of three functions; that is, navigation, communication, and identification. From the satellite system, operation, and control systems to the final and comprehensive realization of integration for the user, it makes the user and the user scheduling commander master the six jurisdictional and scattered questions of Who, When, Where, Why, What, and How? RNSS function can solve the urgent problems as when and where, the identification and communication functions of RNSS can not only answer the commander who the user is, its communication and report functions can also answer questions such as the navigation destination (Where), the mission to be performed (What) and effects of mission (How), and so on.

10.6.1.1 Basic Principles and Methods of Integration

GPS system users intend to realize navigation communication (NAVCOM) integration through the combination of user terminal and the military internet. The Compass navigation system has realized the integration of positioning, timing, and communication under the concept of RDSS. Although Galileo summaries the short points of GPS and GLONASS and formulates the rescue service plan, it does not really solve the problem of prefect integration of navigation, communication, and identification. This target can be better achieved through integration of RNSS and RDSS.

The so-called integration of RNSS and RDSS aims to integrate the business of RNSS and RDSS in the satellite of navigation systems and operation control systems, so that the user can realize the continuous positioning and speed measuring missions autonomously without launching response signals and can conduct the RDSS based position reporting as well as user tracking identification and short message communication as required, so as to realize the dual-mode integration of RNSS and RDSS and application integration of foreign GPS at the user's terminal. Its basic principles and methods are as follows:

1. Arrange RDSS payload and RNSS payload at part of navigation satellites (GEO, IGSO) simultaneously, and the ground control system has RNSS and RDSS signals and signals processing and running abilities.
2. Navigation system and signal format of RNSS and RDSS are unified in the same time system.
3. System GEO satellite RDSS outbound signal and RNSS navigation signal can be both used for user autonomous navigation, and can be used in position reporting and communication service too. That is, the S band signal can be used for RNSS (so-called passive) positioning, and the L band signal of RNSS can be used for communication service.
4. The ground operation and control system RDSS business has the ability to randomly access the user's communication, which can handle short and burst signals and finish the information exchange from the user to the central control system.
5. The user inbound information can bring the user's position to realize position reporting, and can do without the position message. From the RDSS principle, it can figure out the user's position coordinates directly from the answering signal and realize the position reporting "without information" transmitted.

10.6.1.2 Performance Features of RNSS and RDSS Integration

Compared with the integration of RNSS plus communication system (NAVCOM), integration of this dual-navigation system has the following characteristics:

1. Completing the integration of navigation and communication within the navigation system, and this can both enhance the navigation capabilities and avoid the communication distress brought by different communication systems and departmental establishments, with good interoperability.
2. The combination of position reporting link and navigation link provides information transmission link with the safety of GPS P(Y). The user position reporting without position message transmitted can be realized by RDSS principles, with good security.
3. All users within the coverage of satellite beam have random access ability, with the advantages of wide direct communication range, strong communication ability and good real-time performance.
4. The dual-mode receiver has rich positioning methods and high reliability.
5. The benefit-cost of system and the user terminal is high, with relatively low cost.
6. When RDSS is used for realization of two-way timing, its accuracy can be better than 10 ns, and when it is compared with normal GPS one-way timing, its time synchronization accuracy can be improved by 5–10 times.

10.6.2 Realization of Integration with a Foreign System by Multiple-System Information Fusion

Currently, there are two global positioning systems (GPS, GLONASS) that provide civilian service; Galileo in construction also takes the global civilian service as its target, so there may be around three to four global positioning systems in the future. The expediency combination and integrity monitoring reports of these navigation systems constitute the Global Navigation Satellite System (GNSS) expected by the International Civil Aviation Organization.

At present, some of these systems have become quite mature. For example, GPS has reached a 10-m positioning accuracy. But as far as the comprehensive evaluation of accuracy, availability, continuity, and integrity is concerned, GPS neither meets the requirements needed for Class I precision approach, nor does it meet the navigation requirement for a non-precision approach. As can be seen from the current conditions, it is difficult for any independent system to meet the requirements of a global navigation service. The fact is that it is not practical for the International Civil Aviation Organization to build a civilian global satellite navigation system. So, the practical choice is to realize multi-system compatible services to enhance overall navigation performance. Even if there are only two global systems available, there will be 12–22 satellites available within the user's visible range. If there are three systems with the ability to provide service, it can deal with unexpected events when one system is closed and ensure long-term selectivity of the dual system. For the 12–22 selectable satellites, the accuracy of the geometric factor is high, $PDOP \leq 2.0$, thus reducing the UERE requirements for a user's equivalent distance error and meeting the accuracy requirements of the Class I precision approach. As the number of optional satellites increases, the reliability and continuity are also enhanced. It aims to meet all the needs of the Class I precision approach directly through the monitoring of satellite integrity. So, what is the work of system information fusion?

1. *Setting up multi-system comprehensive monitoring station.*
   Measurement and determination toward multi-system navigation signals can be realized on the basis of system monitoring station design by the joint work and compatible designs of GPS monitoring receiver, GLONASS monitoring receiver and Galileo monitoring receiver; it may be an appropriate option to design a compatible monitoring receiver as the compatibility of GPS, Galileo and Compass in system makes it possible.
2. *Realizing the precise orbit determination and modification of system satellite.*
   Use the observational data of the system comprehensive monitoring station, complete the precise orbit determination and modification of the system satellite, and realize the precision modification of foreign satellite orbit on this basis.
3. *Realize the integrity calculation and report of system satellite.*
   This will be described in detail in Section 10.7.4.

10.7 Operation and Control of Multi-System Joint Wide Area Augmentation System

The multi-system joint wide area augmentation system is based on the civil navigation signals of the Compass, GPS, GLONASS, and Galileo for formal operation in the future, to provide a wide area augmentation system that meets the Class I precision approach services for Chinese civilian aviation navigation users. It renders services for China and surrounding areas.

10.7.1 System Composition

System space segment: Compass navigation satellites, GPS, GLONASS, and Galileo navigation satellites.

Operation and control system on the system ground segment: this includes a master control station and 5~7 multi-system monitoring stations.

System application terminal: multi-system compatible user receiver. It receives the navigation satellite signals from Compass, GPS, GLONASS, and Galileo, to achieve positioning, speed measurement, position reporting, and air traffic management.

Figure 10.3 shows a concept map of a multi-system joint wide area augmentation system, which is a joint wide area augmentation system aiming to meet the main satellite navigation system based on China's Compass system. In the Compass system, it must complete operation and control of Compass navigation satellites, and provide accurate correction parameters of the ionosphere. Therefore, for GPS, GLONASS, Galileo, and other navigation satellites, it only provides correction parameters of satellite clock error, correction parameters of satellite ephemeris, and integrity information. The ionosphere correction parameters are obtained from observation of the Compass satellites. So, operation and control on the ground only needs 5–7 monitoring stations. Any of these monitoring stations can inject the parameters into the augmentation satellites selected by the space segment. It can greatly improve the efficiency and reliability of the system.

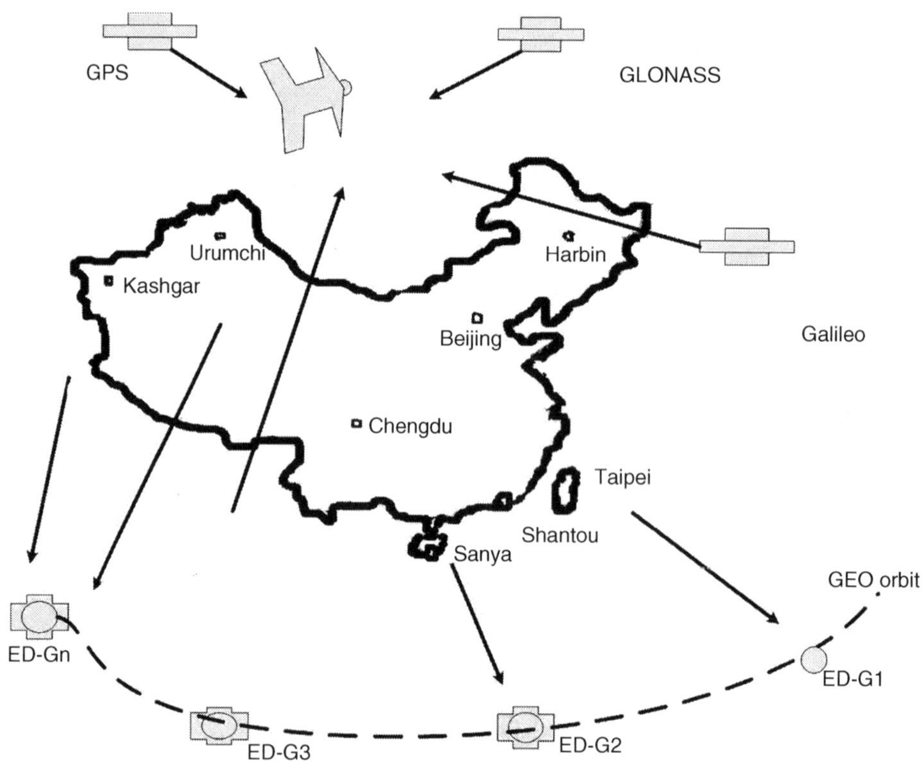

Figure 10.3 Concept of the multi-system joint wide area augmentation system.

10.7.2 System Working Principle

The system treats the GPS, GLONASS, and Galileo satellites, which are monitored as augmentation satellites and to be the user observation satellites, together with the Compass navigation satellites.

The system monitoring station observes the GPS, GLONASS, and Galilean satellites by using monitoring receivers, and provides correction parameters of satellite clock error, and the pseudorange measurement values, carrier phase measured values, and observed values of Doppler integral observations required by precision orbit correction and integrity evaluation, and then sends the observation values to the master control station for calculation and evaluation.

The master control station of the system conducts the calculation of correction values of the clock error, the correction calculation of the precision orbit and the integrity evaluation for the selected satellite based on the health satellites required for meeting the Class I precision approach in China. The selection of these available satellites needs to be dynamic and only needs 8~12 available satellites. These satellites will be injected into the Compass GEO_S satellite through the uplink injection system of the monitoring station. The GEO_S satellite broadcasts monitor satellite clock error and ephemeris correction parameters.

The user receivers within the wide area augmentation coverage area can perform precise navigation positioning according to the available satellites widely distributed by GEO_S satellites and their correction parameters. While the ionosphere correction parameters are corrected by referring to the Compass ionosphere correction parameters.

For satellites that cannot meet the accuracy and integrity, they will not be provided with parameters of clock error correction and ephemeris correction or with unmonitored signs.

The system completes the calculation of multi-system wide area differential correction parameters under the dispatching of master control station. The mission of master control station is to coordinate the master control system of the Compass system, to perform:

1. Selection of wide area augmentation satellites and calculation of calibration parameters.
   - From the GPS, GLONASS, and Galileo satellites, select the satellites of good geometric figure with Compass health satellite and small user equivalent range error to enter into the augmentation satellite constellation.
2. Assignment of satellite monitoring missions to each monitoring station according to the need of available satellite constellations in wide area augmentation systems.
3. Assignment of satellite injection missions to each monitoring station according to GEO_S satellite distribution and working condition in the whole system.

Data exchange between master control station and each monitoring station is performed by the two links of ground communication network and satellite communication network. The mission of wide area differential algorithm is to calculate the three-dimensional ephemeris error, the satellite clock error, and the ionosphere correction parameters of the selected augmentation satellite as the correction parameters broadcasting to users.

By using the dual-band observation data from the monitoring station, the master control station separately processes the three-dimensional ephemeris error of satellite, the satellite clock error, and the ionosphere time delay measured error so as to provide separate ionosphere correction parameters to single-frequency users.

The master control station uses the satellite pseudorange observed by 4~7 known monitoring stations to determine the satellite's position, clock error, and the clock deviation of the monitoring station, and measures the pseudorange from the monitoring station *i* to satellite *j*. After atmospheric error and multipath correction, P_{ij} is:

$$\begin{aligned} P_{ij} &= D_{ij}e_{ij} - B_j + b_i + n_{ij} \\ &= [(R_j + \delta R_j) - S_j]e_{ij} - B_j + b_i + n_{ij} \end{aligned} \tag{10.24}$$

In the formula (10.24), p_{ij} is the measured pseudorange; D_{ij} is the distance vector from monitoring station i to the satellite j; e_{ij} is the unit distance vector from the monitoring station i to the satellite j; R_j is the position of the satellite j calculated in accordance with the GPS (GLONASS......) message; δR_j is the ephemeris error vector of the satellite No. j; S_i is the position of the known monitoring station No. i; B_j is the satellite clock deviation; B_i is the clock deviation at monitoring station; and n_{ij} is the measured noise.

These vectors are shown in Figure 10.4 [15].

For all monitoring stations (I = 1,..., n) and augmentation satellites (GPS or GLONASS..., respectively, defined as j = 1..., m). This is defined as:

$$x = [\delta R^T \;\; B^T \;\; b^T]^T$$

In the formula, $\delta R = [\delta R_1{}^T \delta R_2{}^T \ldots\ldots \delta R_M{}^T]^T$

$B = [B_1 \; B_2 \ldots\ldots B_{n-1}]^T$

$B = [b_1 \; b_2 \ldots\ldots b_m]^T$

If all the measurement formulas (10.24) of all monitoring stations (i = 1, ...n) and the augmentation satellites (j = 1, ...m) are collected and arranged, the following matrix equation can be obtained:

$$\begin{bmatrix} E_1 & -1 & I_1 \\ E_2 & -1 & I_2 \\ \vdots & \vdots & \vdots \\ E_n & -1 & I_n \end{bmatrix} x = D - \begin{bmatrix} E_1 & 0 & 0 & 0 \\ 0 & E_2 & 0 & 0 \\ 0 & 0 & \ddots & 0 \\ 0 & 0 & 0 & E_n \end{bmatrix} \cdot P \tag{10.25}$$

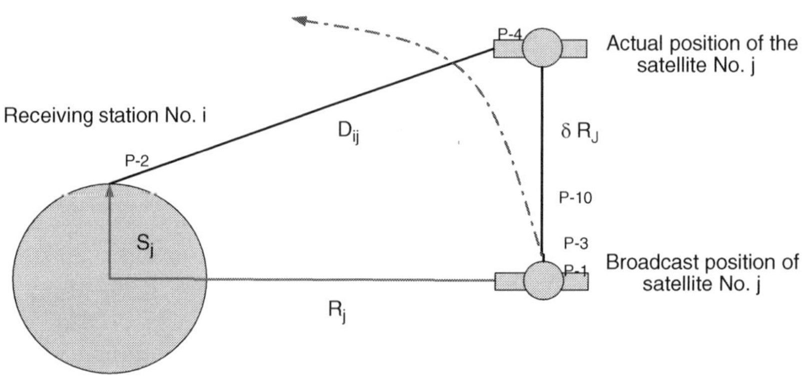

Figure 10.4 Ephemeris error of augmentation satellite.

Where:

$$E_{ij} = \begin{bmatrix} e_{ij}^T & 0 & 0 & 0 \\ 0 & e_{in}^T & 0 & o \\ 0 & 0 & \ddots & 0 \\ 0 & 0 & 0 & e_{im}^T \end{bmatrix} (m \times 3m)$$

$$I = \begin{bmatrix} 1 & 0 & 0 & 0 \\ 0 & 1 & 0 & 0 \\ 0 & 0 & \ddots & 0 \\ 0 & 0 & 0 & 0 \end{bmatrix} (m \times m)$$

Column i:

$$I_i = \begin{bmatrix} 0 & \cdots & 1 & \cdots & 0 \\ 0 & \cdots & 1 & \cdots & 0 \\ \vdots & \cdots & \vdots & \cdots & \vdots \\ 0 & \cdots & 1 & \cdots & 0 \end{bmatrix} [mx(n-1)]i = (\ldots, n-1)$$

$In = 0[mx(n-1)]$ (when $i = n$)

$$\mathrm{D} = [\mathrm{D}_1^{\ \mathrm{T}} \mathrm{D}_2^{\ \mathrm{T}} \ldots\ldots \mathrm{D}_n^{\ \mathrm{T}}]^{\mathrm{T}}$$

$$\mathrm{D}_i = [\rho_i^{\ 1} \rho_i^{\ 2} \ldots\ldots \rho_i^{\ m}]^{\mathrm{T}}$$

$$\mathrm{P} = [\mathrm{P}_1^{\ \mathrm{T}} \mathrm{P}_2^{\ \mathrm{T}} \ldots\ldots \mathrm{P}_n^{\ \mathrm{T}}]^{\mathrm{T}}$$

$$P_i = [(R_1 - S_i)^T (R_2 - S_i)^T \ldots\ldots (R_m - S_i)^T]^T$$

In this formula, the matrix In is set to the 0 matrix because all clock errors are correlated, and are estimated relative to the number n monitoring station clock. If we define the system matrix h and the measurement matrix Z as:

$$H = \begin{bmatrix} E_1 & -I & I_1 \\ E_2 & -I & I_2 \\ \vdots & \vdots & \vdots \\ E_n & -I & I_n \end{bmatrix} \tag{10.26}$$

$$Z = D - \begin{bmatrix} E_1 & 0 & 0 & 0 \\ 0 & E_2 & 0 & 0 \\ 0 & 0 & \ddots & 0 \\ 0 & 0 & 0 & E_n \end{bmatrix} D \tag{10.27}$$

Then the formula (10.25) is converted into:

$$\mathrm{Z} = \mathrm{HX} \tag{10.28}$$

If the motoring station number i cannot observe the satellite number j, then the corresponding elements of the matrix Z and the row vectors of the matrix H in formula (10.28) must be eliminated.

The master control station uses the BLS technique to estimate the three-dimensional ephemeris error vector and clock offset for each GPS satellite in the view of monitoring

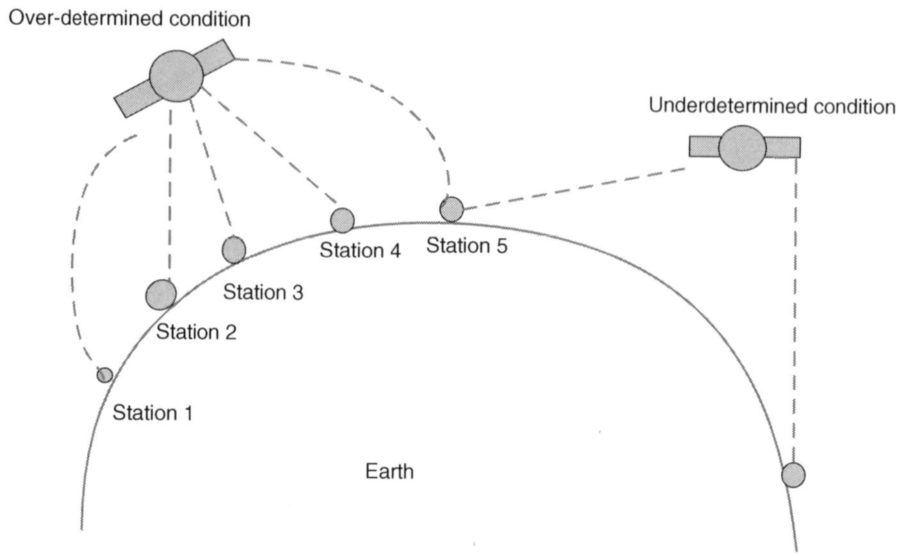

Figure 10.5 Instance diagrams of estimated ephemeris errors under over-determined and underdetermined conditions.

network. If there are more measurement values in the WADGPS network than the unknowns (three-dimensional ephemeris error, satellite clock deviation, and clock offset of monitoring station), then the observational equation for the satellite is specific and the solution is to minimize the measured values' residual sum of squares.

$$X = (H^T H)^{-1} H^T{}_Z \tag{10.29}$$

If the measured value is less than the unknown (Figure 10.5), then the solution is underdetermined. The best estimate is to minimize the two-dimension number of the solutions:

$$X = H^T (HH^T)^{-1} Z \tag{10.30}$$

Under the underdetermined condition, the correction of ephemeris error and clock offset is imprecise, but the user's positioning that uses these correction values is still accurate. Therefore, the most important for the user is the projection of error correction vector on the user to line-of-sight of satellite.

If there are limited monitoring stations in our country, the user near the coastal monitoring station and surrounding area will use the under-determined satellite, resulting in a decrease in accuracy. Therefore, the station shall be established beyond the scope of the user's application.

10.7.3 GEOS Satellite Correction Parameters and Integrity Broadcast Message

The GEO_S satellite is the geostationary satellite of Compass system. Firstly, it is the working satellite of the system, and is also the broadcast satellite that broadcasts ephemeris correction parameters and clock error correction parameters of the GPS, GLONASS, and Galileo satellites. There is a number of GEO_S satellites in the Compass system able to perform mission of broadcasting the wide area differential correction parameters, but the broadcasting missions of several GEO_S satellites shall be optimally allocated, to give the users within China enough available observation satellites. In principle, GEO_S satellites in the eastern region broadcast the correction parameters of GPS or GLONASS (Galileo) satellites that are located in the east of China; GEO_S satellites in the western region broadcast the correction parameters of GPS or GLONASS (Galileo) satellites that are located in the west of China. As the three foreign satellite systems have a large number of observable satellites, up to 33 or more, it's hard to broadcast all the parameters. Therefore, the master control station selects the constellation satellites that enter wide area augmentation system. The principles of its selection are:

1. For regional users, it has good geometric figures that can improve the geometric dilution of precision of users.
2. Good correction accuracy of ephemeris and clock error.
3. Good integrity of satellite.

On this basis, the number of satellites involved in wide area differential can be greatly reduced; for about 12 satellites including the Compass satellite, the capacity of calibration parameters and integrity message shall include:

1. Satellite identification. 6 bits
2. Integrity indication of Class I precision approach. 1 bit
3. The time T corresponding to parameters. 9 bits
4. Clock error corrections. 13 bits
5. Clock speed corrections. 8 bits
6. Satellite $\Delta x \Delta y \Delta z$. $\Delta \dot{x}$ $\Delta \dot{y}$ $\Delta \dot{z}$ correction parameters. $11 \times 3 + 8 \times 3 = 57$ bits
7. Total data bits of each satellite. 94 bits

If the update rate of the correction parameter is: 1 time/3 min.

The integrity indication is the alarm indication: it is arranged in 3 min. The total bit number of message for 12 satellites is 1116 bits. Around three to five GEO_S satellites can ensure the capacity and real-time performance of the message.

10.7.4 Satellite Integrity Monitoring

Satellite integrity monitoring will perform monitoring and judgment by different methods based on satellites of different systems. For the Compass system, the monitoring receiver with wide-narrow correlation receivers equipped by the whole system monitoring station performs the information collection, and carries out system integrity judgment by the master control station based on multi-station monitoring results.

After completing the integrity judgment of the Compass satellite, for the judgment of GPS, GLONASS, or Galileo satellites, this can be completed by the common precision positioning receiver at lower cost than the Compass monitoring receiver, or by the user receiver with good RAIM performance.

At this point, the compatible receiver selects four Compass satellites with qualified integrity test and one foreign satellite (such as: GPS) to be used for integrity monitoring; that is, four Compass satellites and one foreign satellite to constitute five observation equations.

$$\overline{Ru} = R_i - \overline{D}_i (i = 1,\ 2 \ldots\ldots 5) \tag{10.31}$$

Where:

$\overline{Ru}$ is the vector from the center of the Earth to the user.
$\overline{D}_i$ is the vector from user to satellite *i*.
$\overline{R}_i$ is the vector from the center of the Earth to satellite *i*.
i means 1~ 5 satellites.

Since only four satellites are needed to calculate the position of the monitoring receiver, the five equations can form four subsets (each subset excludes a Compass satellite), to obtain four solutions $x_1^j, x_2^j \cdot x_3^j \cdot x_4^j$ of separate receiver position. Comparing with the known position x_0^j of the receiver (x_0^j is the three position parameters of the receiver, $j = 1, 2,$ and 3), we can, respectively, obtain four positional deviations:

$\Delta x_1^j, \Delta x_2^j, \Delta x_3^j, \Delta x_4^j$ When the satellite geometric figures (PDOP value) of four subset solutions are equivalent, the four positional deviations above follow the random characteristic. The total positioning error of the four positioning solutions of the five satellites is:

$$\Delta\overline{x} = \sqrt{\sum_1^4 \Delta X_i^{j2}} \tag{10.32}$$

Decompose $\Delta\overline{x}^j$ into plane position error ΔP and vertical position error ΔH. According to the error requirement of Class I precision approach (CAT-I) from International Civil Aviation Organization:

$$2\Delta p \leq 18.2m$$

$$\Delta p \leq 9.1m$$

$$2\Delta H \leq 7.7m$$

$$\Delta H \leq 3.8m$$

When Δp. ΔH meet the requirements, the GPS satellite can be determined as CAT-1's optional satellite, and make a forecast to the user.

In this calculation, in order to judge $\Delta\overline{x}$ as the maximum value, the HDOP and VDOP values between the satellite and the monitoring receiver in the four subsets of the equation (10.31) shall be lower than the threshold value of CAT-1.

The selected GPS, GLONASS, and other satellites can be determined for their integrity one by one in a similar method, and thus enter into the wide area augmentation constellation. In order to carry out correction parameters and integrity monitoring of the augmentation constellation independently, the monitoring stations that are not involved in satellite clock and satellite orbit monitoring shall be selected when selecting the integrity monitoring station. As shown in Figure 10.6, three monitoring stations as the determination of precision orbit and satellite clock error and three monitoring stations as integrity monitoring can complete the mission. In order to provide enough positive definite satellites to users around China, the number of monitoring stations can be properly increased.

10.7.5 Composition of the Monitoring Station

The monitoring stations and Compass monitoring stations are incorporated and designed uniformly with the basic composition shown in Figure 10.6.

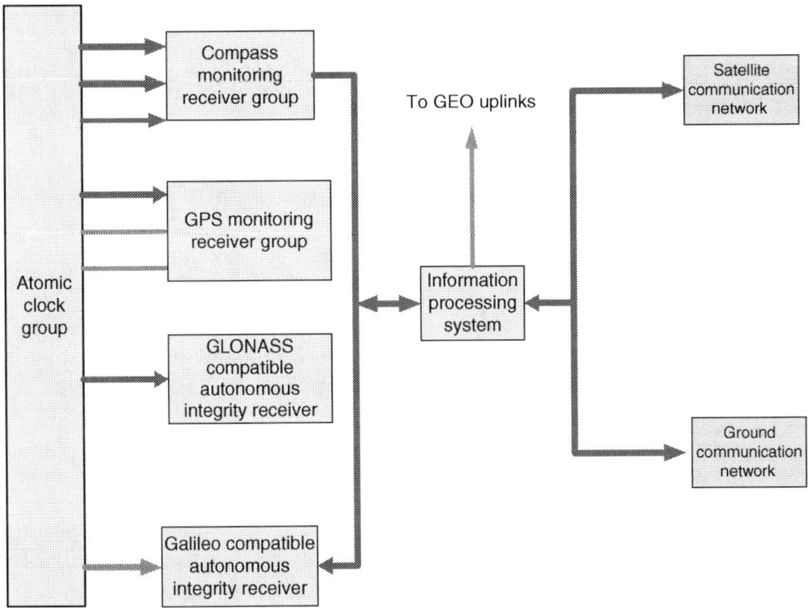

Figure 10.6 Composition of the joint monitoring station.

As shown in Figure 10.6, this consists of the Compass monitoring receiver group, GPS monitoring receiver group, GLONASS receiver, Galileo receiver, information processing and satellite communication network, and the ground communication network.

In Compass monitoring receivers, three receivers work independently and the atomic clock forms a satellite data acquisition system for mutual verification so as to perform the collection and judgment processing of original data required for clock error, ephemeris correction, and integrity judgment of the Compass satellite.

The composition and working principle of GPS monitoring stations are the same as the receivers of Compass monitoring station.

The monitoring of GLONASS and Galilean satellites can be greatly simplified due to the integrity estimation for Compass and GPS satellites. Integrity monitoring of GLONASS and Galileo satellites can be performed by using Compass/GPS/GLONASS and Compass/Galileo compatible receivers, with the principles described in Section 10.4.

The observation and positioning accuracy estimated parameters received by the joint monitoring station are, respectively, classified according to the ephemeris clock error and the integrity judgment information after being processed by the information processing subsystem of this station. For the short timeframes, information such as integrity warning is sent to the master control station by the satellite communication network. The slow variation information can be sent to the master control station by the ground communication network.

The broadcast parameters assigned by the master control station to the monitoring station are injected into the satellite according to the selected GEO Compass satellite uplinks after being received by the information processing sub-system.

10.7.6 Master Control Station Joint Wide Area Differential Software Function

The hardware system of master control station joint wide area differential is uniformly designed with the Compass master control station. In this section, by taking GPS wide area differential software processing as an example, the processing flow of GPS satellite differential integrity on the master control station is introduced.

The application software of central station has six primary function modules: data receiving and processing, GPS satellite precision orbit calculation, GPS satellite correction parameter calculation, integrity calculation and determination, correctness test of data and information, and generation and transmission for broadcast information. There is no need to form the ionosphere broadcast parameters by a separate GPS satellite as a result of composing a wide area differential system with the Compass system. The software function is shown in Figure 10.7 and described next.

1. *Data receiving and processing*
   This function module receives all kinds of inbound data, to dump and allocate.
2. *Ionosphere parameters module*
   Change the ionosphere correction parameters from the Compass information processing system to the correction parameters suitable for the GPS satellite frequency for calculating the precision main orbit and correction parameters of GPS satellite.
3. *GPS satellite precision orbit calculation*
   By using the GPS observation data and referring to the ionosphere correction parameters of the system, determine the precision orbit of GPS satellite, generate the precise ephemeris of GPS satellite, predict the satellite orbit, and generate the next calculation for satellite measurement and control files. This includes the following processes:
   - Correction of the ionosphere effect.
   - GPS satellite orbit determination.
   - Receiver clock error resolution.

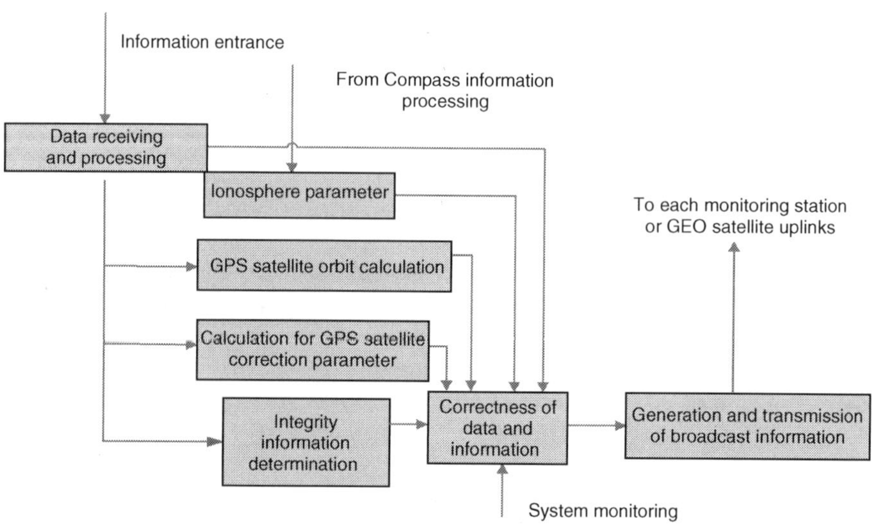

Figure 10.7 Diagram of software processing function for the foreign satellite system.

- Calculation for precise ephemeris of GPS satellites.
- GPS satellite orbit prediction.
- Plan generation.

4. *Satellite corrections calculation*
   Use the received observation data together with the ionosphere correction parameters generated from modules (2) and (3) and the satellite precise ephemeris to calculate the satellite clock error corrections and the ephemeris corrections, including the following processes:
   - Correction of ionosphere effect on observation data.
   - Calculation for slow variable of satellite clock error.
   - Calculation for fast variable of satellite clock error.
   - Calculation of satellite clock error corrections.
   - Calculation for user differential range error (UDRE).
   - Calculation for satellite ephemeris corrections.
   - Satellite ephemeris error (EPRE) calculation.
   - Pseudorange error (RMS) calculation.
5. *Integrity information determination*
   In order to determine the integrity of satellite and differential data, the function block must receive the navigation data of all satellites used for navigation services, when a satellite or differential corrections cannot be used for navigation services, or provide alarm information once satellite differential corrections are not monitored. The function module includes the following processing:
   - Integrity determination of satellite.
   - Integrity determination of differential corrections.
6. *Correctness test of data and information*
   In order to test the correctness of the broad-spectrum information, this function block shall complete the following three tests:
   - Using fitting residual differential to verify the corrections.
   - Conducting the consistency verification by using parallel hardware and software as two sets of data.
   - Conducting contrast verification for GPS original measured data by Compass ionosphere.
7. *Generation and transmission of broadcast information*
   To satisfy the requirements for broadcast information formats and service control, the function block shall achieve the following functions:
   Editing of broadcast information.
   Encryption control of broadcast information.
   Distribution of broadcast information.
8. *In order to complete operation and maintenance of the system, necessary system monitoring and control shall be also provided, including:*
   - Business management.
   - System management.
   - System monitoring and control.
   - System maintenance.

   The internal data flow of the system is as shown in Figure 10.8.
9. *The application software shall consist of the following programs:*
   - Communication program.

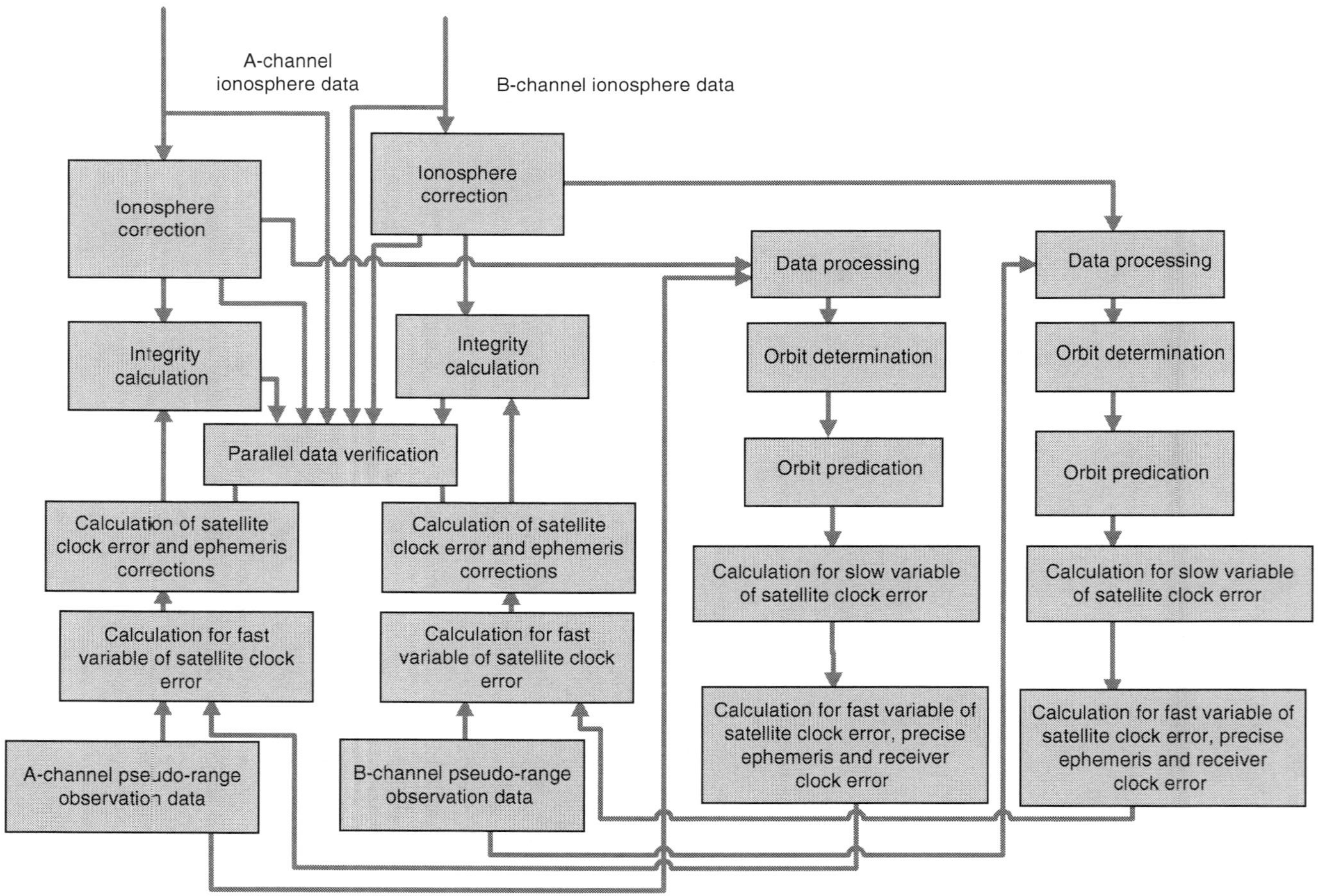

Figure 10.8 Internal data flow of the system.

- Data pre-processing program.
- Precise ephemeris generation program of GPS satellite.
- Ionosphere correction program.
- Estimation program of satellite clock error.
- Correction program of satellite broadcast ephemeris.
- Integrity processing program.
- Data verification program.
- Generation and encryption program of broadcast information.
- Business management program.
- System management program.
- Monitoring display program.

11

Navigation Satellites and Navigation Payload

11.1 Satellite and Navigation Payload History

The beginning of the navigation satellite occurred when the Soviet Union launched the first artificial earth satellite on October 4, 1957. The observation of radio signals sent back to the ground by the satellite made people recognize that the precise time signals from artificial earth satellites at known locations contributed to the navigation of the navigator. Since then, artificial satellites have possessed the same positioning function as the Sun and sidereal table. Moreover, precise satellite time signals can also replace a chronograph on a ship. Ground-based radios can be moved to a small number of satellites to provide global coverage. It is a full-time, all-weather, high-accuracy positioning system. Later, the US Navy launched its first navigation satellite, the Transit satellite in 1958, and started the 621B and TIMATION as satellite navigation experiments simultaneously. In 1973, the US Department of Defense began a GPS research program.

The navigation satellite appeared as "weapon" from the very beginning, and developed rapidly in the Cold War confrontation between the Soviet Union and the USA. The Soviet Union also built цикада and the GLONASS system in the corresponding period. But satellite navigation does not leave well enough alone, as did the atomic bomb. It has the duality of being both weapon and tool, which can enhance military ability and promote economic development; therefore, people like it more and more, and are not worried about the navigation satellites developed in an arms race dragging down the economy, or the oversight of national defense due to the promotion of economic prosperity to develop navigation satellites. It's the best military and civilian representative to date.

The technology of navigation satellites develops toward the direction of relying on the ground less and less to ones where orbital data is obtained by Doppler measurements on the ground, and then ground user positions itself with the Doppler measurements. GPS has developed to a level independent of ground support for 180 days. The improvement at the navigational war level may be a transient state. Now the world is not so peaceful, it may continue for some time. It's also a hot topic of concern to many people. The transfer of application target will lead to the emergence and death of certain technologies. But the pursuit of accuracy, usability, continuity, and good will not stop. The development direction of navigation satellites is clear, long-term, and stable.

GNSS Systems and Engineering: The Chinese Beidou Navigation and Position Location Satellite, First Edition. Shusen Tan.

The navigation payload is the main one related to the navigation accuracy and integrity, and also a direct device for implementing the navigation target. In particular, on-board autonomous integrity and good on-board response of ground integrity channel (GIC) are important development requirements gradually recognized by people; however, the current level is low with weak measures. As seen from the development history of GPS and GLONASS, the satellite navigation is flourishing, but its accuracy, availability, continuity, and integrity have not met the needs of aviation navigation, on which the priority for development shall be put.

GPS developed in BLOCK I, BLOCK II, BLOCK II A, BLOCK II R, and has entered into the BLOCK II F deployment stage. BLOCK I has launched a total of 11 satellites, 10 of which were launched successfully (see Table 11.1) with the first one on February 2, 1978 and the last one on October 9, 1985.

The first phase of satellite launch went through 5 years from the project approval in 1973 to the launch of the first satellite in 1978, but still fails to enter the overall verification of concept of the system; the main obstacle is the traditional and superficial "Catch-22." The two main problems are that: (1) where it's unclear whether the user equipment can work with the satellite, how can the user equipment be verified? (2) How can the satellite be launched without ensuring that the satellites work with the user equipment? The GPS plan was pushed into a situation where it could do nothing. The 621B, of which the configuration is composed of solar-powered transmitters, has to be transformed into a pseudo satellite and arranged in the desert of the test field, thus verifying the cooperative work between the user equipment and the satellite, and make the satellite navigation walk out of the logical impasse. A total of 10 BLOCK I satellites have completed the correctness verification of overall concept of GPS system. The test accuracy of 11 kinds of land, sea, and air carriers is 6–16 m (SEP).

BLOCK II satellites were launched from February 4, 1989, and as of October 1, 1990, a total of nine satellites had been launched. The launch booster of BLOCK II satellite was designed based on the interface of the space shuttle. After the *Challenger* space shuttle crashed, this decision was modified, and the rocket was still a delivery vehicle for GPS satellites. It was found that the BLOCK II satellite was gradually deteriorated for working 14 days after being lunched. Therefore, with the addition of autonomous momentum management capacity, the BLOCK II A satellite can work in 180-day cycle and need not contact with the ground.

The navigation message of BLOCK II/BLOCK II A satellite is completed and loaded by the GPS control system day by day. BLOCK II satellite is equipped with inter-satellite links to complete the distance measurement between each other, and combined with on-board Kalman filter, can provide the autonomous navigation capability for BLOCK II R satellites. Through the distance measurement by inter-satellite links and the error of BLOCK II R satellite Kepler orbit parameters, the BLOCK II R satellite is able to support the navigation accuracy of 16 m spherical probable error (SEP) and need not contact with the ground control system in the same period of 180 days. The comparable error of the BLOCK II A satellite is a kilometer scale at the end of the 180 days.

The GPS system was equipped with initial operational capability (IOC) in December 1993 when the Air Force Space Command announced the start of operation of 24 GPS satellites, which is an important milestone in 20 years of GPS development. The United States Federal Aviation Administration (FAA) announced that GPS can be used for aviation navigation in February 1994.

Table 11.1 Initial GPS launch history.

BLOCK	SVN	PRN	International Identification No.	NASA No.[a)]	Orbit[b)]	Launch Date	Clock[c)]	Availability Date	Withdrawal of Services
BLOCK I									
—	01	04	1978–020A	10684	—	780222	—	780329	850717
—	02	07	1978–047A	10893	—	780513	—	780714	810716
—	03	06	1978–093A	11054	—	781006	—	781113	810518
—	04	08	1978–112A	11141	—	781210	—	780108	811014
—	05	05	1980–011A	11690	—	800209	—	800227	831128
—	06	09	1980–032A	11783	—	800426	—	800516	910306
—	07	11				811218	Failed to launch		
—	08		1983–013A	14189	—	830714	—	830810	930504
—	09	13[d)]	1984–059A	15039	C-1	840613	—	140719	—
—	10	12	1984–097A	15271	A-1	840908	Cs	841003	—
—	11	03	1985–093A	16129	C-4	851009	Rb[e)]	851030	—
—	—	—	—	—	—	—			
BLOCK II									
II-1	14	14	1989–013A	19802	E-1	890214	Cs	890415	—
II-2	13	02	1989–044A	20061	B-3	890610	Cs	890810	—
II-3	16	16	1989–064A	20185	E-3	890818	Cs	891014	—
II-4	19	19	1989–085A	20302	A-4	891021	Cs	891123	—
II-5	17	17	1989–097A	20361	D-3	891211	Cs	900106	—
II-6	18	18	1990–008A	20452	F-3	900124	Cs	900214	—
II-7	20	20	1990–025A	20533	B-2	900326	Cs	900418	—
II-8	21	21	1990–068A	20724	E-2	900802	Cs	900822	—
II-9	15	15	1990–088A	20830	D-3	901001	Cs	901015	—
—	—	—	—	—	—	—	—	—	—
BLOCK IIA									
II-10	23	23	1990–103A	20959	E-4	901126	Cs	901210	—
II-11	24	24	1991–047A	21552	D-1	910704	Cs	910830	—
II-12	25	25	1992–009A	21890	A-2	920223	Cs	920324	—
II-13	28	28	1992–019A	21930	C-2	920410	Rb	920425	—
II-14	26	26	1192–039A	22014	F-2	920707	Cs	920723	—

(Continued)

Table 11.1 (Continued)

BLOCK	SVN	PRN	International Identification No.	NASA No.[a]	Orbit[b]	Launch Date	Clock[c]	Availability Date	Withdrawal of Services
II-15	27	27	1192–058A	22108	A-3	920909	Cs	920930	—
II-16	32	01[f]	1192–079A	22231	F-1	921122	Cs	921211	—
II-17	29	29	1192–089A	22275	F-4	921218	Cs	930105	—
II-18	22	22	1193–007A	22446	B-1	930203	Cs	930404	—
II-19	31	31	1193–017A	22581	C-3	930330	Cs	9304B[g]	—
II-20	07	07	1193–032A	22657	C-4	930513	Cs	930612	—
II-21	09	09	1193–042A	22700	A-1	930626	Cs	930720	—
II-22	05	05	1193–054A	22779	B-4	930830	Cs	930928	—
II-23	04	04	1193–068A	22877	D-4	931026	Cs	931122	—
II-24	—	—	—	—	—	—	—	—	—

a) The NASA catalog number is also known as NORAD or US Space Command grade.
b) If the orbital plane position is not listed, the satellite is no longer in operation.
c) Clock: Rn is rubidium, and Cs is cesium.
d) PRN13 satellites have insufficient power capacity to sustain L1/L2 launch during solar eclipses, so it is possible to turn off the L1/L2 launch by PRN 13 satellites for up to 12 hours per day during this period.
e) The rubidium clock used in PRN03 does not conduct temperature control.
f) The SVN32 satellite had been changed from 32 to 01 on Nov. 28, 1993.
g) The maintenance work of PRN31 satellite failure was completed on June 18, 1993, and the problem of L2 intermittent locking was solved.

GPS provides a precise positioning service accuracy of SEP 16m. Civilian SPS standard positioning accuracy is SEP 25m, providing lower accuracy service of 2drm100m by SA.

The allowable total error (rss) in the space section is approximately 3.5 m (1δ). The reason why rss is selected is that these errors are irrelevant, or statistically independent. The obtained satellite performance was 2.2 ms (1δ) when the rubidium atomic frequency standard (AFS) was adopted, and it was 2.9 m (1δ) when the cesium atomic frequency standard was adopted. Both performance measurements were made 24 h after the last update from the ground. Under normal conditions, the ground updates the satellite every 24 h.

The power at the satellite antenna input end in the space section for L1 I branch is 14.3 dBW, and that of Q is 11.3 dBW, and that of L2 (P (y) code) is 8.1 dBW. The power near the Earth's surface for L1 I (C/A code) is −160.0 dBW, Q [P (y)] is −163.0 dBW, L2 [P (y)] cannot be −166.0 dBW. The satellite is required to work for 7.5 years, and its designed lifetime is 10 years.

Although the construction of GLONASS system began at the same time as the GPS, the first GLONASS satellite was launched on October 12, 1982, and a total of 65 satellites had launched in 14 years as of 1995 (see Table 11.2).

11.2 Navigation Satellite Platform

As with other satellite platforms for communication, and so on, the navigation satellite platform includes measurement, control, propulsion, thermal control, structural and power supply, and other subsystems that offer the wireless transmission channel between the ground station and the satellite, attitude, and orbit control, and the torque required to produce the attitude and orbit control, satellite thermal control, structural support, and power, respectively. As the navigation satellites need to maintain aa high-accuracy orbit and stable time and signal transmission delay, the control and propulsion sub-system needs to be equipped with a stable and independent attitude and orbit system. The control of satellite attitude has little influence on the orbit and it is preferable to have an accurate force vector measurement in the process of orbit control. The thermal control subsystem requires stable and accurate operating temperature for payload and atomic clock frequency on the satellite. The orientation and stabilization system of GLONASS satellite is designed with the active three-axis directional stabilization system with control flywheel and jet unloading in order to ensure the satellite to the sun and orientation initially, the vertical axis of the satellite pointing to the Earth's core and orbital correction engine thrust orientation. When the satellite shifts to predetermined orbit structure point (system or operating point), which is completed through satellite correction and the attitude-adjustments power plant, the cylindrical airtight container, also the main bearing structure of the satellite, should be used to create necessary operating temperature conditions for satellite equipment.

Taking the GLONASS satellite as an example, the satellite platform has the following characteristics:

Orientation and stabilization systems
Orientation accuracy (degrees):
Longitudinal axis 0.5–1.0

Table 11.2 GLONASS unfolding phase.

No.	Cosmos	Launch Time	Start Date	Status	Group No.
1	1413	821012	821110	840330	1
2	1490	830810	830902	851029	2
3	1491	830801	830831	880609	2
4	1519	831229	840107	880128	3
5	1520	831229	840115	860910	3
6	1554	840519	840605	860916	4
7	1555	840519	840609	870917	4
8	1593	040904	840922	851128	5
9	1594	840904	840922	860916	5
10	1650	850518	850606	851128	6
11	1651	850518	850604	870917	6
12	1710	851225	860117	890306	7
13	1711	851225	860120	870917	7
14	1778	860916	861017	890705	8
15	1779	860916	861017	881024	9
16	1780	860916	861017	881012	8
17	1838	870424	—	a_3	9
18	1839	870424	—	a_3	9
19	1840	870424	—	a_3	9 10
20	1883	870916	871010	880606	10
21	1884	870916	871009	880820	10
22	1885	870916	871005	890307	11
23	1917	880217	—	a_3	11
24	1918	880217	—	a_3	11
25	1919	880217	—	a_3	12
26	1946	880521	880601	900510	12
27	1947	880521	880603	910918	12
28	1948	880521	880603	910918	13
29	1970	880916	880920	900521	13
—	1971	880916	880928	890830	13
30	1972	880916	881003	920812	14
31	1987	890110	890201	940203	14
32	1988	890110	890201	940203	14
33	1989[A]	890110	Geodetic survey	—	15
—	2022	890531	890704	900123	15
34	2023	890531	890615	891108	15
35	2024[A]	890531	Geodetic survey	—	16
36	2079	900519	900620	940817	16
37	2080	900519	900617	940827	16
38	2081	900519	900611	930120	17
39	2109	191208	910101	940610	17
40	2110	901208	901229	940120	17
41	2111	910404	901228	960815	18
42	2139	910404	910428	941114	18
43	2140	910404	910428	930604	18
44	2141	920130	910504	920616	19
45	2177	920130	920224	930629	19
46	2178	920130	20222	—	19
47	2179	920130	920218	—	20
48	2204	920730	920819	970805	20

Table 11.2 (Continued)

No.	Cosmos	Launch Time	Start Date	Status	Group No.
49	2205	920730	920819	940827	20
50	2206	920730	920825	960826	20
51	2234	930217	930314	94017	21
52	2235	930217	930825	—	21
53	2236	930217	930314	970823	21
54	2275	940411	940904	+	22
55	2276	940411	940518	990909	22
56	2277	940411	940516	970829	22
57	2287	940811	940907	990203	23
58	2288	940811	940904	+	23
59	2289	940811	940907	+	23
60	2294	941120	941211	+	24
61	2295	941120	941215	970727	24
62	2296	941120	941216	+	24
63	2307	950307	950330	+	25
64	2308	950307	950405	+	25
65	2309	950307	950405	971226	25
66	2316	950724	950826	+	26
67	2317	950724	950822	+	26
68	2318	950724	950822	+	26
69	2323	951214	960107	+	27
70	2324	951214	990426	+	27
71	2325	951214	960118	+	27
72	2362	981230	990129	+	28
73	2363	981230	990129	+	28
74	2364	981230	990218	+	28

Notes:

a) The dates in the table use Moscow time (UTC + 3 h + 00 min)

b) What's listed in status is the date when the satellite was stopped use. "+" means operation in orbit.

c) Russian President commanded GLONASS for official start on Sep. 14, 1993.

d) By the end of 1995, there were 24 working satellites and GLONASS had operated in the whole system.

Solar cell 5
Force vector 5–11

Correction system
Engine thrust (N):
Correction 5
Stability 0.1
Total impulse (N) 90,000

Power supply system
Working power of solar battery (W) 1250
Battery working capacity (Ah, Wh) 45, 1260
Output voltage (V) 27 ± 1

Temperature control system
Temperature range (°C)
Frequency standard (15–30) ± 1

Sealed cabin 0–40
Engine assembly 5–40
Energy consumption (W) 36
Service life (years) 3–5
Continuous working time (day) 1415–1485
Weight (kg) (structure only) 237
Energy consumption (average day and night, W) 1000
GPS-related indicators are increased slightly.

11.3 Navigation Payload Requirements

The number of transmitted operating frequencies of payload for navigation satellites is greater than 2, and power level will allow the standard antenna power level reaching the ground to be about −160 dBW or so. And the corresponding noise of the transmitter is required to be minimized. The group delay variation of the two L1-band transmitter links is consistent, preferably within 1 ns, allowing the user to obtain accurate ionosphere correction.

The gain flatness of transmitter within the operating frequency shall be preferably controlled within 1 dB so that the code-related processing function of user receiver has better symmetry.

It's a requirement that some navigation satellites have the ability to receive uplink signals on ground, and to enable high-accuracy pseudo-range measurements so as to provide independent time synchronization accuracy for the system. The GLONASS satellites also carry a laser reflector, allowing the ground control system to perform high-accuracy laser range measurements.

The most advanced navigation satellite payload is also equipped with inter-satellite links to allow inter-satellite distance measurements and to complete the ephemeris error estimates by on-board Kalman filter, and to realize 180-day autonomous navigation at last. And it has the ability to restart and load the on-board software system.

The integrity of on-board payload will be dominated by automatic detection of satellite, so as to accurately inform the user in time. In particular, there shall be a high-accuracy atomic time frequency system of which the long-term stability shall be not less than 1×10^{-13}/day.

The reliability of the satellite payload is ensured by the redundancy of device and components. The high-reliability components (transistors, integrated circuits, relays, etc.) verified by flight is also used. The failure of each component should not reduce the performance of system components, and cannot be transmitted to the entire system.

The single-point failure will be minimized and a detailed reliability analysis should be made on all satellite units in order to determine the effectiveness of these compromise proposals.

11.4 GPS Satellite Navigation Payload

BLOCK II R is the most advanced in GPS satellite navigation payload [8], and it is an alternative to BLOCK II and BLOCK II A. In 1978, Martin Marietta (a GE Astronomical

Space Subsidiary at that time) and the ITT Aerospace Communications Department were mainly responsible for the development.

The GPS navigation signal is generated and transmitted by the Total Navigation Payload (TNP). The heart of TNP is AFS that provides precise time and frequency standard for GPS. AFS consists of two rubidium atomic frequency standards and one cesium atomic frequency standard.

L-band transmission system consists of three transmitter chains, namely three frequencies of L1, L2, and L3, with L1, L2 for navigation, and L3 for nuclear explosion detection.

The important role of TNP of BLOCK II R satellite is to maintain the stability of time system (TKS) and autonomous navigation (Auto NAV) capabilities, which is a huge leap forward compared with previous satellites.

The center of the BLOCK II R satellite is a cube with each side of 6 feet (2 m) long, the unfolded solar panels are about 30 feet (9.14 m), and the takeoff weight of satellite is 4480 pounds (2.24 ton). Its weight in orbit is about 2370 pounds (1.2 ton). The satellite is launched with a Delta launch vehicle. The satellite is equipped with 16 boosters and a set of standard bus devices: telemetry, tracking and command (TT&C), payload control equipment (PCE), satellite processing units (SPU), and attitude reference systems.

The designed life of satellite is 10 years, with the storage time of 4 years, and autonomous navigation time of 180 days.

11.4.1 Atomic Frequency Standard

1. Atomic frequency standard requirements are based on that: The satellite clock maintains the accuracy of 6 ns relative to GPS time.
2. The reliability of the R_b GPS BLOCK II R atomic frequency standard is 0.750 (at the end of 7.5 years) with power consumption of 15 W and weight of 14 lbs. The reliability of the C_S GPS BLOCK II R atomic frequency standard is 0.775 (at the end of 7.5 years) with power consumption of 26 W, weight of 22 lbs.

11.4.2 On-Board Processing

The mission of on-board processing is to: produce navigation information, complete ephemeris calculation and data encryption, generate P code and C/A code, monitor the health of the payload, and provide clock error correction.

On-board processing is completed in the mission data unit (MDU) that provides navigation data storage when the uplink load is made via the satellite TT&C subsystem within the control section. It is assigned to the L-band system as required, loaded in the P code or C/A code to be launched into the space. The MDU will modify these navigation signals when necessary to implement the SA policy.

The MDU is capable of autonomous operation for up to 180 days without the need to receive updated navigation data from the ground control system, which is achieved by processing inter-satellite pseudo-range measurement data and inter-satellite exchange data by way of using this information to complete the satellite ephemeris and clock correction parameter calculation, and send it to the user.

MDU software program, complex, and huge, must complete the navigation information and clock correction parameter processing originally completed by the original ground control system CS, and also include integrity monitoring, curve fitting of navigation parameters for autonomous operation for 180 days. User ranging accuracy URA estimation, navigation datamation, availability of options, Universal Time Coordinated (UTC) control, and assisted offset recovery. The whole mission processing (MP) software is written in the Ada language.

Space programs are reprogrammable and can be completely reprogrammed according to ground commands. After a cold start, the processor directs the program within the PROM. The work program is transferred by use of the S-band uplink data link through this program. The PROM program possesses adequate diagnostic capability to verify whether the processor, memory, and data interfaces required for upstream loading and execution of the flight program run properly.

In addition to the uplink loading program, it can also do part of the uplink load to facilitate change. The MDU software also provides control of the timekeeping system. The short-term stability of the system clock relative to GPS time is maintained by the stability of the voltage-controlled VCXO, while the long-term stability is maintained by coupling the system clock to a more stable reference clock. The phase of the system clock is compared with the phase of the reference clock by use of a phase meter, and then the phase difference is converted into a ΔF command for adjustment.

11.4.3 Wave Band System

The primary functions of the wave band sub-system are:

1. Producing C/A code, P code. And the PN code is formed at 1.023 Mbps and 10.23 Mbps, respectively.
2. Performing data modulation. The module 2 addition operation is made for 50-bit data and P-code, C/A code.
3. The encoded output phase is modulated on the L1 carrier at 1575.42 MHz and on the L2 carrier at 1227.6 MHz. The data is sent out at these two frequencies to allow high-accuracy ionosphere delay correction for the user.
4. L1 solid-state high-gain amplifier outputs the level of 50 W. L2 level output is 10 W.
5. Combining L1 and L2 in the duplexer. When L3 is installed, it will be a triplexer.
6. The duplexer delivers the combined radio frequency (L1 and L2) power to the antenna system that is a phased array composed of helical antennas, producing shaped beams covering the whole Earth.

The shaped beam of the antenna is as shown in Figure 11.1. The path loss is minimal when the satellite is directly overhead (90° elevation angle), and is maximum at the edge (the satellite is on the horizon) of the Earth's coverage area. The path length difference between these two extreme cases is about 500 km, so the resulting path loss difference is 2.1 db. The shaped beam is designed as shown in Figure 11.1, symmetrical about the axis from the satellite to the center of the Earth, so as to uniformly illuminate the Earth's surface. At a viewing angle larger than 28°, the radio-frequency energy radiated from the antenna is almost zero and the total radiant energy within 28° is the maximum.

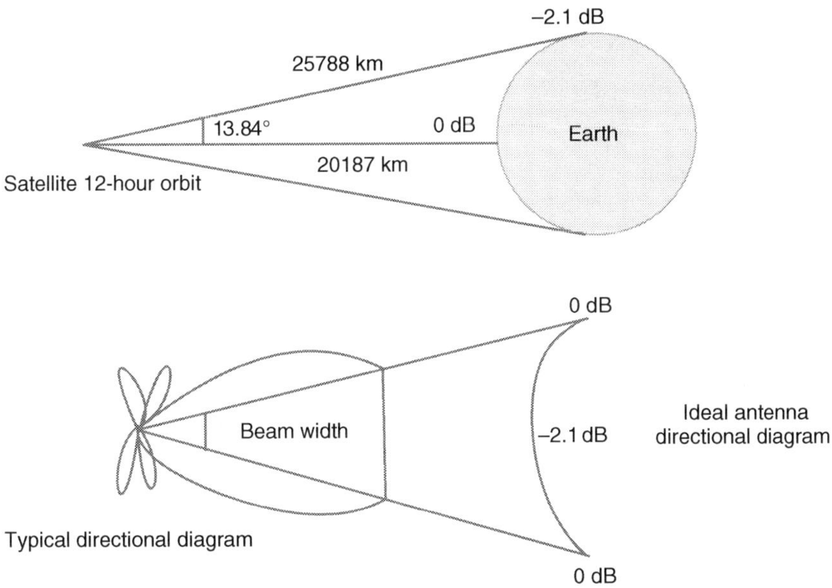

Figure 11.1 Antenna beam width and path loss values [15].

11.4.4 Horizontal Link

All GPS BLOCK IIR satellites contain a Horizontal Link Transponder Unit (CTDU); it has data communication and ranging mode for inter-satellite distance measurement and intercommunication in order to realize the ability to independently work for 6 months, while ensuring the URE less than 6 m.

CTDU is a time division multiple access (TDMA) frequency hopping spread spectrum communication system, operating in the UHF band, and it contains a 5 Mbps pseudo-random code and a 108 W RF output power.

The primary functions of the CTDU include transmitting data between satellites by use of frequency hopping carrier waves, conducting one-way ranging measurements between satellites by use of two kinds of carrier frequencies, and measuring the internal (CTDU) time delay. The secondary functions include interception and identification of horizontal link transmission; symbol recovery, timing of frequency hopping and frames; and generating the intercepted synchronous head in the transmission mode. The horizontal link transponder and data unit block diagram are as shown in Figure 11.2. The radio frequency converter obtains received frequency hopping signals and makes it down to the intermediate frequency, assisted by the digital frequency synthesizer. The dedicated digital signal processor conducts digital conversion on the intermediate frequency signals and completes the detection, tracking, and demodulation of received signals.

11.4.5 Autonomous Navigation

The navigation messages of BLOCK I and BLOCK II GPS satellite broadcast are loaded by the ground control system CS every day. The clock correction parameters and ephemeris parameters in the message are predicted based on the current estimate of CS.

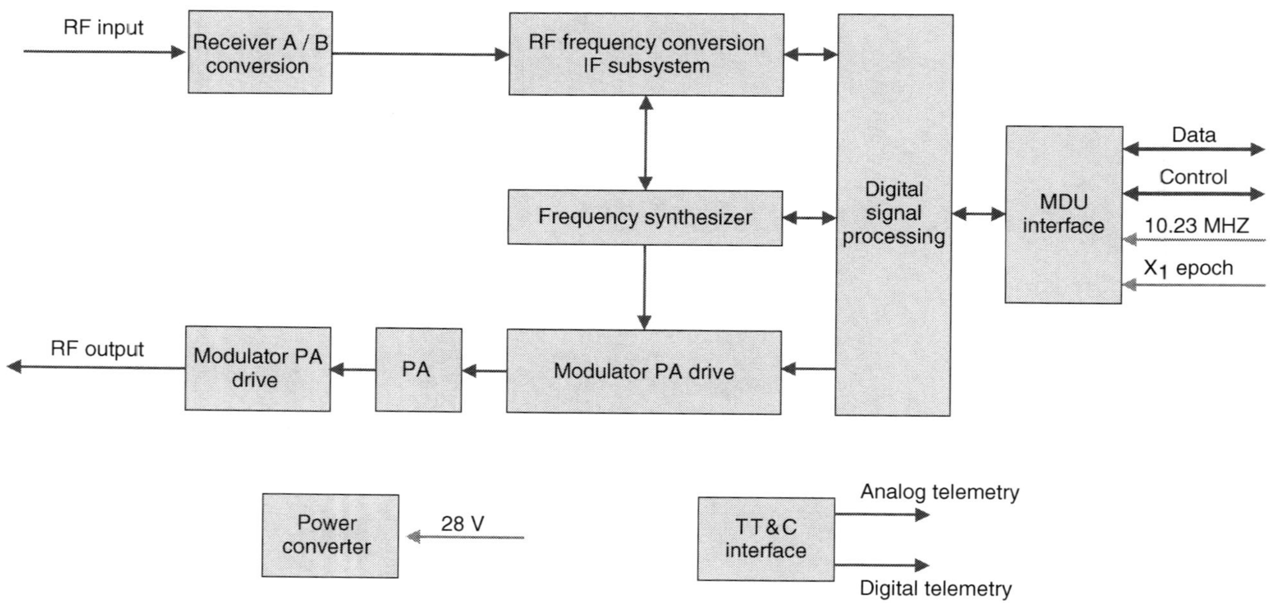

Figure 11.2 Horizontal link transponder and data unit block diagram.

In addition to this mode of operation, the BLOCK II R satellite is also able to independently estimate the ephemeris and clock error and generate navigation messages by satellite borne functions, which is called autonomous navigation or AutoNav. Its development is driven by the following four factors.

1. *Durability*
   The ground control CS is the most vulnerable part of the system. With continuous operation of 180 days by AutoNav and the ability to meet accuracy, it will reduce vulnerability, improve durability, and perhaps eliminate any detection station in the future.
2. *Reducing uplink load requirements*
   After using AutoNav, the data loaded to satellite uplink by CS is reduced, which slows the demand for instantaneity of uplink stations.
3. *Integrity*
   The horizontal link ranging provides an independent benchmark, so that satellites can directly compare the ephemeris and clock difference.
4. *Accuracy*
   AutoNav can shorten the ephemeris and clock difference forecast to every 4 h from every 24 h by CS, which undoubtedly improves the accuracy of using ephemeris and clock difference.

AutoNav estimates ephemeris and clocks based on inter-satellite pseudo-range measurements provided by the UHF horizontal link, and also exchanges information. In this way, the satellite performs clock skew and pseudo-range measurements between each pair of satellites, and each satellite updates its clock and ephemeris Kalman filter with such measurements.

The UHF horizontal link used in AutoNav is in successor compatibility with the horizontal link used as relay NDS on BLOCK II where the horizontal link also performs the corresponding NDS function. As a result, the AutoNav function is completed with a minimum of additional on-board devices. However, the need of successor compatibility with BLOCK II is a serious constraint to the design of the system. AutoNav must use the inherent link protocol and cannot interfere with the BLOCK II satellites located in the track at the same time. In such protocol, each satellite is assigned a time slot of 1.5 s, which is located in a TDMA frame of 36 s. In each frame, the satellite is launched in its assigned time slot and received in the other 23 time slots. AutoNav is completed in one cycle consisting of several TDMA frames.

The block diagram of AutoNav is shown in Figure 11.3. CTDU completes the horizontal link communication and ranging function. The data from the CTDU is sent to the processor, and the processor performs other functions in the diagram. The results of the processor require the formatted navigation messages and send them to the L-band system.

At the beginning of each cycle, all satellites broadcast the ranging signal in the same TDMA frame. Two frequencies are used for correction of the delay generated by the high ionic layer. In the subsequent frames, they broadcast their pseudo-range measurements, thus enabling bidirectional ranging function. At the same time, other parameters such as ephemeris are broadcasted in the output message. These derived measurements are received by the clock component and ephemeris Kalman filter.

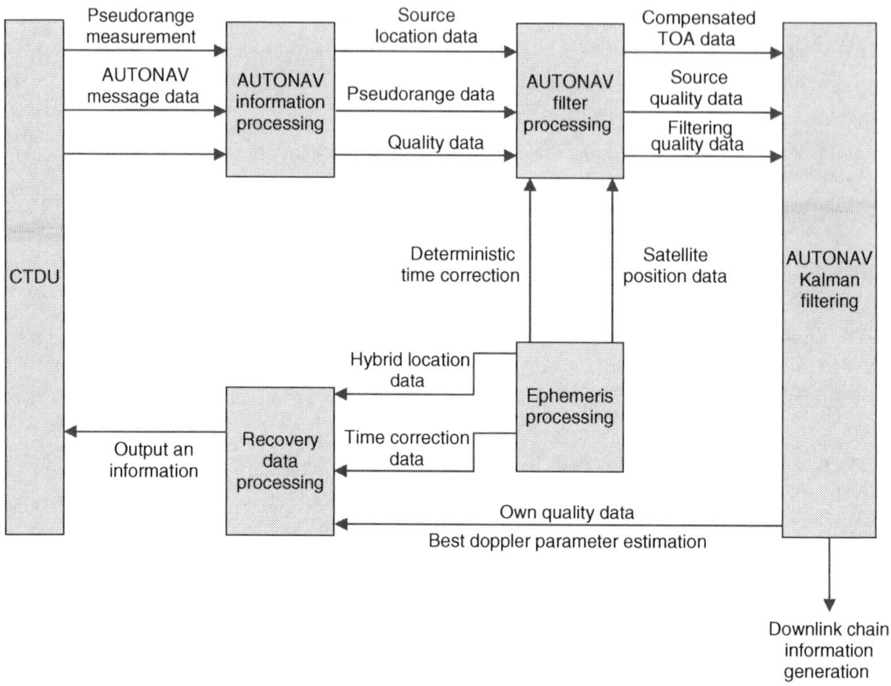

Figure 11.3 AUTONAV processing block diagram.

AutoNav can meet the system's accuracy and integrity requirements with 4–7 BLOCK II R satellites operation in orbit. During the period when BLOCK II R and BLOCK II are mixed for use, the user receiver does not have to distinguish AutoNav data from non-AutoNav data.

Obviously, the key element of GPS is the exact time, and the frequency calibration (FS) is the precise holder of GPS. The key to improving the performance of AutoNav is the high-accuracy atomic clock.

11.5 GLONASS Navigation Satellite and Navigation Payload

The GLONASS navigation satellite has experienced two generational developments, and the first generation is the current GLONASS satellite in operating orbit. The second generation GLONASS-M satellite is in the testing stage and through downlink test will replace the first-generation network satellite.

11.5.1 Functions of the GLONASS Navigation Satellite

The GLONASS navigation satellite is a basic component of the GLONASS system and its on-board equipment mainly completes the following functions:

1. Transmitting two types of high-stability radio navigation signals.
2. Receiving, keeping, and transmitting navigation information.
3. Forming, coding, keeping, and transmitting high-accuracy time signals.

4. Forwarding or transmitting the satellite radio monitoring signals and the correction parameters of the on-board time system.
5. Receiving, replying, decoding, and correcting one-time instructions.
6. Receiving, storing, and processing the control program of satellite in-orbit functioning pattern.
7. Forming and transmitting the telemetry data of on-board equipment.
8. Receiving and processing correction codes and on-board time scale phase codes.
9. Urgently calling the ground control system about the threshold alarm of important parameters.
10. Monitoring and analyzing the state of on-board equipment, and forming control commands.

The GLONASS system presets two application modes: normal (standard) mode and analysis mode. Under the normal mode, the navigation satellite is in the in-orbit working condition, and transmits radio navigation signals. Under the analysis mode, navigation satellite will be out of network, and conducts the analysis monitoring.

11.5.2 Satellite Composition

On-board equipment includes navigation transmitter, time system, control combination system, orientation and stable system, correction system, power system, thermal control system, on-board filling and environmental protection equipment, structural components, and cable network.

In order to ensure the reliability, most of the on-board equipment adopts redundant backups.

11.5.2.1 Navigation Transmitter

The mission is to form and transmit the high-stability radio navigation signal at two frequency bands of L_1 and L_2. The signals of the L_1 frequency band contain ranging code, on-board time scale, and navigation data (ephemeris, time correction parameters, frequencies, and phase correction). The L_2 frequency navigation signals only contain ranging code, and only have the function of correcting ionosphere suppressed transmission delay.

The on-board navigation transmitter includes the navigation signal formation equipment and antenna feed system. Navigation signal formation is completed by the independent module, the main equipment of which, the navigation signal formation devices, are placed in closed devices.

In order to monitor the state of on-board equipment, the amplitude and signal telemetric transducers are provided, and the signals that the transducers detect are input into the telemetric system. The amplitude transducer responds to the amplifier high-power output voltage level. The combination and switch of instrument equipment are not only conducted according to the commands, but also operate automatically according to the state analysis results of transducers.

The function of antenna feed system is to directionally transmit the navigation signals at two frequency bands of L_1 and L_2. It includes transmitters (12 pieces), distribution units (4 pieces), distributor, and cable.

The structure of antenna feed system is a grid shape, composed of two groups of spiral arrays. There are four in the center group and eight in the peripheral group, distributed on the ring with a diameter of 850 mm.

Major technical characteristics of on-board transmitter are as follows:

Frequency range (MHz)
  L_1 channel 1597–1621
  L_2 channel 1241–1261
Transmission gain (dB)
  L_1 channel center 10
  Within angle of ±15° 12
  Within angle of ±19° 8
  L_2 channel center 9
  Within angle of ±15° 11
  Within angle of ±19° 9
Output power (W)
  L_1 channel 64
  L_2 channel 10
Effective transmitted power (W)
  L_1 channel 30
  L_2 channel 21
Energy consumption (W) 530

11.5.2.2 Time System

The mission of the on-board time system is to continuously output highly-stabilized time-frequency signals and form standard timing signals. It ensures generation of standard frequency accuracy signals and synchronization pulse signals: f_1 is sine wave, the rest is pulse signal. The synchronizer also forms the time interval signal. The time serial number is a 32-bit binary sequence code with a frequency of 100 Hz.

The on-board time system includes: atomic frequency standard (3 sets), synchronous frequency, and a time scale forming device.

Atomic frequency standard is composed of quartz-crystal oscillator, atomic ray tube, and automatic frequency tuning system. The quartz-crystal oscillator produces the oscillation frequency of 5 MHz, and atomic ray tube and automatic frequency tuning system form high-stability frequency signals of 5 MHz.

Synchronous frequency and time scale forming device includes a sinusoidal amplifier and a forming device. Atomic frequency standard outputs high-stability frequency signals of 5 MHz, and sinusoidal signal amplifiers send them respectively to external application users.

On-board timing system operation is in the standard or on-duty mode. On-duty mode, namely test mode, ensures the frequency output and the output of quartz-crystal oscillator. After completing the examination for satellite equipment, it enters into the standard mode according to the ground monitoring and control subsystem commands.

If the on-board time scale exceeds the confirmed standard, the correction can be adopted according to the ground monitoring and control subsystem commands. If not corrected to the rated range, the signals that "Phase is not credible" shall be sent to users. The conversion from normal mode to on-duty mode is conducted according to the satellite control commands.

Major technical parameters of satellite clock are as follows:

Nominal value of output frequency (MHz) 5
The maintained accuracy of time scale (ns) 20
Daily stability $1 - 5 \times 10 - 13$
Within the scope of ±20 KHz, in comparison with the fundamental component of 5 MHz, the decrement of signal number spectrum component exceeds 90 db.
Weight (kg) 20.7

11.5.2.3 Control Combination System

Control combination system includes an on-board instruction system, an on-board computing processor, an on-board telemetric system, and a control combiner.

The functions of on-board instruction equipment are: using the antenna and visible light to measure the satellite orbit parameters in cooperation with the monitoring and control system; controlling the satellite system to carry out the one-time instructions; processing the time program and outputting navigation information; receiving, forming and sending on-board time scale signals; sending telemetry information; and transmitting the information of switching the ground equipment.

The on-board instruction system equipment is composed of low-frequency components, high-frequency components, password protection, antenna feed system, and a set of optical corner reflectors. The antenna feed system contains weak directional receiving and transmitting antenna and directional transmitting antenna. The directional antenna is used under the condition that the satellite is in the normal state.

The missions of on-board computing and processing are: to store and process navigation information, form navigation frames, and transmit them to the navigation transmitter; to monitor the state of navigation transmitter and implement the switch; to store time program codes and phase step codes and send them to the measuring system; and to receive, form, and send the call signals for the ground control integrated facilities. On-board computing and processing equipment sends out the automatic testing information with message operating characteristics along with the telemetric channel; and under the failure condition, it forms an "on-board computing and processing report" and sends it through the navigation channel, and stops using the satellite, so as to analyze the failure condition at the ground station.

Technical indexes of control combination system are as follows:

On-board computing and processing
Speed of information record (H) 1000
Set-up time after power supply (min) 3
Processing information interval (day) 30
Maximum processing speed (program/s) 2
Instruction system equipment
Frequency range cm
Time correction accuracy (ms) 0–2
Receiving speed (Hz)
One-time instruction 100
Time program 1000
Control combiner
Total measuring error (m) 3

Antenna radio band 0.3
Visible light wave band
Number 1000
Number of telemetry channel 512

Satellite control integrated equipment ensures connection with the satellite after entering into the orbit, and powers up the rocket apparatus after the satellite separating from the carrier rocket. It distributes power among all electric equipment; when correcting and unloading control wheels, the subsystems make the combination operation and under the condition of orientation disorders or power failure, it controls the satellite system.

11.5.2.4 Orientation and Stable System as well as Its Auxiliary Equipment

The missions of orientation and stable system are: to damp and stabilize the satellite and ensure the initial to-Sun and to-Earth orientation; the vertical axis of satellite points to the Earth's core and the solar cell points to the Sun orientation; and the driving force vector of orbit correction engine goes along with velocity vector orientation (during the operation of correction impulse).

The GLONASS satellite is equipped with the control flywheel and air injection unloading system, and initiative triaxial orientation and stabilization system. The orientation and stable system includes sensitive elements (angular velocity measurement module, to-Sun and to-Earth directional instruments, Sun attitude instrument of yaw channel, geomagnetometer), and actuating elements (electromechanical actuator, solar cell array drive unit, electromagnetic device, stable flywheel, power unit, and control combination).

The orientation and stable system operates in the following modes: damping stabilization, initial to-Sun orientation, rotation around horizontal axis, initial to-Earth orientation, to-Earth orientation, and orientation at correction time.

Satellite solar cell is composed of four plane rectangular plates, each two of which constitute a wing, and symmetrically distributes relative to the stable plane (XOY). Solar cell wings rotate around the axes (OZ) perpendicular to the stable plane, and point to the Sun when running normally. The transfer of each satellite to the predetermined orbit structure point (system or working point) is conducted with the satellite correction and attitude-adjusting power units (two symmetric modules).

11.6 Galileo Navigation Satellite and Alternative Schemes of Navigation Payload

Galileo is under development and construction, and according to the comparison studies of European global navigation satellite system (GNSS-2), there are two kinds of navigation satellites: MEO satellite and GEO satellite. There are three kinds of different proposals for satellite schemes [17]. A block diagram of MEO and GEO satellites is shown in Figure 11.4.

11.6.1 Satellite Definition

The MEO satellite platform is designed according to the main technical requirements:

- About 30 satellites constitute the constellations, distributed on three orbit planes;

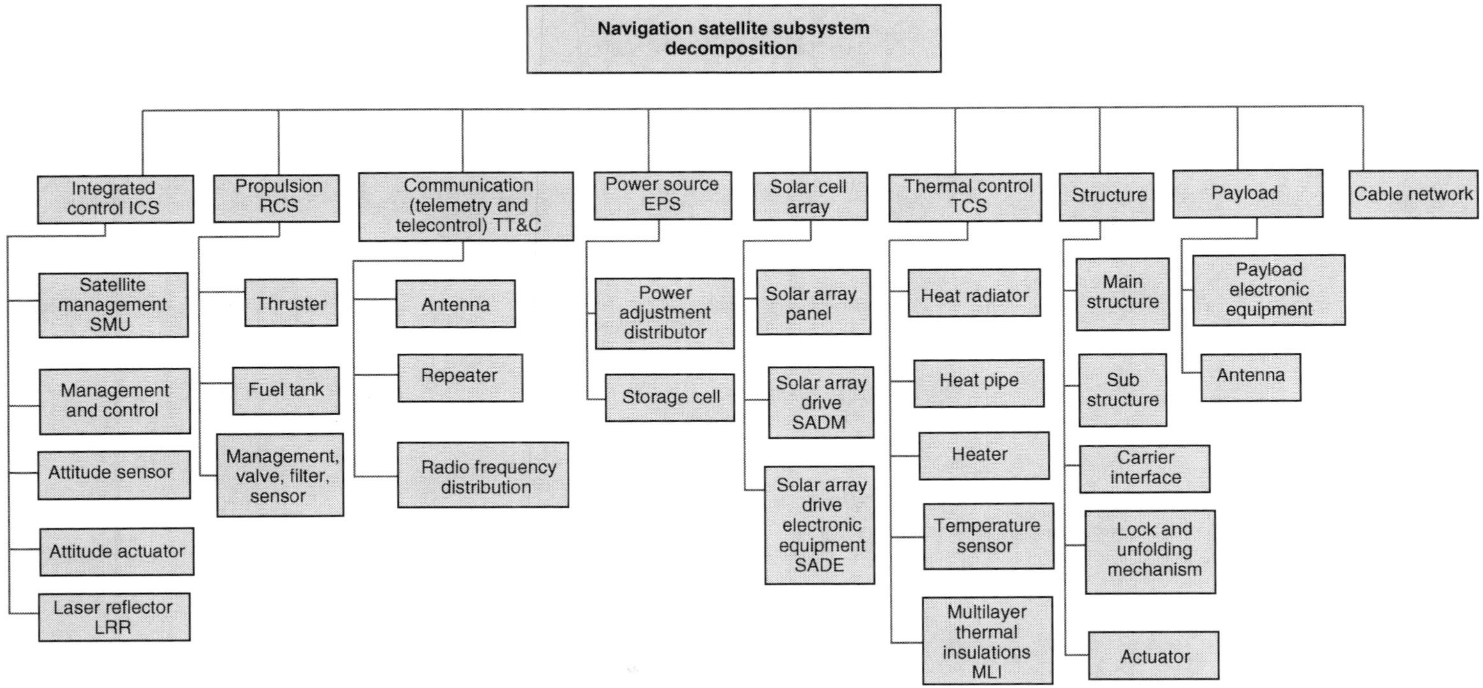

Figure 11.4 Block diagram of Galileo MEO/GEO satellite subsystems.

- Orbital elevation is about 24,000 km, and angle of tilt is larger than 50°;
- For navigation payload scheme 4, the weight is 73 kg, the power is 454 W (without certainty degree and allowance, with the addition of 71 W increment adjustment to alternate the payload scheme 3);
- Antenna aperture: 1.3 m
- Communication (TT&C) load, 56 kg/500 W without allowance.
- Occupation area of payload electronic equipment: 1.33 m^2
- Alternative rockets:
  - Proton M-Breeze+ long integer constant, transmitting quality of 4880 kg
  - Soyuz ST + Fregat 1300 kg
  - Ariane 5 Esv 4000 kg
- Reliability (satellite, including payload) 0.8 (10 years)
- Lifetime (15 years)

Except for carrying navigation, communication payloads, the GEO satellite has European geostationary navigation coverage business (EGNOS) capacities.

Navigation payload of GEO satellite will increase power, antenna aperture increases to 1.9 m, and is required to have generalities with MEO and considered to use the same launching rockets as MEO.

The payload of the Galileo navigation satellite has three different kinds of signal structure, two kinds of different satellites (MEO and GEO), and there are the alternative payloads from different companies.

In order to ensure the system availability, the payload must have a certain reliability, and the measures to achieve high reliability include:

Simplifying the structure of payload (e.g., minimizing the number of load)
Improving the functional integration (e.g., minimizing the number of equipment)

11.6.2 MEO Satellite Configuration

According to the satellite system, the following requirements of top-level design shall be met:

1. Attitude control modes: three-axis stabilization and yaw control;
2. Payload in-orbit operation attitude: antenna pointing to nadir;
3. The field of view and orientation requirement met by platform sensor and actuator;
4. Total space coverage of TT&C antenna;
5. All illumination periods, solar array to-Sun orientation shall be optimized and simplified.

The rotation axis of solar array is perpendicular to the yaw axis of satellite.

1. *Integrated control and data management system (ICS)*
   This subsystem is used for the satellite management. The functions are:
   - Attitude and orbit control (AOCS)

   Obtaining the satellite attitude data through ICS sensor, and completing the attitude adjustment by the actuator;
2. *Data handling (DH)*
   Including: telecontrol management and processing of the telecontrol commands from the ground, and after the confirmation, allocating them to different users of satellites to achieve the aims of telecontrol management;

Telecontrol management. Collect the telemetry data of all on-board devices, send them to the ground by the TT&C subsystem. The sending rate is constant, to adapt to more detailed telemetry report information under the abnormal condition.
On-board reference time. Keep the time information and allocate them to on-board users.
Data storage and retrieval. Through data storage, select the communication and transmit telemetry information to the ground.
Arrangements and implementation of automatic activities, including the arrangements for instruction encryption and other autonomous works, fault detection, isolation, and recovery.

3. *Attitude and orbit control system (AOCS)*
   The mission is to keep the payload pointing to the Earth, and to make the satellite rotate under yaw control, solar array rotate around its axis, and solar panel constantly perpendicular to solar incident angle.
4. *Propulsion system*
   The mission is to provide thrust and torque, complete satellite speed adjustment, orbit acquisition, position holding, and shift to the abandoned orbit after the termination of satellite life.
   The main working modes include:
   - Stable ignition. It is used for the position holding along the track or vertical orbits, basically one time each year, and working for a few minutes each time. It is also used for the reposition of auxiliary satellites and the disposal of end-of-life satellites.
   - Non-modulation condition. When maneuvering, carry out the open-loop thrust control.
   - Modulation condition. It is used for auxiliary attitude control.
5. *TT&C subsystem*
   Telemetry, tracking, and control systems (TT&C). Complete on-board satellite telecontrol commands and the reception of satellite navigation data, send satellite status information and data to the ground station, and respond to the distance measurement signals of the ground station. The main equipment includes:
   - two sets of hot stand-by telecontrol receivers;
   - two sets of hot stand-by receivers of navigation data;
   - two sets of cold stand-by transmitters;
   - two sets of omnidirectional antennas, respectively, on the $\pm Z$ axes.
     - two sets of responders at the S frequency band, used for the telecontrol reception and telemetry transmission.
     - two sets of dedicated receivers at the S frequency band used for receiving navigation data.

   Both the two sets of omnidirectional antennas can receive the ground telecontrol signals at the transfer orbit, emergency situation, and on-station stage.
   Navigation data are received by omnidirectional antenna. Both the two sets of transponders at S frequency band can receive and transmit the telemetry data flow transmitted by the telecontrol base band. Both telemetry and telecontrol devices adopt ESA subcontract standards. The functions are shown in Figure 11.5.
6. *Power subsystem (EPS)*
   Its function is meeting the power requirements of on-board users.

It provides 42.5 V full-modulation busbar voltage.
EPS basic compositions:
- two solar array wings
- two solar array drive mechanisms (SADM)
- one solar array electronic instrument (SADE)
- two groups of 46A nickel-hydrogen storage batteries (respectively composed of 23 monomers)
- one power adjustment distributor (PCDU)

7. *Solar array unit*
   Two solar array wings providing the primary power supply for all on-satellite devices and storage batteries. Two solar array drive units (SADA) guarantee the solar array will follow the Sun during the mission.
   Performance indexes are as follows:
   - Power offered at the end of life −2000 W
   - Total area of solar cell −12.5 m^2
   - Designed life 15 years
   - Total mass of solar wing 42.0 kg
   - Unfolding mode Rigid unfolding
   - Solar array tracking Around pitch axis
8. *Thermal control subsystem*
   Make the temperature, humidity stability, and homogeneity of the devices in its charge in the ordered mission stage and lifespan be within the coordinated range.
9. *Structure*
   The primary functions are:
   - Bearing the static force and dynamic load of all states;
   - Providing mechanical support for all subsystems;
   - Providing the interfaces between satellites and dispatchers;
   - Providing necessary heat-dissipation radiation surface;
   - Providing operation windows for different components;
   - Providing stable size, and ensuring the alignment of satellite regulation;
   - Providing the shield against space radiation for electronic components.
10. *Cable network*
   Provides the internal connections between electrical equipment.

11.6.3 Payload of MEO Satellite Navigation

Navigation payload includes the processing for navigation, time, orbit, and integrity. The principle is as shown in Figure 11.5. The following information are automatically produced and transmitted:

1. Signals of two different kinds of frequencies of OAS open service.
2. Signals of 1–2 kinds of frequencies of CAS controlled service.
3. Signals of two different kinds of frequencies of CAS2.

It has not been determined whether OAS and CAS use the same two frequencies or each business uses the independent frequency.

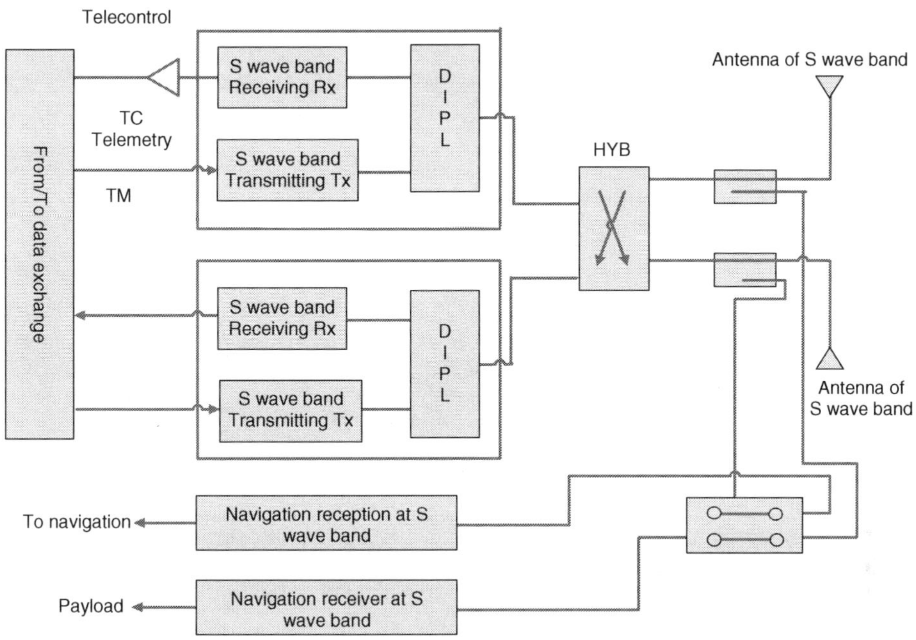

Figure 11.5 Block diagram of TT&C system (I).

All frequency signals are produced according to the on-board atomic frequency standard.

Navigation data are sent out from the different navigation data tables of NAV processor, and different navigation data is shown in Table 11.3.

Each table may not include only one information entity, and each information entity has an effective time and a release time. NAV processed table automatically selects latest data with effectiveness meeting the requirements, and then produces NAV information.

The data in satellite table are provided by two channels: One channel is provided by orbit and synchronization processor through global area network and global uplink station; and the other is provided by integrity processing through regional network, regional uplink interface, global uplink interface, and uplink station.

11.6.4 GEO Satellite and Navigation Load

GEO satellite has two kinds of alternative schemes including small-sized GEO satellite and medium-sized GEO satellite, provided by different companies, and the function of its platform payload is the same as that of MEO satellite.

Directly in-orbit GEO satellite position keeps: within 15 years controlled at: 0.1° from north to south and 0.1° from east to west; the satellite position of geostationary transfer orbit scheme keeps: within 15 years controlled at 7° from north to south and 0.1° from east to west. In March 2005, it was known that Galileo adopted the scheme of a (27+3) MEO satellite constellation instead of the GEO satellite scheme. How to realize integrity monitoring and prediction is worthy of attention.

Table 11.3 Navigation data grouping.

	Feed Arrangement
User integrity data	Integrity processing facilities
User clock parameters	Orbit and synchronization processor
User space accuracy signals	Orbit and synchronization processor
User orbit parameters	Orbit and synchronization processor
User precise orbit simulation parameters	Orbit and synchronization processor
Other user NAV information: • Satellite outage warning • Constellation available mode • Difference between Galileo system time and UTC • Time difference between Galileo and GPS	Orbit and synchronization processor/navigation control center

11.7 Compass Satellite Navigation and Payload

The Compass satellite is a part of the space segment of China's Compass Navigation System, and it is composed of three basic forms of satellites, and, respectively, applies to GEO, IGSO, and MEO orbits.

The Compass Navigation Satellite consists of satellite platform and payload. Satellite platform consists of TT&C, data management, attitude and orbit control, propulsion, thermal control, structure and power supply, and other subsystems. Payload includes navigation subsystem and antenna subsystem. A GEO satellite also contains an RDSS payload. Thereby, the Compass satellite creates the precondition for providing integrated navigation, communication, and time services.

The Compass Navigation Satellite designs respectively the navigation signals of several coarse codes and precise ranging codes at three frequency bands of 1559~1610 MHz, 1215~1300 MHz, and 1164~1215 MHz. There are two service modes of open service and authorization service. Compass satellite navigation frequencies launched after 2012 will use the frequency band of 2483.5~2500 MHz, and there is the satellite-to-satellite link at the Ka frequency band to a provide semi-autonomous satellite navigation test.

The satellite platform adopts the mature technology of China's communication satellites, with a designed life of 10 years, and an in-orbit life of 8 years.

The radiation power of payload of a Compass satellite can make the strength of the ground receiving signals greater than −160 dBW. The time and frequency subsystem on satellite has the unified frequency source of navigation load generated by atomic frequency standard, which realizes a no gap switch of atomic frequency standards, and also has the function of satellite laser ranging.

Compass navigation satellite has the on-board information processing capability, the function of completing the measurement, demodulation, decoding, classification of uplink input signals, the generation and encryption of navigation message, and so on, the function of reloading partial program, and the function of autonomous monitoring for satellite integrity.

11.8 Comparison and Development Direction of Navigation Payload

By comparing the satellite navigation payload of the previously mentioned several satellite navigation systems, there are the following common characteristics and development tendencies.

1. There shall be rich working frequency bands and modulation coding signals at L frequency band as far as possible. GPS, respectively, has three navigation signals including 1575 ± 12 MHz, 1227 ± 12 MHz, and 1176 ± 12 MHz. Through QPSK and BOC modulation segmentation, the service for different users is to be realized at the same frequency.
2. Navigation satellite develops to the direction of autonomous navigation (AutoNAV) capability, and GPS BLOCKIIR satellite can keep the rated navigation accuracy within 180 days under the condition of being separated from the support of the ground control system, which is a significant measure to enhance the reliability during wartime.
3. Three pillars of autonomous navigation are: high-performance atomic clock, satellite-to-satellite link distance measurement and communication, and on-board information processing. The way to overcome the poor long-term stability of the atomic clock is also to develop satellite-to-satellite link technology.
4. The system integration level should be improved to enhance the satellite reliability.

12

Satellite Navigation User Receiver

Satellite navigation user receiver is an application terminal of the satellite navigation system. It consists of a navigation receiver and a navigation positioning processor. The receiver completes the receiving of navigation signals and pseudorange measurement; while the navigation positioning processor completes the calculation of navigation positioning, as shown in Figure 12.1.

The main characteristics of the satellite navigation user receiver are signal acquisition, tracking, pseudorange measurement, and positioning treatment under relative high speed motion between satellite and receiver, so it is necessary to analyze the motion characteristics.

12.1 Relative Motion Characteristics Between the User and the Satellite

Because of the high-speed motion of the satellite relative to Earth, and the high-speed motion of the user-carrier relative to the ground, there is always high-speed relative motion between satellite navigation signals and navigation receiver for the user, whether so-called static users or dynamic users. The pseudorange measurement and positioning treatment of the user receiver is subject to dynamic effects. As the signal accumulation time of pseudorange measurement is in the tens of milliseconds, there are the following misunderstandings in pseudorange engineering measurement:

1. Simply using the following formula to design the accuracy of pseudorange measurement

   $$\sigma_{DLL}^2 = \frac{B_L d}{C/N_0}\left[1 + \frac{1}{T\,(C/N_0)(1-d)}\right]$$

   This ignores the fact that the signal is often limited bandwidth in engineering implementation. The filter with overlarge bandwidth will often cause the decline of C/No. In fact, the error in receiver measurement contains dynamic stress error term and random error term, but the formula is only the random error term, which only accounts for a small part of the error in receiver measurement. In the GPS P code receiver, the random error is 0.2 m, the fixed bias term that contains dynamic stress and channel delay is 0.5 m, so the key point of design is to try hard to reduce the error of fixed bias term such as dynamic stress, and so on.

GNSS Systems and Engineering: The Chinese Beidou Navigation and Position Location Satellite, First Edition. Shusen Tan.

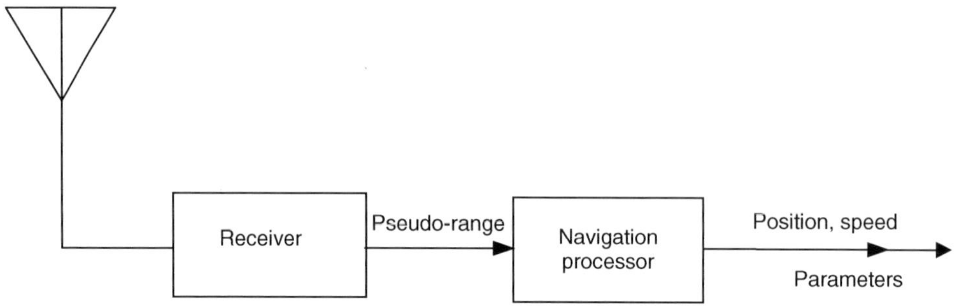

Figure 12.1 Satellite navigation user receiver.

2. The selective basis for signal accumulation time T is insufficient. So, the rather narrow loop bandwidth B_L (0.01 Hz) is selected, which causes locking losing frequently in the loop.
3. Pseudorange measurement is deemed to be the time difference between satellite transmission signal restored from the received signal and local time scale; and it is only related to the resolution ratio of two signals and the leading edge random jitter.
4. Ignoring the accurate reduction at pseudorange measurement time.
5. Confusing pseudorange smoothing with signal accumulation time.
6. Confusing the carrier smoothing with carrier auxiliary code ring tracking.
7. Improving the measuring accuracy of pseudorange measurement with wide correlation technology and ignoring the premise of wide correlation application.

In order to improve the user receiver's pseudorange measurement accuracy fundamentally, and reduce the effect of fixed bias terms, it is necessary to analyze the relative motion characteristics between user and satellite. So, we shall first analyze the motion characteristics within 100 ms of signal accumulation time.

By taking Beijing and Sanya as the static user's observation stations, dynamic analysis is conducted within 100 ms towards the motion states of 1–5 GEO satellites, 6–8 IGSO satellites, and 9–12 MEO satellites of the Compass system. Assuming that the relative radial motions between user and satellite all have first-order dynamic characteristics, the variance, maximum difference, minimum difference, distance variability, and elevating angle separately compared with the actual radial motion curve are calculated; please refer to Table 12.1 and Table 12.2 for data.

The following conclusion can be drawn after comparison of the data in Table 12.1 and Table 12.2.

1. Distance variability of MEO satellite is larger than that of GEO satellite and IGSO satellite, and their maximum radial distance variability is:
   320 m/S for a GEO satellite;
   284 m/S for a IGSO satellite;
   857 m/S for a MEO satellite;
2. The radial displacement caused by the maximum acceleration within 100 ms:
   0.0022 m for a GEO satellite;
   0.0029 m for a IGSO satellite;
   0.0114 m for a MEO satellite;

Table 12.1 Analog computation of relative motion dynamic characteristics for Beijing static users.

Time	Satellite No.	Variance (m)	Maximum Difference (m)	Minimum Difference (m)	Inter-Radial Distance variance (m/100 ms)	Elevating Angle (°)	Remarks
0.00	1	0.0012	0.0019	−0.0013	16.2721	37.5154	
12.00	1	0.0012	0.0020	−0.0013	15.7598	37.5451	
23.92	1	0.0012	0.0019	−0.0012	16.2861	37.5091	
0.00	2	0.0010	0.0017	−0.0011	−22.5963	30.7253	
12.01	2	0.0010	0.0017	−0.0011	−23.3484	30.7195	
23.92	2	0.0010	0.0016	−0.0011	−22.5900	30.7300	
0.00	3	0.0013	0.0022	−0.0014	−3.6133	43.2898	
12.00	3	0.0014	0.0022	−0.0014	−4.2538	43.3162	
23.92	3	0.0013	0.0022	−0.0014	−3.6068	43.2924	
0.00	4	0.0008	0.0014	−0.0009	26.9392	25.6164	
12.00	4	0.0009	0.0014	−0.0009	26.4794	25.6358	
23.92	4	0.0008	0.0014	−0.0009	26.9531	25.5950	
0.00	5	0.0006	0.0009	−0.0006	−31.2707	16.0445	
12.00	5	0.0006	0.0010	−0.0006	−32.0434	16.0252	
23.92	5	0.0006	0.0009	−0.0006	−31.2645	16.0594	
0.00	6	0.0012	0.0020	−0.0014	28.4966	42.2586	Starting
2.88	6	0.0003	0.0005	−0.0003	26.2581	5.0871	Variation point
8.95	6	0.0003	0.0005	−0.0003	−25.1459	5.0351	Rising
15.39	6	0.0018	0.0030	−0.0020	−7.1820	77.6669	Maximum
17.24	6	0.0018	0.0029	−0.0019	0.0028	73.2143	Variation point
23.92	6	0.0012	0.0019	−0.0003	28.4501	42.4729	Ending
0.00	7	0.0018	0.0029	−0.0019	−4.9962	76.7520	Starting
1.50	7	0.0017	0.0028	−0.0019	0.0022	72.8499	Variation point
10.86	7	0.0003	0.0005	−0.0003	26.4348	5.0309	Falling
16.93	7	0.0003	0.0006	−0.0004	−24.5857	5.0877	Rising
23.38	7	0.0018	0.0029	−0.0019	−7.4700	77.5352	Maximum
23.92	7	0.0018	0.0029	−0.0019	−5.0862	76.7753	Ending
0.98	8	0.0003	0.0005	−0.0003	−24.9753	5.0480	Rising
7.44	8	0.0018	0.0029	−0.0020	−7.8648	77.4136	Maximum
9.51	8	0.0018	0.0029	−0.0019	−0.0093	72.8132	Variation point
18.83	8	0.0003	0.0005	−0.0004	26.8016	5.0525	Falling
4.63	9	0.0006	0.0010	−0.0007	−81.7136	5.0414	Rising
8.12	9	0.0068	0.0110	−0.0074	−4.9687	76.6756	Maximum
8.27	9	0.0068	0.0110	−0.0073	−0.0765	76.2521	Variation point
12.60	9	0.0005	0.0009	−0.0060	84.8301	5.0538	Falling
2.79	10	0.0010	0.0017	−0.0011	−67.7333	5.0403	Rising
5.84	10	0.0070	0.0113	−0.0076	−0.0127	82.6389	Variation point

(Continued)

Table 12.1 (Continued)

Time	Satellite No.	Variance (m)	Maximum Difference (m)	Minimum Difference (m)	Inter-Radial Distance variance (m/100 ms)	Elevating Angle (°)	Remarks
5.96	10	0.0070	0.0114	−0.0075	4.5548	83.3790	Maximum
9.86	10	0.0006	0.0010	−0.0006	82.5068	5.0632	Falling
2.87	11	0.0014	0.0022	−0.0015	−49.4577	5.0705	Rising
3.73	11	0.0021	0.0035	−0.0023	−27.1524	9.5778	Maximum
4.64	11	0.0018	0.0029	−0.0020	−5.0609	5.0342	Falling
10.84	11	0.0010	0.0016	−0.0011	−68.6558	5.0217	Rising
13.94	11	0.0071	0.0114	−0.0076	0.0643	84.0441	Variation point
14.02	11	0.0070	0.0114	−0.0077	3.0101	84.4649	Maximum
17.94	11	0.0006	0.0009	−0.0006	83.1778	5.0801	Falling
0.58	12	0.0008	0.0013	0.0008	−76.8364	5.1158	Rising
2.44	12	0.0037	0.0060	−0.0040	−29.7995	26.7257	Maximum
3.45	12	0.0033	0.0053	−0.0036	−0.0775	19.7573	Variation point
4.44	12	0.0018	0.0029	−0.0020	15.3017	5.0362	Falling
9.32	12	0.0014	0.0022	−0.0015	−47.6817	5.0011	Rising
11.83	12	0.0062	0.0100	−0.0067	0.0035	58.4634	Variation point
12.25	12	0.0063	0.0102	−0.0068	16.7331	61.3441	Maximum
15.32	12	0.0005	0.0009	−0.0006	85.7298	5.0995	Falling

Notes:
a) Rising: After the satellite is out of the horizon and the elevating angle is over 5°;
b) Variation point: The distance changes from large-small to small-large, or vice versa;
c) Maximum: The maximal height;
d) Falling: The elevating angle of the satellite falls below 5°.

3. The maximum radial distance variability and maximum radial displacement increase with latitude of the observation station, and the data at the Beijing Station is greater than that at the Sanya Station.
4. The radial distance variability of the low elevating angle is large. The high elevating angle has large radial displacement and poor linearity.
5. For fixed observation station, the maximum radial distance error caused by MEO navigation satellite's nonlinear effect within 100 ms is 0.0114 m. For the accuracy of 0.1 m pseudorange measurement, it can be ignored. There shall be enough pseudorange measurement accuracy only considering the first-order dynamic effect.
6. If the user's motion trajectory can be regarded as the first-order dynamic effect within the signal's accumulation time of receiver non-coherent delay locked loop (DLL) correlation processing, the relative motion between the user and the satellite is also consistent with the first-order dynamic effect. When only considering the effects of relative motion speed on the pseudorange measurement, the acceleration terms can be ignored. All vehicles, ships, and high-speed aircraft can satisfy this condition. Therefore, even under high dynamic conditions, the relative motion between the user and the satellite within the integration period can be regarded as constant motion.

Table 12.2 Analog computation of relative motion dynamic characteristics for Sanya static users.

Time	Satellite No.	Variance (m)	Maximum Difference (m)	Minimum Difference (m)	Inter-Radial Distance Variance (m/100 ms)	Elevating Angle (°)	Remarks
0.00	1	0.0014	0.0022	−0.0015	25.3382	49.3213	
12.00	1	0.0014	0.0023	−0.0015	25.1954	49.3186	
23.92	1	0.0014	0.0022	−0.0014	25.3537	49.3076	
0.00	2	0.0014	0.0022	−0.0015	−24.6390	49.8542	
12.00	2	0.0014	0.0023	−0.0015	−25.0849	49.8485	
23.92	2	0.0014	0.0023	−0.0015	−24.6317	49.8596	
0.00	3	0.0017	0.0028	−0.0019	0.8282	68.2821	
12.00	3	0.0018	0.0028	−0.0019	0.5396	68.3150	
23.92	3	0.0017	0.0028	−0.0019	0.8347	68.2848	
0.00	4	0.0009	−0.0014	−0.0010	37.0985	29.7325	
12.00	4	0.0009	−0.0014	−0.0010	36.9982	29.7117	
23.92	4	0.0009	−0.0014	−0.0009	37.1125	29.6996	
0.00	5	0.0008	0.0014	−0.0009	−37.0829	29.1015	
12.00	5	0.0009	0.0014	−0.0010	−37.5766	29.0845	
23.92	5	0.0008	0.0014	−0.0009	−37.0740	29.1203	
0.00	6	0.0016	0.0027	−0.0018	18.7212	65.2587	Starting
6.09	6	0.0004	0.0007	−0.0005	4.5086	7.4358	Minimum
6.63	6	0.0005	0.0008	−0.0005	0.0289	8.0846	Variation point
13.35	6	0.0019	0.0030	−0.0021	−0.4677	89.0786	Maximum
23.92	6	0.0016	0.0027	−0.0018	18.6468	65.4230	Ending
0.00	7	0.0016	0.0025	−0.0017	8.9005	57.2108	Starting
14.07	7	0.0005	0.0008	−0.0005	5.0143	7.4418	Minimum
14.68	7	0.0005	0.0008	−0.0005	0.0230	8.2676	Variation point
21.33	7	0.0018	0.0030	−0.0020	−0.4394	89.0220	Maximum
23.92	7	0.0016	0.0025	−0.0016	8.8733	57.3704	Ending
0.00	8	0.0006	0.0011	−0.0007	−9.4574	15.1896	Starting
5.38	8	0.0019	0.0030	−0.0020	−1.1089	88.8063	Maximum
5.57	8	0.0018	0.0030	−0.0020	−0.0113	86.9784	Variation point
22.05	8	0.0004	0.0007	−0.0005	4.9329	7.4787	Minimum
22.64	8	0.0005	0.0008	−0.0005	−0.0003	8.2601	Variation point
23.92	8	0.0006	0.0010	−0.0007	−9.3595	15.0495	Ending
3.58	9	0.0007	0.0012	−0.0008	−78.8652	5.1498	Rising
6.78	9	0.0066	0.0107	−0.0071	−14.7900	68.9748	Maximum
7.25	9	0.0065	0.0105	−0.0070	−0.0116	65.9471	Variation point
12.87	9	0.0008	0.0013	−0.0009	75.1575	5.0353	Falling
1.60	10	0.0013	0.0021	−0.0014	−55.5881	5.0747	Rising
4.82	10	0.0070	0.0113	−0.0075	−0.0460	80.2019	Variation point

(Continued)

Table 12.2 (Continued)

Time	Satellite No.	Variance (m)	Maximum Difference (m)	Minimum Difference (m)	Inter-Radial Distance Variance (m/100 ms)	Elevating Angle (°)	Remarks
4.99	10	0.0070	0.0113	−0.0076	6.7046	81.1675	Maximum
8.86	10	0.0008	0.0014	−0.0009	71.9669	5.0301	Falling
9.66	11	0.0012	0.0020	−0.0013	−56.5745	5.1119	Rising
12.92	11	0.0070	0.0114	−0.0076	−0.1407	82.1061	Variation point
13.05	11	0.0070	0.0114	−0.0076	4.9960	83.1675	Maximum
16.97	11	0.0009	0.0114	−0.0009	71.7428	5.0301	Falling
1.37	12	0.0007	0.0011	−0.0007	−80.7873	5.0818	Rising
3.30	12	0.0035	0.0056	−0.0037	−38.8747	25.3603	Maximum
5.92	12	0.0018	0.0029	−0.0019	−5.7564	5.0141	Falling
7.85	12	0.0017	0.0027	−0.0018	−22.9039	5.0455	Rising
10.65	12	0.0055	0.0089	−0.0059	0.0242	47.5716	Variation point
11.43	12	0.0058	0.0093	−0.0062	25.7152	54.0183	Maximum
14.29	12	0.0006	0.0009	−0.0006	84.8147	5.0415	Falling

12.2 Pseudorange Measurement and Error Analysis

The navigation receiver uses non-coherent DLL to complete the acquisition, tracking of the navigation signals, and recovers satellite navigation signals.

The definition of pseudorange is discussed in Section 7.2.1 in Chapter 7. It indicates the product by multiplying the sum of relative time difference between the satellite and the user at a given time and the transmission delay of signal from the satellite to the user by the speed of light, as shown in formula (7.1). It contains the two quantitative definitions, one is the time when the pseudorange occurs, and the other is the size of pseudorange. The time when the pseudorange occurs should be the time when the pseudorange measurement signal is transmitted from the satellite. The relative geometrical relationship is shown in Figure 12.2. In the signal accumulation time, both the satellite from t_{sini}-t_{senq} and the user from t_{sini}-t_{sent} can be regarded as high-speed linear motion. Because of the relative motion between satellite and user, when the rate of spreading code (chip) launched by the satellite reaches the receiver, its width changes and is not the same as the chip width of the local receiver. It is also different from the chip width launched by the satellite. The pseudorange ρ_{ini} (t) at the initial moment of the signal accumulation is not equal to the pseudorange ρ_{ent} (t) at the end of the signal accumulation. Many documents define the measurement value ρ_{ent} (t) at local time t_{rent} as the pseudorange measured at the pseudo t_{rent} time. Since the satellite begins to transmit measurement signals at t_{sini} to t_{rent}, and since the user starts to accumulate the signals at t_{sini} to t_{rent}, as the relative motion between the satellite and the user can be seen as a linear relationship, when the carrier loop is used for assisting and tracing the code ring and it is in good condition, the chip width of the user received pseudo-code is not changed, and the symmetry of the user-related properties is essentially unchanged. The pseudorange measurements under relative motion can be equal to those on fixed spatial

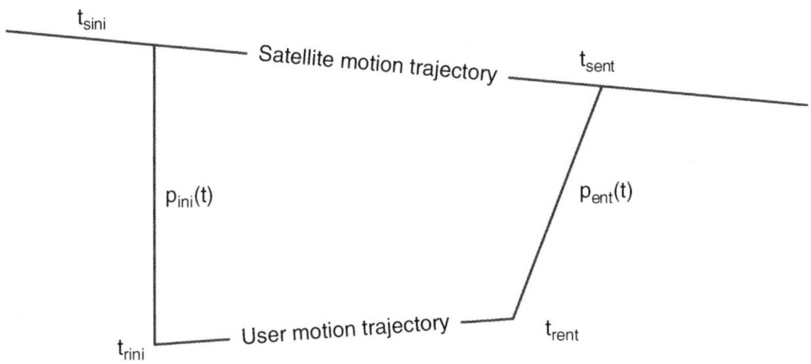

Figure 12.2 Geometrical relationship of pseudorange.

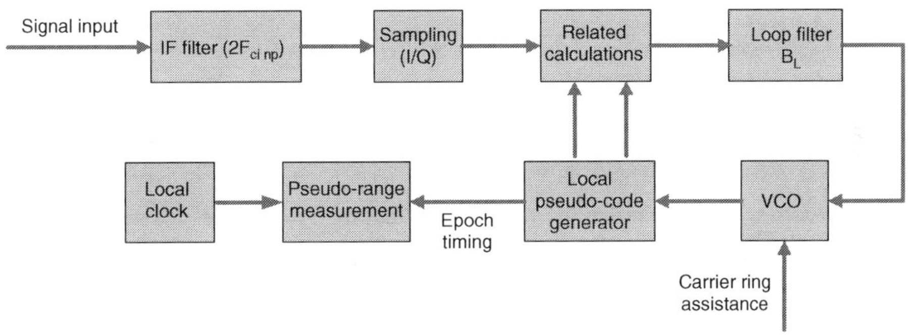

Figure 12.3 Typical DLL delay locked loop.

distances. The signal accumulation between t_{rini} and t_{rent} does not affect the accuracy of pseudorange measurement. So, it can still be regarded as the pseudorange measurement at t_{rent}.

A typical DLL delay locked loop with the premise of high-precision ranging is as shown in Figure 12.3. The design of pseudorange accuracy and selection of loop parameters will be discussed next.

12.2.1 Types and Characteristics of Pseudorange Error

From the perspective of the definition of pseudorange, investigate the recovery process of the user receiver towards the satellite signal, the influence of noise on the accuracy of ranging, the stability of channel delay differential, and so on, and its pseudorange measurement error can be divided into three parts: random error, fixed deviation (systematic error), and dynamic stress error. The random error follows the Gaussian distribution; it is caused by the system thermal noise and expressed with standard deviation RMS. The fixed deviation is caused jointly by the satellite signal transmission channel delay and the user receiving channel delay, this deviation can be weakened by accurate calibration and ultimately affect the ranging accuracy and positioning accuracy in the form of system residuals. In actual engineering application, calibration deduction is usually not made; instead, it is to control long-term delay stability of launch channel and reception channel separately, and to list the channel fixed delay in the difference between satellite clock and user clock separately.

12.2.2 Dynamic Stress Error

The premise for accurate pseudorange measurement is that the user receiver can restore pseudorange phase of the transmitted signal accurately. From the perspective of the theoretical analysis, because of dynamic effect, the chip width of the received code is no longer that of the launching code. The width is not constant, so the received pseudo-code and local pseudo-code cannot be aligned one-to-one. Only the code phase at the phase alignment moment can accurately reflect the pseudorange. In other words, there is dynamic stress error left in the measured pseudorange under the condition of incorrect alignment. The residual error varies with the relative motion between the satellite and the user receiver. This error cannot be weakened by statistical methods, nor can it be eliminated with accurate calibration, so it becomes part of the pseudorange measurement: that is, pseudorange error:

$$\delta_{DLL} = \delta_{n\,DLL} + \delta_{d\,DLL} \tag{12.1}$$

Wherein, $\delta_{n\,DLL}$ is the random error caused by noise, and $\delta_{d\,DLL}$ is the dynamic stress error.

In engineering practice, when the constant speed reaches 3000 km/s, the width of code element changes only for 1 ns and can be ignored in general.

12.2.3 Pseudorange Random Error

As mentioned previously, the satellite navigation receiver uses a non-coherent delay locked loop (DLL) to recover the received satellite signal. As figured out by A.J. Van Dierendonck in 1992, when the bandwidths of navigation signal and receiver are infinite, the code loop tracking noise variance calculated with early power phase detection algorithm is

$$\delta^2_{nDLL} = \frac{B_c d}{c/N_0}\left[1 + \frac{1}{T(c/N_0)(1-d)}\right] \tag{12.2}$$

In engineering practice, neither the bandwidth of signal nor that of the receiver can be infinite, so there is large difference between the previous variance and engineering practice.

J.W. Betz [18] models the navigation signal transmission channel and the receiver channel bandwidth as rectangular filters, and obtains the variance approximation formula of thermal noise by analysis:

$$\delta^2_{nDLL} = \begin{cases} \dfrac{B_L(1-0.5B_cT)}{2(C/N_0)}2d\left[1 + \dfrac{1}{T(C/N_0)(1-d)}\right] & bd \geq \dfrac{\pi}{2} \\[2ex] \dfrac{B_L(1-0.5B_cT)}{2(C/N_0)}\left[\dfrac{1}{b} + \dfrac{b}{\pi-1}\left(2d - \dfrac{1}{b}\right)^2\right]\left[1 + \dfrac{1}{T(C/N_0)(1-d)}\right] & \dfrac{1}{2} \leq bd \leq \dfrac{\pi}{2} \\[2ex] \dfrac{B_L(1-0.5B_cT)}{2(C/N_0)}\left(\dfrac{1}{b}\right)\left[1 + \dfrac{1}{T(C/N_0)}\right] & 0 \leq bd \leq \dfrac{1}{2} \end{cases} \tag{12.3}$$

Wherein, $\delta^2_{n\,DLL}$ is the noise variance normalized with width of code element (chip), and δ_{nDLL} is the standard deviation (1σ);

$\delta_{n\,DLL}$ multiplying code width is the noise standard deviation in second (time unit), and then multiplying speed of light is the noise standard deviation in meter (distance unit).
b is the signal bandwidth after normalized with the width of code element. In normal satellite navigation user receiver, $b = 2$;
d is the correlation distance after normalized with the width of code element, $1/d$ means the number of correlators that can be scheduled between the advanced codes and the immediate codes that are copied locally;
B_L is the bandwidth of code tracing loop, in Hz;
T is the signal accumulation time, in s;
C/N_0 is the carrier-to-noise ratio of the receiver, in Hz.

The navigation frequency is quite crowded, so except for considering the carrier-to-noise ratio of launching signals in engineering application, the degradation of the carrier-to-noise ratio of the received signal by other navigation signals in the same frequency band shall also be considered.

The first line of the formula (12.3) is applicable for conditions when $b \geq 4$ and the wide correlation $d > \frac{1}{2}$; the second line is applicable for conditions when $b \leq 2$ and the wide correlation $d = \frac{1}{2}$; and the third line is applicable for conditions when $b > 4$ and the narrow correlation $d < \frac{1}{8}$.

Generally speaking, the random error can be reduced by reducing the loop bandwidth, extending the signal accumulation time and narrow correlation concurrent processing. But it is easy to lose the lock of the loop by reducing the loop bandwidth that will reduce signal acquisition and tracking ability; and the extending of signal accumulation time will increase the dynamic stress error.

The preconditions for narrow correlation processing are $b \geq 4$ and the total number of narrow correlation device in concurrent processing is $1/d = 16$. The minimum mean variance ratio of narrow correlation and wide correlation is 1:2; with the increase of actual bandwidth of signals, the number of narrow correlation can be decreased. This method is widely used for determining whether there is distortion in the satellite navigation signals (i.e., determining the integrity of navigation signals). Since the actual occupied bandwidth of the navigation signal is twice of the chip rate, when $b \geq 4$, it will certainly receive other signals at the same time, thus causing deterioration of the carrier-to-noise ratio. But this determination method for navigation signal integrity is quite complex, so multiple concurrent processing narrow correlation devices must be used for determination of the integrity. The narrow correlation processing is then used only for the undistorted signal to improve pseudorange measurement accuracy. Table 12.3 describes pseudorange measurement accuracy with wide and narrow correlations for the two code element widths and for different signal bandwidths.

It can be seen that the cost of improving the ranging accuracy with narrow correlation technology is quite huge. Correctness of the result cannot be determined with only one narrow correlator. The results of the code phase tracking can be confirmed after the navigation signal is determined to be without distortion using 16 narrow correlators. The fewer the narrow correlators, the greater the possibility of misjudgment. Through comparison of 16 narrow correlators with one wide correlator, the ratio of complexity is 16:1, and the ratio of ranging error is 1:2. So the navigation user receiver usually adopts wide correlator, not the narrow one. That is $b = 2, d = 1/2, B_L \leq 1.0$ Hz, $T \leq 0.02$ s, substitute

Table 12.3 Pseudorange measurement accuracy for the two code element widths and different correlation space.

Conditions		Measurement Accuracy (cm)	
		$C/N_0 = 38$ dB Hz	$C/N_0 = 41$ dB Hz
$T_c = 10.23$ Mbps $B_L = 1$ Hz $T = 0.02$s	$d = 1/2$ Signal bandwidth $B_S = 80$ MHz	26	18.3
	$b = 8, d = 1/16$ Signal bandwidth $B_S = 80$ MHz Total number of correlators $n = 32$	10	6.5
	$d = 1/2$ Signal bandwidth $B_S = 40$ MHz	26	20
	$b = 4, d = 1/8$ Signal bandwidth $B_S = 40$ MHz Total number of correlator $n = 16$	13	10
$T_c = 2.046$ Mbps $B_L = 0.05$ Hz T = 0.02 s	$d = 1/2$ Signal bandwidth $B_S = 40$ MH	29	20
	B_S=40 MHz $n = 16$	14	10

them into (12.3), and the result is as follows.

$$\sigma^2_{nDLL} \approx \frac{0.73B_L}{2(C/N_0)}\left[1 + \frac{2}{T(C/N_0)}\right]$$

In order to acquire the pseudorange measurement accuracy of 0.3 m, pseudorange smoothing technology can be adopted.

12.2.4 Pseudorange Smoothing Technology

Pseudorange smoothing technology is with the linear characteristic based on the motion speed of the satellite relative to the receiver within the smoothing time $T_A = 20{\sim}100$ ms. When there are n pseudorange measurement values $\rho_1, \rho_2 \ldots \rho_n$ with the equal interval, the pseudorange value after linear smoothing is

$$\rho_A(t_A) = \frac{1}{n}\sum_1^m \rho_n$$

The pseudorange value the moment after smoothing is

$$t_A = \frac{1}{n}\sum_1^n t_n$$

The pseudorange error after smoothing is

$$\delta_{ADLL} = \frac{1}{\sqrt{n}}\delta_{nDLL}$$

Given $T = 2$ ms, $\delta_{nDLL} = 0.6$ m. Given $T_A = 20$ ms and $n = 10$, $\delta_{ADLL} = \frac{1}{\sqrt{10}}0.6 = 0.2$ (m).

12.3 Positioning and Filtering Processing

Section 7.2.2 introduces navigation positioning solving equation and positioning accuracy estimation. The navigation filter can be used in order to further improve positioning accuracy and velocity measurement accuracy. The function of positioning and filter processing is to convert pseudorange observation values into user position estimation values and there are two methods of conversion.

The principle of the first method is shown in Figure 12.4: first, convert a group of pseudorange observation values into unfiltered positions through position calculator; then reduce the noise error in position estimation with α/β tracker to obtain the filtering position. The principle of the second method is shown in Figure 12.5: convert a group of pseudorange observation values into filtered positions uniaxially with a Kalman filter.

12.3.1 α/β Tracker

The first conversion method uses position algorithm and α/β tracker filtering [18]. As mentioned previously, the satellite signals received by the receiver contain noises; assuming that the noise has the mean value of zero and follows the Gaussian distribution, it will cause pseudorange error and then cause error in unfiltered positions. α/β tracker shall eliminate the position data errors as much as possible.

The α/β tracker filters out the position estimation values that the calculator inputs at the end position of each update period. The length of update period Tup is identical to the position update period of the receiver. Assuming position estimation value adopts geocentric Cartesian coordinate system, three α/β trackers shall filter out three position variables of X, Y, and Z separately. By taking the filtered single position variable as an example, the α/β tracker is introduced.

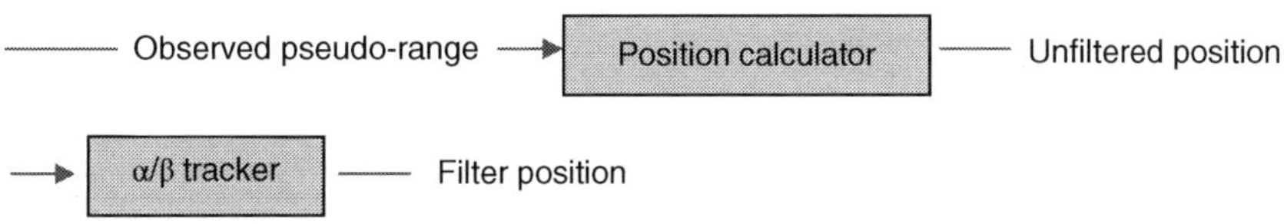

Figure 12.4 Navigation processor with α/β Tracker.

Figure 12.5 Navigation processor with Kalman tracker.

Working Principle of α/β Tracker

At the end of the kth update period, if α/β tracker receives unfiltered position X_k from position calculator, it shall be compare with the predicted position $\hat{X}_k$ of the tracker to obtain the filtered position $\overline{X}_k$ by calculation. It is the weighted average of predicted position and unfiltered observation position, and its relationship is shown as follows.

$$\overline{X}_k = (1-\alpha)\hat{X}_k + \alpha X_k \tag{12.4}$$

This formula is the relationship between filtered position, predicted position and unfiltered position, and the weighting coefficient α shall be valued within [0, 1]. If $\alpha = 0$, the filtered position equals the predicted position; as α gets larger, the weight of unfiltered position gets larger, and the weight of the predicted position gets smaller; and when $\alpha = 1$, the filtered position equals the unfiltered position.

How the tracker obtains the filtering estimation value $\overline{\dot{X}}_k$ of user's velocity will be introduced next. First of all, the unfiltered observation velocity $\dot{X}_k$ is calculated according to the following formula.

$$\dot{X}_k = (\overline{X}_k - \overline{X}_{k-1}) / \mathrm{Tup} \tag{12.5}$$

Then, the tracker conducts weighted summation toward unfiltered observation velocity $\dot{X}_k$ and predicted velocity $\hat{\dot{X}}_k$ to figure out the filter velocity estimation value $\overline{\dot{X}}_k$.

$$\overline{\dot{X}}_k = (1-\beta)\hat{\dot{X}}_k + \beta\dot{X}_k \tag{12.6}$$

Next, the tracking filter shall use the user's current velocity, position filtering estimation value, and second-order model of user's motion to predict user's future velocity and position. The second-order model of user's motion is to assume that the user's velocity is a first-order constant (acceleration being zero). So the tracker can set the predicted velocity in the future as the current filtering speed.

$$\hat{\dot{X}}_{k+1} = \overline{\dot{X}}_k \tag{12.7}$$

The user's position predicted by the tracker is

$$\hat{X}_{k+1} = \overline{X}_k + \mathrm{Tup}\,\overline{\dot{X}}_k \tag{12.8}$$

Up until now, the tracker can receive the unfiltered observation position of another filter. Formula (12.4)–(12.8) can be simultaneous and be simplified into three commonly used formulas of the α/β tracker.

$$\overline{X}_k = \hat{X}_k + \alpha(X_k - \hat{X}_k) \tag{12.9}$$

$$\overline{\dot{X}}_k = \hat{\dot{X}}_{k-1} + \beta / \mathrm{Tup}(X_k - \hat{X}_k) \tag{12.10}$$

$$\hat{X}_{k+1} = \overline{X}_k + \mathrm{Tup}\,\overline{\dot{X}}_k \tag{12.11}$$

α and β determine the response attenuation of the tracker and can be changed independently. The response of the tracker can reach "optimization" and the commonly used formula put forward by T.R. Benedict and G.W. Bordner is shown in (12.12).

$$\beta = \alpha^2(2-\alpha) \tag{12.12}$$

A mild under-damping tracking response is given.

The α/β tracker is a time-discontinuous and low-pass filter with the gain of 1. Values of α and β determine the bandwidth of filter, and the error of α/β determines the damping of filter, as the α and β decrease, the bandwidth decreases as well; as the α and β increase, the bandwidth increases too. When α and βchange from 0 to 1, the bandwidth changes from 0 Hz to (1/2 Tup) Hz. So, by decreasing α and β, the noise component in the filtered position data can be reduced.

The cost paid for reducing α and β can be compensated by increasing the time constant of tracker. As the tracker is a second-order model based on the user's motion, the user's acceleration will cause error in the user's position estimation. The time required to eliminate the error is proportional to the time constant of tracker. If the user has a uniform acceleration, the tracking estimation value of the user's position in the turning period shall have errors constantly. For the uniform acceleration, the error shall be constant too and proportional to the time constant of tracker. The reduction of α and β shall result in the increase of deviation error caused by acceleration in filtered position data.

In the article of *α/β Tracking Error of Orbit Targets* by Sinsky, the noise reduction formula caused by the tracker and the deviation formula caused by linear uniform acceleration are deduced.

Noise reduction amount represents the ratio of filtering output signal variance to non-filtered input signal variance, and is shown in the following formula

$$\begin{aligned} K_{\bar{x}} &= \sigma_{\bar{x}}^2/\sigma_x^2 \\ &= \frac{2\alpha^2 + \beta(2 - 3\alpha)}{4 - \beta - 2\alpha} \end{aligned} \tag{12.13}$$

Wherein, $\sigma_{\bar{x}}^2$ is the filtering output signal variance; σ_x^2 is non-filtered input signal variance; $K_{\bar{x}}$ is the variance reduction rate; when $\alpha = 1$, $K_{\bar{x}}$ shall be constant as 1. That is because when $\alpha = 1$, the filtered position is the unfiltered position, and the noise of data is not reduced.

During the period of linear uniform acceleration, the formula of deviation caused by acceleration is shown here.

$$|\varepsilon_{\bar{x}}| = (1 - \alpha)\, a\, \mathrm{Tup}^2/\beta \tag{12.14}$$

In this formula, a is the acceleration of the carrier, when $\alpha = 1$, for all β that is not zero, $|\varepsilon_{\bar{x}}|$ equals zero. When the filtering data equals to the unfiltered data, the filter cannot be the reason of the deviation caused by acceleration. It can be proved that this error applies to dimensional isotropic acceleration conditions. Table 13.4 lists the corresponding relation between the deviation caused by variance reduction rate, acceleration and α when the ship acceleration is 0.13 m/s^2 and the update period is 5.4 s.

Table 12.4 shows that if α is quite small, so is the variance reduction rate, the deviation caused by the acceleration becomes large. If α is quite large, the deviation caused by the acceleration is small, but the variance reduction rate approaches 1. So, when α is selected, two factors shall be considered comprehensively so as to reach a satisfying comprehensive effect at the specified noise and acceleration conditions. In order to measure the overall positioning accuracy and define a standard function J_x, it equals to the sum of the error caused by 2σ noise and the error caused by the acceleration. This is based on the worst case where both errors are present simultaneously. Thus,

$$J_{\bar{x}} = 2\,\sigma_{\bar{x}} + \left|\sum \bar{x}\right| \tag{12.15}$$

Table 12.4 Functional relationship between α and the deviation error caused by variance reduction rate and acceleration.

α	0.1	0.2	0.3	0.4	0.5	0.6	0.7	0.8	0.9	1.0
$k_{\bar{x}}$	0.08	0.15	0.24	0.32	0.41	0.51	0.61	0.71	0.84	1.0
$\lvert\varepsilon_{\bar{x}}\rvert$	648	136	50	23	11	5.9	3.0	1.4	0.51	0.0

It can be figured out from formula (12.13) that

$$\sigma_{\bar{x}} = \sigma_x \sqrt{k_{\bar{x}}} \tag{12.16}$$

In the formula, σ_x equals the product of multiplying chip error σ_c of the receiver by horizontal dilution of precision (HDOP), that is $\sigma_x = \sigma_c(\text{HDOP})$.

Here the code element error is deemed as the user equivalent distance error, and the propagation error of ionosphere, and so on is ignored.

α/β Filter of Switch Type

In order to ensure the high-precision positioning when there is no acceleration and no large error when there is acceleration, the most suitable solution is to use the variable α tracker. The α/β tracker of switch type is one of such variable α filters. When no acceleration is detected by the tracker, the filtered position data is provided to the user. And when acceleration is detected by the tracker, the non-filtered position data is provided to the user. In actual operation, the tracker shall choose to provide the filtered data and the non-filtered data at a fixed α value. Because when α value is fixed, the predicted position output of the tracker is used for detecting the acceleration of the ship. This predicted position shall be subtracted from the non-filtered position, as is shown in the following formula.

$$\delta_{X,k} = X_k - \hat{X}_k \tag{12.17}$$

When there is no acceleration, the average value of difference $\delta_{X,k}$ is 0; if there is acceleration, it is not zero, and its value equals to the acceleration deviation caused by the tracker. When the difference goes through a low-pass filter, the acceleration can be detected. Adopting a method similar to α/β tracker, the filtered position deviation $\bar{\delta}_{X,k}$ can be shown as follows.

$$\bar{\delta}_{X,k} = (1-\gamma)\,\delta_{X,k-1} + \gamma\,\delta_{X,k} \tag{12.18}$$

γ is a low-pass filter constant, from (12.19), it can be used for the α/β tracker of switch type.

$$\begin{aligned}
&\text{If } \sqrt{\bar{\delta}^2_{X,k} + \bar{\delta}^2_{Y,k} + \bar{\delta}^2_{Z,k}} < TR, \text{ then} \\
&X_k^{out} = \bar{X}_k \\
&\text{If } \sqrt{\bar{\delta}^2_{X,k} + \bar{\delta}^2_{Y,k} + \bar{\delta}^2_{Z,k}} \geq TR, \text{ then} \\
&X_k^{out} = X_k
\end{aligned} \tag{12.19}$$

Wherein, TR is the threshold.

If the sum of low-pass filter outputs on three axes is smaller than the threshold, the filtered position data shall be output to the user; if it is greater than or equal to the threshold, the unfiltered position value shall be output to the user.

Advantages of using α/β filter of switch type are: during the period when the user has no acceleration, it allows the use of small α value and keeps the ability of reducing noise of normal α/β filter unchanged. During the period with acceleration, the position accuracy acquired by the user will decrease, but it is still acceptable.

12.3.2 Kalman Filter

The other method to convert pseudorange observation values into filtered position estimation is to make the pseudorange observation value pass through the Kalman filter [18]. The Kalman filter is a minimum error covariance calculator, that is, when the filtered noise is Gaussian noise and its average value is zero, there is the minimum variance towards the error in user position; when the noise is not Gaussian noise, the Kalman filter is an optimal linear filter. The Kalman filter is implemented by vectors and matrixes rather than scalars in α/β tracker. The first vector in Kalman filter is the user state vector

$$\mathrm{X}_{\mathrm{K}} = [\overline{X}_{\mathrm{K}}\overline{\dot{X}}_{\mathrm{K}}\overline{Y}_{\mathrm{K}}\overline{\dot{Y}}_{\mathrm{K}}\overline{Z}_{\mathrm{K}}\overline{\dot{Z}}_{\mathrm{K}}\overline{T}_{\mathrm{K}}\overline{\dot{T}}_{\mathrm{K}}]^{\mathrm{T}} \tag{12.20}$$

In the formula, the superscript ^{T} represents transposition. State vector fully describes the user's filtered position, velocity and clock error rate. This is a vector formula for each component measured separately with α/β filter.

The next matrix is the transition matrix

$$\varphi = \begin{bmatrix} 1 & \Delta T & 0 & 0 & 0 & 0 & 0 & 0 \\ & 1 & 0 & 0 & 0 & 0 & 0 & 0 \\ & & 1 & \Delta T & 0 & 0 & 0 & 0 \\ & & & 1 & 0 & 0 & 0 & 0 \\ & & & & 1 & \Delta T & 0 & 0 \\ & & & & & 1 & 0 & 0 \\ & & & & & & 1 & \Delta T \end{bmatrix} \tag{12.21}$$

Here, ΔT is the time interval between pseudorange observed values; and the predicted state vector at the next time interval can be obtained by multiplying the transition matrix by state vector.

$$\hat{X}_{\mathrm{K+1}} = \varphi \mathrm{X}_{\mathrm{K}} \tag{12.22}$$

Wherein, $\hat{X}_{\mathrm{K+1}} = [\hat{X}_{\mathrm{K+1}}\ \hat{\dot{X}}_{\mathrm{K+1}}\hat{\mathrm{Y}}_{\mathrm{K+1}}\hat{\dot{Y}}_{\mathrm{K+1K+1}}\ \hat{\dot{Z}}_{\mathrm{K+1}}\hat{T}_{\mathrm{K+1}}\ \hat{\dot{T}}_{\mathrm{K+1}}]^{\mathrm{T}}$

Through calculation for formula (12.22), it can be seen that

$$\hat{X}_{\mathrm{K+1}} = X_{\mathrm{K}} + \Delta \mathrm{T} X_{\mathrm{K}} \tag{12.23}$$

And

$$\hat{X}_{\mathrm{K+1}} = \hat{X}_{\mathrm{K}} \tag{12.24}$$

Kalman filter and α/β tracking adopt the same second-order user's motion model. So the prediction equation adopted by (12.23) and (12.24) is nearly the same with that which α/β tracker adopts, with only a slight difference, which is that the α/β tracker uses the update period T up of the receiver as the interval between K^{th} and $K+1^{th}$ state, while the Kalman filter uses the time interval ΔT between pseudorange observation values as the time interval of adjacent state. The reason is that before the computer filters out the user's position, the position calculation and the combination of α/β tracker waits for a complete set of pseudorange observation values. On the contrary, Kalman filter estimates a new user position on receiving a new pseudorange observation value.

The filter calculates the predicted distance $\hat{P}_K$ toward the satellite according to predictive state vector $\hat{X}_K$ and known position of the satellite of which pseudorange is to be measured. When the Kalman filter receives the observed pseudorange (P_K) from the receiver, it calculates the difference between the predicted and observed pseudoranges, and this difference is residual $(P^R)_6$; usually called

$$P^R = (P_K - \hat{P}_K) \tag{12.25}$$

According to magnitude of this residual, the filter balances its predicted state matrix and builds a new state matrix, as shown in the following formula

$$X_K = \hat{X}_K + K_0 P^R \tag{12.26}$$

K_0 is the Kalman gain vector and its form is

$$K_0 = [K_{0X}\dot{K}_{0X}K_{0y}\dot{K}_{0y}K_{0Z}\dot{K}_{0Z}K_{0T}\dot{K}_{0T}]^T \tag{12.27}$$

In the following equation formulas, the residuals of $(X_K\text{-}\hat{X}_K)$ and (12.26) are equivalent. In α/β, $(\overline{\dot{X}}_{(K-1)})$ is equivalent to the predicted user's velocity $\hat{\dot{X}}_K$.

Kalman gain vector converts the pseudorange residual into predictive state vector changes. To realize this conversation, two operations are conducted separately, one is coordinate transformation, and the other is weighting.

Coordinate transformation operation enables the Kalman filter to convert the pseudorange residual into predictive state vector changes, the pseudorange residual is the one-dimensional vector from the user to the satellite line of sight, state vector X, Y, Z, and the eight-dimensional vector of the T-axis, this motion is shown in Figure 12.6.

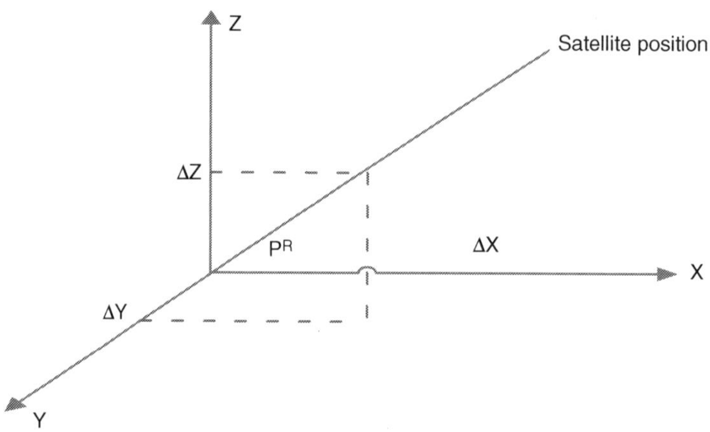

Figure 12.6 Conversion from pseudorange to the user state coordinate.

For gain residual (P^R), there are:

$$\Delta X = P^R COS\theta_X \tag{12.28}$$

$$\Delta Y = P^R COS\theta_Y \tag{12.29}$$

$$\Delta Z = P^R COS\theta_Z \tag{12.30}$$

The angles between the sight vector from satellite to the user and the positive X-, Y-, and Z-axes are θ_X, θ_Y, and θ_Z, respectively.

The coordinate transformation of the clock error term is equal to 1, because the change in residual is directly dependent on the same variation as the clock error.

The Kalman gain vector can also conduct the acceleration calculation similar to α/β of α/β tracker. The variation of the predicted state vector for a given residual can be determined through this calculation. For example, when α is 0.3 and residual is +2, X_K can be calculated by adding 0.6 into $\hat{X}_K$, and the other terms of the Kalman gain vector can be performed in the same manner. If every item of Kalman gain vector is rewritten as the product of the weighted term and the coordinate transformation term, then we can get

$$K_0 = [W_X COS\theta_X \dot{W}_X COS\theta_X W_Y COS\theta_Y \dot{W}_Y COS\theta_Y W_Z COS\theta_Z \dot{W}_Z COS\theta_Z W_T \dot{W}_T]^T \tag{12.31}$$

W is the weighted term, and COS is the coordinate transformation term.

The weighted term determines the magnitude of the change in predicted state vector for a given residual. This residual is the difference between the predicted pseudorange and the observed pseudorange, and is caused by processed noise and measurement noise.

Figure 12.7 explains the relationship between processed noise and measurement noise and its influence on the predicted pseudorange, the actual pseudorange, and the observed pseudorange.

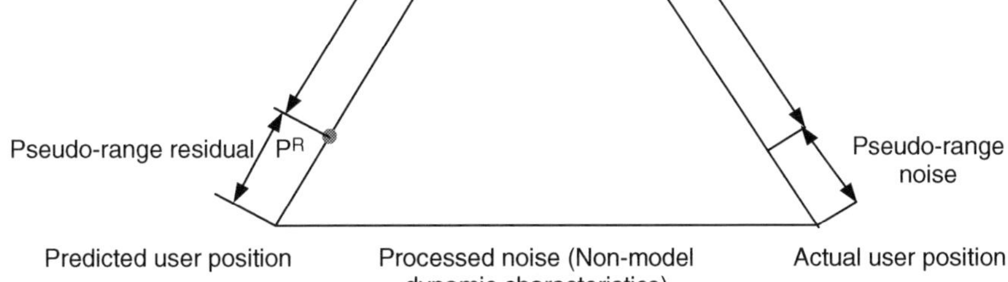

Figure 12.7 Relationship between processed noise and measurement noise and its influence on the predicted, the actual, and the observed pseudoranges.

Any difference between the predicted pseudorange and the actual pseudorange is caused by processed noise. The figure shows that the velocity of satellite motion and signal propagation through the atmosphere is non-modeled and is ignored. So the processed noise equals to the user's non-model motion characteristic. In predicting the user's state, a zero-acceleration model is used for user motion and user clock error, so what impacts the processed noise first is user acceleration and clock error rate. Any difference between the actual pseudorange and the predicted pseudorange is caused by measurement noise. The measurement noise equals the noise error of the receiver.

The output of the Kalman filter is the best possible estimate of the user state. Due to processed noise, the user state is different from the user's predicted state; therefore, the filter should process as much noise as possible. And the user state can be estimated more accurately in this way. However, the measurement noise can cause wrong estimation toward the user state; the Kalman filter should filter out as much noise as possible. Unfortunately, both measurement noise and processed noise present as noise of the observed pseudorange. If the weighting factor of Kalman gain vector is small, it makes the filter ignore the measurement noise, thus ignoring processed noise. If the weighting factor of the Kalman gain vector is large, a large amount of processed noise and measurement noise will pass through the filter. Therefore, if the measurement noise is larger than the processed noise, the accuracy gained by filtering out the measurement noise will be better than that lost in filtering out the processed noise, and the weighting factor shall be very small. However, if the processed noise is much larger than the measurement noise, the precision obtained by processed noise will be better than the accuracy lost in measurement noise, and the weighting factor is large. Kalman gain vector weighting factor is calculated by compromising between processed noise of the filter and filtering out the measurement noise.

In order to calculate Kalman gain vector completely, we must define several new matrices and the first is measurement noise matrix R. Assume that the timing receiver is adopted, it can only generate one pseudorange observation value at one time, then R becomes a scalar and items in R are the expected variance of measurement noise, called measurement noise covariance. This paper will adopt measurement noise covariance and it equals to expected chip variance (σ_{cn}).

The next matrix to be defined is processed noise matrix Q. The terms that make up the matrix are called the processed noise covariance, and they represent the expected covariance of the individual bit positions in the user state. Assume that there is no orthogonal correlation between each noise of the user state, and then all diagonal terms of the matrix are 0, at this time Q turns to

$$Q = \begin{bmatrix} \delta_X^{\ 2} & & & & & & & 0 \\ & \delta_{\dot{X}}^{\ 2} & & & & & & \\ & & \delta_Y^{\ 2} & & & & & \\ & & & \delta_{\dot{Y}}^{\ 2} & & & & \\ & & & & \delta_Z^{\ 2} & & & \\ & & & & & \delta_{\dot{Z}}^{\ 2} & & \\ & & & & & & \delta_T^{\ 2} & \\ 0 & & & & & & & \delta_{\dot{T}}^{\ 2} \end{bmatrix} \tag{12.32}$$

These $б_x^{\ 2}$ are the expected variances of measurement noise at the X coordinate of the user. Each other item has the similar definition.

For this research, $б_x^2$, $б_y^2$, $б_z^2$, and $бT^2$ are set to zero, the premise of this setting is that the processed noise in these terms is caused by (X.Y.Z.T) processed noise. Processed noise covariance of the velocity terms equals to the product of the pseudorange sampling period (ΔT) multiplied by the square of the maximum expected acceleration on the relevant axis. This product equals to the maximum expected non-model change for any velocity term during a sample period. The relative size of Q and R determines the weighting property of the Kalman gain vector.

The last matrix to be defined is matrix H, according to the assumption that we adopt timing receiver, H turns into a vector, and it needs the new position of the user or the satellite to redefine each pseudorange observation value:

$$H = [\cos\theta_x \quad \cos\theta_y \quad \cos\theta_z \quad 0 \quad 1 \quad 0] \tag{12.33}$$

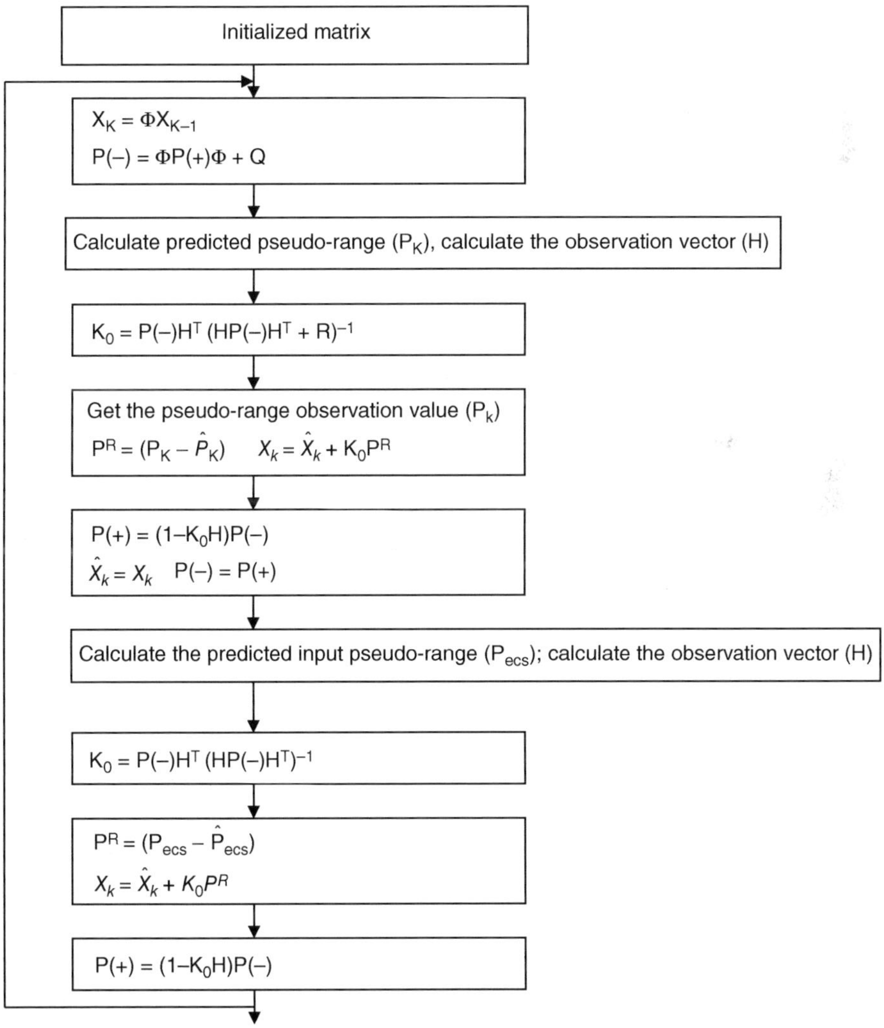

Figure 12.8 Flow chart of Kalman filter implementation.

Observation matrix determines the conversion characteristics of Kalman vector gain. Before incorporation of each successive pseudorange observation value into the state vector, the following formula can be adopted to recalculate the Kalman vector gain

$$P(-) = \Phi R(+)\Phi^{T} + Q$$

$$P(-) = \Phi R(+)\phi^{T} + Q \tag{12.34}$$

$$K_0 = P(-)H^{T}(HP(-)H^{T} + R)^{-1} \tag{12.35}$$

$$P(+) = (1 - K_0 H)P(-) \tag{12.36}$$

$P_{(-)}$ and $P_{(+)}$ are virtual matrixes for intermediate calculation. When the Kalman filter is initialized, the initial value of P(+) is Q, the actual implementation of the Kalman filter is shown in Figure 12.8.

As both the Kalman filter and the α/β tracker use the second-order user motion model in predicting the user position, as long as there is user acceleration, the output of the Kalman filter will have deviation. When R increases or Q decreases, the filter will cause the prediction state vector R to increase so as to reduce the residual weight calculated from the pseudorange observation values. So, the output noise decreases and the output deviation will increase; vice versa, when Q increases or R decreases, the output noise will increase and output deviation will decrease. However, for any given amount of noise reduction, the deviation caused by the acceleration contained in the output of Kalman filter may be smaller than the output of standard α/β tracker. The reason is that the Kalman filter input pseudorange observations are input one by one upon receiving, that is, inputting one upon receiving one; while α/β tracker inputs are non-filtered position data, and are input after receiving a complete set of pseudorange observations. So the Kalman filter can detect user acceleration with only a third of α/β tracking time. Thus, the update time of the filter is effectively shortened by two-thirds. Because the deviation caused by acceleration is proportional to the square of the filter update time, the deviation can be reduced to one-ninth, and that is why the Kalman filter can maintain a good amount of noise reduction, and will not have a significant deviation caused by acceleration.

References

1 Tan, S. *Current Situation and Development of Satellite Application (Vol. 1): Current Situation and Development of Navigation Satellite Application System*, Ed. J. Xu, China Science and Technology Press, 2001.

2 Ruth Blatter, M.A. *Radio Determination Satellite Service and Standard*. Trans. K. Tong, W. Lu, F. Ye, et al., National Defence Industry Press, Beijing, China, 1989.

3 Xu, Q. *Technology and Military Application of GPS*, The People's Liberation Army Press, Beijing, China, 1997.

4 Panetii, A. and Maggi, E. GalileoSat Team, *GNSS-2 Comparative System Studies Phase 2: Technical Note 3.5.2.a Satellites Definition – Alenia Aerospazio*, Doc. No. GNSS2-P2-SS-231, TN 3.5.2.a, Trans. Chunfa Liao, China Academy of Space Technology, Beijing, China, 2001

5 Liu, J. *Satellite Navigation Positioning Principle and Method of GPS*, Science Press, Beijing, China, 2003.

6 GNSS2 CSS Team, *GNSS-2 Comparative System Study Phase 2: GALILEO System and Segment Justification Files – Volume I System Level Justification*, Doc. No. GNSS2-P2-SYS-501, Trans. A. Zhou, China Academy of Space Technology, Beijing, China, 2001.

7 Aigle, F., Ammour, N., and Salgado, G. *GNSS-2 Comparative System Study Phase 2: Integrity Function Trade-offs*, Doc. No. GNSS2-P2-sys-202, TN 3.2.3, Trans. Y. Li, X. Li, Y. Deng, et al., China Academy of Space Technology, Beijing, China, 2001.

8 Parkinson, B.W., Spilker, J.J., and Axlrad, P. *GPS Theory and Application*, Trans. H. Li, Q. Liu, and H. Xie, Xi'an Institute of Navigation Technology, Xi'an, China, 1999.

9 Massatt, P. and Zeitzew, M. The GPS constellation design – Current and projected, *Proceedings of the 1998 National Technical Meeting of the Institute of Navigation, "NAVIGATION 2000,"* Long Beach, CA, January 21–23, pp. 435–445, 1998.

10 Study Team, *GNSS-2 Comparative System Studies Phase 2: Final Assessment and Conclusions*, Doc. No. GNSS2-P2-man-201, TN9, Trans. N. Wang, China Academy of Space Technology, Beijing, China, 2001.

11 Shen, Y. *Spread Spectrum*, National Defence Industry Press, Beijing, China, 1995.

12 Preisig, C. *GPS Marine Receiver Characteristic Analysis*, Trans. S. Tan, Surveying Technology (Supplement VI), Xi'an Surveying Institute, Xi'an, China, January 1988.

13 Chenebault, J., Provenzano, J.P., and Richard, F. INES an innovative European navigation system relying on a low earth orbit constellation, *Proceedings of the 1998 National Technical Meeting of the Institute of Navigation, "NAVIGATION 2000,"* Long Beach, CA, January 21–23, pp. 225–238, 1998.

GNSS Systems and Engineering: The Chinese Beidou Navigation and Position Location Satellite, First Edition. Shusen Tan.

14 Parkinson, B.W. A history of satellite navigation, Navigation, *Journal of the Institute of Navigation*, Vol. 42, No. 3, pp. 111–126, 1995.

15 He, X. and Chen, Y. Analysis of the performance of a GPS receiver with inertial velocity aided, *Proceedings of the 1998 National Technical Meeting of the Institute of Navigation, "NAVIGATION 2000,"* Long Beach, CA, January 21–23, pp. 175–182, 1998.

16 Parkinson, B. and Spilker, J.J. (Eds) *Global Positioning System, Theory and Applications, Vol. I*, American Institute of Aeronautics and Astronautics, Inc. Washington DC, 1996.

17 Parkinson, B. and Spilker, J.J. (Eds) *Global Positioning System, Theory and Applications, Vol. II*, American Institute of Aeronautics and Astronautics, Inc. Washington DC, 1996.

18 Betz, J.W. and Kolodziejske, K.R. Extended theory of early-late code tracking for a bandlimited GPS receiver, *Navigation, Journal of the Institute of Navigation*, Vol. 47, No. 3, pp. 211–226, 2000.

Further Reading

1 Han, L. and Sun, Q. *Military Digital Map*, The People's Liberation Army Press, Beijing, China, 1997.

2 Zhang, S. *Measurement Positioning Theory and Application of GPS Satellite*, National University of Defence Technology Press, Changsha, China, 1996.

3 Parkinson, B.W. *NAVSTAR GPS*, Trans. G. Qu, et al., Surveying and Mapping Press, Beijing, China, 1983.

4 Laramas, C. *GNSS-2 Comparative System Studies Phase 2: Technical Note 3.5.2.a Satellites Definition*, Doc. No. Project Ref (GNSS2-P2-SS-211) and TN 3.5.2.a, Trans. Y. Zhu, China Academy of Space Technology, Beijing, China, 2001.

5 Jocic, L., DiEsposti, R., Kastenholz, C., et al., Architecting GPS modernization, *Proceedings of the 1998 National Technical Meeting of the Institute of Navigation, "NAVIGATION 2000,"* Long Beach, CA, January 21–23, pp. 363–370, 1998.

6 Chang, J.M. A rapid alignment technique without S-turn and accelerometer, *Proceedings of the 1998 National Technical Meeting of the Institute of Navigation, "NAVIGATION 2000,"* Long Beach, CA, January 21–23, pp. 569–574, 1998.

7 Issler, J.-L., Fourcade, J., Lestarguit, L., et al., High reduction of acquisition and tracking thresholds of GPS spaceborne receivers, *Proceedings of the 1998 National Technical Meeting of the Institute of Navigation, "NAVIGATION 2000,"* Long Beach, CA, January 21–23, 1998, pp. 123–131.

8 Upton, D.M., Upadhyay, T.N., Marchese J., et al., Commercial-off-the-shelf (COTS) GPS interference canceller and test results, *Proceedings of the 1998 National Technical Meeting of the Institute of Navigation, "NAVIGATION 2000,"* Long Beach, CA, January 21–23, 1998, pp. 319–325.

9 Lin, L. Antenna nulling technique for GPS adaptive antenna, *Proceedings of the 1998 National Technical Meeting of the Institute of Navigation, "NAVIGATION 2000,"* Long Beach, CA, January 21–23, pp. 309–318, 1998.

10 Moore, T.A. Use of the GPS aided inertial navigation system in the navy standard missile for the BMDO/Navy LEAP technology demonstration program, *Proceedings of the 8th International Technical Meeting of the Satellite Division of the Institute of Navigation, ION GPS-95*, Palm Springs, CA, *September* 12–15, pp. 104–112, 1995.

11 Estrada, V., Yoo, K., and Moore, T. GPS-Aided Inertial Navigation System (GAINS) Evaluation Methodology for Airborne Missile Applications. Proceedings of the 1998 National Technical Meeting of the Institute of Navigation, "NAVIGATION 2000," Long Beach, CA, January 21–23, pp. 79–86, 1998.

12 Brown, R.G. and Huang, P.Y.C. *Introduction to Random Signals and Applied Kalman Fitering*, 2nd Edn, John Wiley & Sons, Inc., New York, 1992.
13 Ornedo, R.S. and Farnsworth, K.A. GPS and radar aided inertial navigation system for missile system applications, *Proceedings of the 1998 National Technical Meeting of the Institute of Navigation, "NAVIGATION 2000,"* Long Beach, CA, January 21–23, pp. 549–556, 1998.

Index

GNSS Systems and Engineering: The Chinese Beidou Navigation and Position Location Satellite, First Edition. Shusen Tan.

Made in the USA
Columbia, SC
17 September 2026

c6791c2f-3bf4-4de8-938b-5ef18c2cf8e1R01